Digital Representations of the Real World

HOW TO CAPTURE, MODEL, AND RENDER VISUAL REALITY

EDITED BY

Marcus A. Magnor
Oliver Grau
Olga Sorkine-Hornung
Christian Theobalt

CRC Press
Taylor & Francis Group
Boca Raton London New York

CRC Press is an imprint of the
Taylor & Francis Group, an **informa** business

CRC Press
Taylor & Francis Group
6000 Broken Sound Parkway NW, Suite 300
Boca Raton, FL 33487-2742

CRC Press is an imprint of Taylor & Francis Group, an Informa business

Printed and bound in India by Replika Press Pvt. Ltd.

Printed on acid-free paper
Version Date: 20141204

International Standard Book Number-13: 978-1-4822-4381-9 (Hardback)

Visit the Taylor & Francis Web site at
http://www.taylorandfrancis.com

and the CRC Press Web site at
http://www.crcpress.com

Contents

Foreword

Over the last two decades, the realism achievable using computer graphics has increased to the point where it is now impossible, in visual effects applications, to distinguish computer-generated imagery from reality. Over the same time period, our ability to capture and re-synthesize the real world inside the computer has kept pace. These techniques enable us to quickly and accurately capture 3D objects and scenes with a high degree of geometric and photometric fidelity.

Examples of such capture systems include active range scanning, most notably affordable real-time depth cameras such as Kinect™. They also include passive image-based modeling algorithms, which take as input collections of regular RGB images or videos and produce 3D shape and appearance models. Recent examples of such systems include the research Photo Tourism system, the consumer-level Photosynth Web service, as well as 3D image-based capture systems such as 123D® Catch.

This book, *Digital Representations of the Real World: How to Capture, Model, and Render Visual Reality*, contains a comprehensive compendium of the myriad techniques that enable us to capture, model, and render the world with a high degree of realism. It reviews the variety of sensors, such as regular cameras, wide-angle omnidirectional cameras, active range scanners, and plenoptic (multi-viewpoint) cameras, used to capture 3D scenes, as well as fundamental algorithms, such as 3D structure and motion recovery and stereo correspondence, used to process this sensed imagery.

The book also describes 3D modeling techniques, including both generic object models such as 3D meshes, and more domain-specific models such as human shape and motion models, needed to efficiently capture and manipulate 3D scenes. Finally, it describes how these shape and appearance models can be rendered in a way that meets both speed (e.g., real-time interactivity) and realism requirements, often using techniques such as image- and video-based rendering and incorporating modern models of visual perception and fidelity.

The scope and breadth of the techniques and systems used to capture, model, and render realistic simulacra of 3D scenes are quite daunting and can be a challenge for newcomers. This book provides an excellent introduction to and survey of this diverse field, written by some of the foremost researchers and practitioners in the field. Whether you are a novice to this exciting and challenging area, or an experienced veteran working in this field, you are sure to discover a wealth of useful and inspiring information in these pages.

Please dive in and enjoy!

Richard Szeliski
Microsoft Research

Preface

Marcus Magnor, Oliver Grau, Olga Sorkine-Hornung, and Christian Theobalt

Reality: The final frontier. Since the early beginnings of computer graphics, creating authentic models of real-world objects and achieving visual realism have been major goals in graphics research. Over the years, ingenious ways have been devised to represent real objects digitally, to efficiently simulate and emulate the laws of optics and physics, and to re-create perceptually authentic appearance. Ever-increasing CPU and GPU performance paved the way, up to the point where the memory and computational power available today afford genuine visual realism.

With visual realism within reach of modern hard- and software, intriguing new computer graphics applications have become possible. By combining computer graphics methods with video acquisition technology and computer vision algorithms, real-world events can now be interactively explored and experienced from an arbitrary perspective, almost like a video game. At the same time, the pursuit of visual realism has created new challenges. Higher visual realism can be achieved only from more detailed and accurate scene models. Consequently, the modeling process has become the limiting factor in attaining visual realism. Following the traditional paradigm, the manual creation of digital models consisting of 3D object geometry and texture, surface reflectance characteristics and scene illumination, character motion and emotion is a very labor-intensive, tedious process. The cost of conventionally creating models of sufficient complexity to engage the full potential of modern graphics hard- and software increasingly threatens to stall further progress in computer graphics.

To overcome this bottleneck, an increasing number of researchers and engineers worldwide have started to investigate alternative approaches in how to create digital models directly and automatically from real-world objects and scenes, with encouraging results: By now, entire cities are being digitized using panorama video footage, 3D scanners, and GPS; from CAD data and measured surface reflectance characteristics, highly realistic digital mock-ups of prototypes are being created, e.g., for the automotive industry; algorithms are being developed to create stereoscopic movies from standard, monocular footage; and live TV sports broadcasts are being augmented in real-time with computer graphics annotations. Other graphics application areas that work on merging the real with virtual worlds are special effects production for movies and computer games. In their goal to construct convincing virtual environments and digital actors, special effects production companies heavily rely on techniques to capture models from the real world. Still, a lot of time must be spent on manual post-processing and modeling. As an alternative approach, the computer graphics and vision communities are working on image- and video-based scene reconstruction approaches that can capture richer and more complex models of objects, humans, and entire complex scenes.

The trend toward model capture from real-world examples is also being pushed by new sensor technologies becoming available at mass-market prices. Microsoft's Kinect™ depth cameras, Lytro's light field cameras, Point Grey's Ladybug™ omni-directional cameras, and other companies' products offer unprecedented, novel ways to capture the appearance as well as other attributes of real-world objects and events. Finally, the pervasiveness of smartphones containing video chips, GPS, orientation sensors, and more gadgetry may in the near future lead to new real-world capture paradigms based on swarms of networked handheld devices.

Robust methods to unobtrusively capture comprehensive digital models of the real-world are one important part for attaining visual realism in computer graphics. Still, model reconstruction from real-world captured data remains, in general, an ill-posed problem that is prone to errors and failure cases. Insight into our human visual perception, however, allows for developing new model-adaptive, perception-aware rendering approaches that are able to perceptually mask and conceal modeling error–induced visual artifacts. Investigating how to best integrate new capture modalities, reconstruction approaches, and visual perception into the computer graphics pipeline, or how to alter the traditional graphics pipeline to make optimal use of the many new possibilities, has become a top priority in computer graphics.

The following 23 chapters present the state-of-the-art of how to create visual realism in computer graphics from the real world. A total of 48 authors from all over the world have joined up to compile a comprehensive overview, covering in 5 parts the entire pipeline from acquisition, reconstruction, and modeling to realistic rendering and applications. While editing the book, we tried to strike a balance between a general, comprehensive introduction to this exciting new research area and a practical guide that shows how to get started on re-implementing and using many of the most frequently encountered methods. We hope that it will be helpful to graduate students as well as researchers in academia and industry who are working in computer graphics, computer vision, multimedia, or image communications and who want to start their own research experiments in the challenging new field of real-world visual computing.

For MATLAB® and Simulink® product information, please contact:
The MathWorks, Inc.
3 Apple Hill Drive
Natick, MA, 01760-2098 USA
Tel: 508-647-7000
Fax: 508-647-7001
E-mail: info@mathworks.com
Web: www.mathworks.com

Contributors

Johannes Behr Fraunhofer IGD

Philippe Bekaert Hasselt University

Tamy Boubekeur Telecom ParisTech–CNRS LTCI–Institut Mines-Telecom, Paris, France

Edmond Boyer INRIA Grenoble Rhône-Alpes

Dan Casas University of Surrey–Centre for Vision, Speech and Signal Processing

Darren Cosker Department of Computer Science, University of Bath

Carsten Dachsbacher Karlsruhe Institute of Technology

Qionghai Dai Tsinghua University

Robert Dawes BBC Research and Development

Edilson de Aguiar CEUNES/UFES in São Mateus

Enrique Dunn The University of North Carolina at Chapel Hill, USA

Elmar Eisemann Delft University of Technology

Martin Eisemann TU Braunschweig, University of Technology

Peter Eisert Humboldt Universität zu Berlin, Fraunhofer HHI

Dieter W. Fellner Fraunhofer IGD / TU Darmstadt

Jan-Michael Frahm University of North Carolina at Chapel Hill, USA

Martin Fuchs University of Stuttgart

Juergen Gall Bonn University

Leonid German Institut für Informationsverarbeitung, Leibniz Universität Hannover

Bastian Goldlücke University of Konstanz, Department of Computer and Information Science

Oliver Grau Intel Visual Computing Institute

Volker Helzle Institute of Animation, Visual Effects and Digital Postproduction at Filmakademie Baden-Wuerttemberg

Anna Hilsmann Humboldt-Universität zu Berlin, Fraunhofer HHI

Adrian Hilton University of Surrey – Centre for Vision, Speech and Signal Processing

Peng Huang University of Surrey–Centre for Vision, Speech and Signal Processing

Muhannad Ismaël Université de Reims Champagne Ardenne

Jan Kautz NVIDIA Corporation

Oliver Klehm MPI Informatik

Thomas Knop Stargate Germany

Andreas Kolb University Siegen

Hendrik P. A. Lensch University of Tübingen

Max Limper Fraunhofer IGD/TU Darmstadt

Christian Lipski Metaio GmbH

Yebin Liu Tsinghua University, China

Céline Loscos University of Reims Champagne-Ardenne

Laurent Lucas Université de Reims Champagne-Ardenne

Fabrizio Pece ETH Zurich, Department of Computer Science

Yannick Remion Université de Reims Champagne Ardenne, France

Lorenz Rogge TU Braunschweig, University of Technology

Bodo Rosenhahn Institut für Informationsverarbeitung, Leibniz Universität Hannover

Kai Ruhl TU Braunschweig, University of Technology

Holly Rushmeier Yale University, USA

Michael Stengel TU Braunschweig, University of Technology

Graham Thomas BBC Research and Development

Kiran Varanasi Technicolor Research

Sven Wanner Heidelberg Collaboratory for Image Processing

Ruigang Yang University of Kentucky, USA

Jiejie Zhu SRI International

Stefanie Wuhrer Cluster of Excellence on Multimodal Computing and Interaction, Saarland University

Image Credits

Figure 1.3 Images courtesy of Philippe Bekaert at Hasselt University, Belgium, EU “2020 3D Media” project

Figure 2.2 Images courtesy of BBC, UK

Figure 3.2 Images courtesy of Philippe Bekaert at Hasselt University, Belgium, iMinds “explorative television” project

Figure 3.3 Image courtesy of Philippe Bekaert at Hasselt University, Belgium

Figure 3.4 Image courtesy of Philippe Bekaert at Hasselt University and Eric Joris at CREW vzw, “ICoSOLE” and “DreamSpace” EU projects

Figure 4.2 Image credits [Raposo et al. 13]

Figure 4.3 Image credits [Butler et al. 12] - accompanying video

Figure 4.4 Image courtesy of [Kolb et al. 10], Eurographics Association, 2010

Figure 4.5 Image courtesy of Left pmdtechnologies GmbH

Figure 4.7 Image courtesy of [Kolb et al. 10], Eurographics Association, 2010

Figure 4.8 Image courtesy of [Lefloch et al. 13], SPIE, 2013

Figure 9.2 Image courtesy of [Nießner et al. 13], ACM 2013

Figure 9.3 Image courtesy of [Keller et al. 13], IEEE 2013

Figure 13.2 Figures adapted from [de Aguiar et al. 08a]

Figure 13.5 Figures adapted from [Neumann et al. 13b]

Figure 15.1 Image courtesy of Michael Stengel

Figure 15.2 Image courtesy of Michael Stengel

Figure 15.3 Image courtesy of Michael Stengel

Figure 15.4 Image courtesy of Michael Stengel

Figure 15.5 Image courtesy of Mirko Sattler, Ralf Sarlette, and Reinhard Klein

Figure 15.6 Image courtesy of Michael Stengel

Figure 15.7 Image courtesy of Anna Hilsmann

Figure 17.1 3D Model "Iggy" by Dan Vulanovic was used for this image under the CC-Attribution 3.0 license

Figure 17.2 3D Model "Iggy" by Dan Vulanovic was used for this image under the CC-Attribution 3.0 license

Figure 17.3 3D Model "Iggy" by Dan Vulanovic was used for this image under the CC-Attribution 3.0 license

Figure 17.4 3D Model "Iggy" by Dan Vulanovic was used for this image under the CC-Attribution 3.0 license

Figure 17.5 3D Model "Iggy" by Dan Vulanovic was used for this image under the CC-Attribution 3.0 license

Figure 17.6 3D Model "Iggy" by Dan Vulanovic was used for this image under the CC-Attribution 3.0 license

Figure 17.7 3D Model "Iggy" by Dan Vulanovic was used for this image under the CC-Attribution 3.0 license

Figure 17.8 3D Model "Iggy" by Dan Vulanovic was used for this image under the CC-Attribution 3.0 license

Figure 17.9 3D Model "Iggy" by Dan Vulanovic was used for this image under the CC-Attribution 3.0 license

Figure 17.10 3D Model "Iggy" by Dan Vulanovic was used for this image under the CC-Attribution 3.0 license

Figure 17.11 3D Model "Iggy" by Dan Vulanovic was used for this image under the CC-Attribution 3.0 license

Figure 20.2 Image courtesy of Filmakademie Baden-Württemberg, The Gathering 2011

Figure 21.5 Images courtesy of Philippe Bekaert and Tom Mertens, Hasselt University, EU "2020 3D Media" project

Figure 21.6 Images courtesy of Philippe Bekaert at Hasselt University, Belgium, iMinds "explorative television" project

Figure 22.3 Image courtesy of [Schwartz et al. 11a], Eurographics Association 2013 / 2011

Figure 22.5 Image courtesy of [Lavoué et al. 13], ACM 2013

Figure 22.7 Image courtesy of [Schwartz et al. 11a], Eurographics Association 2011

Figure 23.1 Image courtesy of Filmakademie Baden-Württemberg, Jahre Leben 2013

Figure 23.3 Image courtesy of Filmakademie Baden-Württemberg, Dark Matter 2014

The Editors

Marcus Magnor is professor of computer science at Technische Universität (TU) Braunschweig, Germany, where he is chair of the computer graphics lab. He also holds an appointment as adjunct professor in the Physics and Astronomy Department at the University of New Mexico, USA. He earned his BA (1995) and MS (1997) in physics from Würzburg University and the University of New Mexico, respectively, and his PhD (2000) in electrical engineering from Erlangen University. After his postdoctoral time at Stanford University, he joined the Max-Planck-Institut Informatik in Saarbrücken as Independent Research Group leader. He completed his habilitation in 2005 and received the venia legendi for computer science from Saarland University. His research interests center around the natural phenomenon of images, from their formation, acquisition, and analysis to image synthesis, display, perception, and cognition. Areas of research include, but are not limited to, computer graphics, vision, visual perception, image processing, computational photography, astrophysics, imaging, optics, visual analytics, and visualization. He is the recipient of an ERC Starting Grant as well as being a Fulbright scholar, an elected member of the Braunschweigische Wissenschaftliche Gesellschaft, and laureate of the Wissenschaftspreis Niedersachsen.

Oliver Grau joined Intel as associate director of operations of the Intel Visual Computing Institute in Germany in October 2012. He earned a PhD from the University of Hannover, Germany, in 1999. Prior to Intel he worked for BBC R&D in the UK on innovative tools for visual media production. Since 2013 he has been a visiting professor at University of Surrey, UK. Oliver's research interests are in the intersection of computer vision and computer graphics techniques. His prior work included immersive virtual production systems, stereoscopic video production tools, free-viewpoint visualization of sport scenes, and Web-delivery of free-viewpoint experiences. More recent research interests include visual computing for new user experiences and digital content creation tools. Dr. Grau has a long track history of leading interdisciplinary work in more than 10 major collaborative projects, between academic and industrial partners. He has published a number of scientific papers and holds several patents.

Olga Sorkine-Hornung is an associate professor of computer science at ETH Zurich, where she leads the interactive geometry lab at the Institute of Visual Computing. Prior to joining ETH she was an assistant professor at the Courant Institute of Mathematical Sciences, New York University (2008–2011). She earned her BSc in mathematics and computer science and PhD in computer science from Tel Aviv University (2000, 2006). Following her studies, she received the Alexander von Humboldt Foundation Fellowship and spent two years as a postdoc at the Technical University of Berlin. Professor Dr. Sorkine-Hornung is interested in theoretical foundations and practical algorithms for digital content creation tasks, such as shape representation and editing, artistic modeling techniques, computer animation, and digital image manipulation. She also works on fundamental problems in digital geometry processing, including reconstruction, parameterization, filtering, and compression of geometric data. Professor Dr. Sorkine-Hornung received the EUROGRAPHICS Young Researcher Award (2008), the ACM SIGGRAPH Significant New Researcher Award (2011), the ERC Starting Grant (2012), the ETH Latsis Prize (2012), and the Intel Early Career Faculty Award (2013).

Christian Theobalt is a professor of computer science and the head of the research group "Graphics, Vision, & Video" at the Max-Planck-Institute for Informatics, Saarbrücken, Germany. From 2007 until 2009 he was a visiting assistant professor at Stanford University. He earned his MSc degree in artificial intelligence from the University of Edinburgh, Scotland, and his Diplom (MS) degree in computer science from Saarland University, in 2000 and 2001, respectively. In 2005, he earned his PhD (Dr.-Ing.) from Saarland University and the Max Planck Institute for Informatics. His research lies on the boundary between computer vision and computer graphics. For instance, he works on 4D scene reconstruction, marker-less motion capture, machine learning for graphics and vision, and new sensors for 3D acquisition. Dr. Theobalt has received several awards: The Otto Hahn Medal of the Max–Planck Society (2007), the EUROGRAPHICS Young Researcher Award (2009), the German Pattern Recognition Award (2012), and an ERC Starting Grant (2013). He is also a co-founder of the Captury (www.thecaptury.com).

Acknowledgments

This book is the result of work by experts from the fields of computer graphics, computer vision, and visual media production. We are deeply indebted to all contributing authors and thank everyone for the considerable time and effort they have devoted to this project. The idea for this book came about in the fall of 2013 at the Dagstuhl seminar on Real-World Visual Computing. Schloss Dagstuhl, the Leibniz Center for Informatics, situated in the peaceful and picturesquely forested hills of the northern Saarland in the westernmost part of Germany, offers computer scientists from all over the world the unique opportunity to get together for a full week to present their latest research, discuss and exchange novel ideas, and to get to know each other on a personal level. We thank the staff of Schloss Dagstuhl for their heartwarming hospitality as well as for the superb cuisine that kept everyone's body, soul, and mind together (or as the Saarland natives say: "Hauptsach gudd gess").

The content presented in this book constitutes mostly fundamental research that is available for everyone to read, re-implement, and use for their own purposes, free of charge. This is possible only because of publicly funded research. We gratefully acknowledge the support from all the funding agencies who invested in the research the results of which are presented in this book, in particular the German Science Foundation (DFG), the Swiss National Science Foundation (SNF), and the European Research Council (ERC).

We thank all the people at CRC Press who have helped us in getting this book written, edited, proofread, printed, and published in such a short time. We would explicitly like to thank Sarah Chow and Joselyn Banks-Kyle for their great help and support.

While all of the above were necessary ingredients to make this book happen, there is one person without whom the book would not have come into existence. Felix Klose was the good soul of our project. He prepared the LaTeX templates, set up the project's Wiki pages, reminded authors of deadlines, collected permission forms, made sure all chapter files were compiled, and much more. Felix, thank you for your commitment and perseverance!

The Editors

Marcus Magnor, Oliver Grau, Olga Sorkine-Hornung, Christian Theobalt

Part I

Acquiring the Real World

1

Camera Sensor Pipeline

Jan Kautz, Hendrik P.A. Lensch, Céline Loscos, and Philippe Bekaert

1.1 Introduction

The very first step of most real-world visual computing applications is the acquisition of images or video. However, acquiring meaningful image and video data is surprisingly challenging. This stems from the fact that real-world cameras and sensors are far from perfect concerning sampling or measuring and a substantial amount of processing needs to be applied before the data can be used. The different sensor types will be discussed, in particular with regard to how they affect the quality of data, how measurement noise affects image acquisition, and how to create dense color samples from sparse samples as they are acquired by most cameras. In order to characterize a given camera, it has to be calibrated in terms of radiometry and color as well as lens and geometric calibration. Applying all these steps results in well-calibrated and meaningful images.

1.2 Sensor Technology

Most digital imaging sensors operate based on the inner photoelectric effect. In the depletion area of the p-n junction of a photo diode an incoming photon of sufficient energy will create an electron-hole pair producing a photo current. In principle, every photon of sufficient energy (wavelength) can contribute to the effect, but the specific *quantum efficiency* depends on the wavelength. For a silicon photo diode the response covers the visible and the near infrared spectrum (400-1000nm).

The photoelectric effect produces a current that is linearly proportional to the radiant power, i.e., it can be used for physical measurements for all practical considerations inside a camera. Only at very strong illumination non-linear effects might occur [Anisimov et al. 77].

The light sensitive part of a pixel on a sensor typically is a photo diode. Besides photo diodes cameras might otherwise employ photo transistors with the added benefit of preventing further exposure by electronically closing the gate.

The charge collected during exposure needs to be stored, amplified and finally converted to a digital value. The design of a sensor allocates resources for these steps. The well capacity indicates how many electrons can be accumulated in one exposure, the scale indicates how much the photo current is amplified, and the bit rate is correlated to the number of discernible intensity values.

There are two fundamentally different approaches on how the charge is transfered to the A/D unit. In charge coupled devices (CCD) the charge is transported pixel by pixel to the end of each row, and the pixels in the last row are successively piped through a single amplifier and converter. Transfer between pixels can be carried out with hardly any loss. The main benefit of a CCD is that the information gathered by all pixels is basically processed by the same amplifier and converter. They undergo the same transformation. As a draw-back, a CCD can only read out rectangular regions. The speed of a CCD is also limited by the frequency of the A/D unit. As quality is typically decreasing with speed, reading off a multi-megapixel CCD at highest quality can take up to a couple of seconds. The process can be accelerated by providing multiple A/D units and splitting the sensor plane into tabs. This, on the other hand, leads to difficult to control conversion settings which are independent for each tab. In video cameras often interlaced read-out is used to provide higher frame rate. Each frame contains only half the rows, specifically every other row. Two subsequent frames alternate between the two sets of rows. The process of de-interlacing then performs a spatio-temporal interpolation between these to half-frames.

The second approach often employed is based on CMOS technology with the ability to group more electronic processing close to each pixel. Similar to random access memory, pixels can be addressed per row or individually. From just reading off a few pixels one can for example quickly sample an image histogram. Each pixel is equipped with a small amplifier which in the early days led to rather noisy CMOS images as each pixel is amplified individually. The additional electronics per pixel reduces the space available for the photo-sensitive part, lowering the fill factor. A lens on top of each pixel can counteract this loss in fill factor. Benefits of CMOS sensors are the flexibility of addressing and lower production cost.

More exotic sensors include for example back-illuminated CCDs where the support structure is thinned and the illumination is provided from the back-side avoiding photons being blocked by the electronic wires. For light sensitive applications this approach is typically combined with electron multiplying CCD (EMCCD) that employ solid-state impact ionization to multiply the number of generated photo-electrons. On the CMOS side

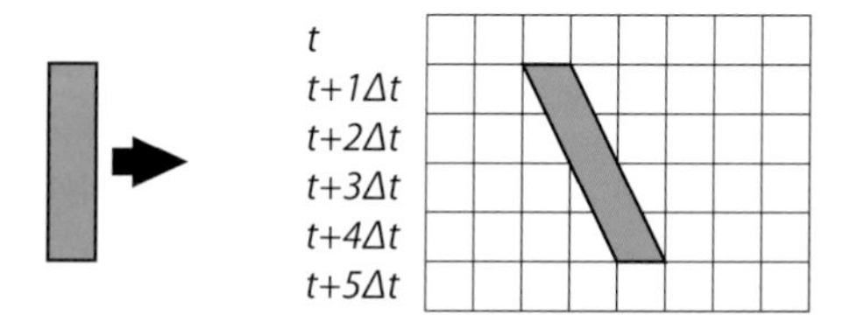

Figure 1.1: Rolling Shutter. In a rolling-shutter sensor each row will start exposure at a slightly different time. This results in distortion of moving parts. A vertical line moved to the right will be sheared.

so-called scientific CMOS sensors provide high-quality imaging by a more elaborate design of the per-pixel amplifiers.

Another important factor on sensors is how the entire image is read off the chip. Some CCD sensors provide a shielded area for storing the accumulated charge of one exposure while the sensor still is illuminated. As this electronic shutter transfer is synchronous one obtains the same global shutter for all pixels.

This is in contrast to the cheaper and faster rolling shutter most often found in CMOS chips where some rows are read out while others are still being exposed (Figure 1.1). In order to guarantee the same exposure duration, the exposure and read out of the rows is staggered. As a consequence the different rows will capture the scene at different moments in time. Special care is necessary when employing rolling shutter cameras for 3D reconstruction in dynamic environments as each camera (each row) potentially captures a different slice of the space time volume. Even though two cameras expose synchronously the same scene feature might be recorded at different times depending on its position in the respective camera image.

1.3 Noise

Inherent to digital imaging are a number of noise sources that affect every captured image. Reibel et al. [Reibel et al. 03] discerns two major classes: *temporal* and *non-temporal* noise.

The temporal noise sources vary with the scene brightness, and the temperature of the sensor. A fundamental limit to the accuracy of photographic measurements is the photon shot noise. Any source of light creates photons according to a temporal Poisson random process, i.e., the rate at which photons arrive at the sensor fluctuates. The variance of the photon shot noise is linearly correlated to the light intensity. Therefore, the standard deviation and at the same time the signal-to-noise ratio increases with the square root of the signal (SNR $= N/\sqrt(N) = \sqrt(N)$ for N photons). Similarly, heat can knock electrons loose in the silicon, producing a so-called dark current. The effect is independent of the actual signal, but the dark current can limit the maximum exposure duration when exceeding the well capacity. Dark current enters the subsequent amplification and A/D conversion step. Thus, these electrons are indistinguishable from photo electrons. Cooling the camera reduces the effect of the dark current shot noise as the noise doubles every $5 - 8°$ Celsius. Another temporal source of noise is the amplification and conversion step where thermal noise and a frequency-dependent flicker noise in the amplifier as well as quantization in the digitization step degrade the signal.

Non-temporal noise occurs due to static defects of the sensor. Due to slight irregularities, the area of the photo-sensitive part might vary and the properties of the per-pixel electronics might differ. The amount of dark current varies from pixel to pixel, resulting in a fixed pattern per-pixel bias independent of the signal. Similarly, the effect of photo-response non-uniformity corresponds to the amplifier gain being different per pixel. Some pixels reach saturation earlier than others, a problem mainly found in CMOS sensors. Finally, the actual amplification might not be perfectly linear, corrupting the direct linear relationship between photons and electrons. For HDR imaging, therefore, the actual photon transfer curve needs to be estimated (see Section 1.5).

The individual noise sources co-occur all at the same time during image capture and cannot always be disentangled. If accurate photometric calibration is required, cooling and taking a number of calibration images can improve image quality and allows to quantify the potential variance [Granados et al. 10, Hasinoff et al. 10]. Most common is to capture and average a series of dark frames with the same exposure time as the intended shot but leaving the cover on the lens. This way, the dark current and its spatial non-uniformity can be characterized. The variance of the readout noise can be captured by a bias frame, an image of zero exposure time. In order to quantify the photo-response non-uniformity, i.e., the per-pixel bias, a flat field is needed, a picture taken without a lens where each pixel receives exactly the same exposure. A practical difficulty is to ensure a really uniform illumination on the sensor. Perfect would be a large homogeneous area light source such as a monitor with added diffusor or a quite distant point light source. In a similar way a flat field captured with the lens can correct for vignetting. Considering all these measures, Granados et al. [Granados et al. 10] developed a noise-optimal pipeline for combining multi-exposure photos into a single HDR image.

Figure 1.2: Example of the common Bayer color filter array (CFA).

1.4 Demosaicing and Noise Reduction

Most digital cameras are single sensor cameras, i.e., only a single sensor measures the incoming light. However, a pixel in a CCD or CMOS sensor cannot sense the wavelength of the incoming light, but only its power. To enable color imaging, a color filter array (CFA) is overlaid on the sensor pixels: each pixel now senses only light within a specific wavelength range, typically corresponding to red, green, and blue wavelengths. The most common pattern is the Bayer pattern, with one red pixel, two green pixels, and one blue pixel in each 2×2 block of pixels (Figure 1.2). The use of a CFA leads to colors being sensed sparsely and missing color information needs to be filled in. This process is commonly called demosaicing, and many different techniques have been proposed over the years [Li et al. 08].

The simplest method is to simply take all the samples for a given color channel and to bilinearly interpolate from the nearest neighbors [Longere et al. 02]. As one might expect, this yields artifacts across edges and in areas with high-frequency texture content, since correlation between color channels is not taken into account. For instance, if there is a strong discontinuity between two neighboring green pixels, there is a high chance that there is a discontinuity also in the red and blue channels, but simple per-channel bilinear interpolation cannot reproduce this.

Quality can be increased with gradient-based methods, which typically estimate a local gradient direction followed by filtering along estimated edge directions and not across, thus avoiding the issues discussed above. The well-known Malvar–He–Cutler demosaicer (the default method in MATLAB®) falls into this category [Malvar et al. 04]. It still performs bilinear interpolation, but corrects it with a local gradient estimate using a 5×5 pixel window. This yields much improved results but can still lead to "zippering" artifacts, i.e., a visible high-frequency pixel pattern along high-frequency edges.

The best quality can be achieved by exploiting image self-similarity [Zhang et al. 11]. Instead of trying to estimate local image features across sparsely sampled color channels, self-similarity is used to derive the missing information. The LDI-NLM algorithm (and the very similar LDI-NAT), works as follows [Zhang et al. 11]. First, a standard directional interpolation method as described above is used to create an initial estimate of the green-channel. The green channel is then enhanced by running non-local means (NLM) [Buades et al. 05] on it. NLM will find similar patches for each pixel and compute a weighted average of those patches, which in turn is likely to improve the interpolated samples as additional data is being used. Following the reconstruction of the green channel, an initial estimate of the R and B channels are created (using information from the now complete green channel). Then, NLM is again run on the initial red and green channels. The LDI-NAT version proceeds similarly but uses soft thresholding in a sparse transform domain (similar to the BM3D denoising algorithm [Dabov et al. 07]). LDI-NLM and LDI-NAT achieve excellent results and outperform most other methods.

Noise Reduction It is important to note that these demosaicing methods assume noise-free input data. Of course, this is not usually the case. Applying these methods to noisy input data, however, often emphasizes color noise. Subsequent denoising (e.g., using the state-of-the-art BM3D denoiser [Dabov et al. 07]) of the demosaiced images is then necessary. Joint demosaicing and denoising is possible, but only little research has been conducted in this area to date [Chatterjee et al. 11].

1.5 Radiometry and Color

Sensing Radiance

As described in Section 1.2, the A/D unit converts the charge of each pixel to a digital value. This conversion is directly proportional to the charge, i.e., linear in the number of photoelectrons that have reached the sensor pixel (discounting noise). Most professional cameras allow the user to access this raw data, i.e., without any post-processing such as white-balancing, gamma correction, noise reduction, and so forth. If the raw data cannot be accessed on a particular camera, it is still possible to calibrate the response curve of the camera.

Color

Different sensors use different color filter arrays and different manufacturing processes, which leads to device-dependent color measurements. To output physically meaningful and device-independent color coordinates, such as CIEXYZ or CIELAB, the camera must be calibrated. This process is often called device characterization and requires two components [Johnson 02]: 1) calibration: determining the device's color space; and 2) characterization: finding a mapping between the device color space and the device-independent color space, e.g., CIE tristimulus values.

Suppose a color target is being captured. In discretized form, the trichromatic response value $[R, G, B]$ of a specific pixel on the sensor is given as the sum of the product of the spectral power distribution (irradiance) of the light source $P(\lambda)$, the surface reflectance of the imaged object $S(\lambda)$, and the spectral sensitivies of the color filters $D_{r/g/b}(\lambda)$:

$$R = \sum_{\lambda} P(\lambda)S(\lambda)D_r(\lambda)\Delta\lambda, \tag{1.1}$$

$$G = \sum_{\lambda} P(\lambda)S(\lambda)D_g(\lambda)\Delta\lambda, \tag{1.2}$$

$$B = \sum_{\lambda} P(\lambda)S(\lambda)D_b(\lambda)\Delta\lambda, \tag{1.3}$$

where the summation is over the visible spectrum. Now this is very similar to the computation of device-independent color values, such as CIEXYZ:

$$X = \sum_{\lambda} P(\lambda)S(\lambda)\bar{x}(\lambda)\Delta\lambda, \tag{1.4}$$

$$Y = \sum_{\lambda} P(\lambda)S(\lambda)\bar{y}(\lambda)\Delta\lambda, \tag{1.5}$$

$$Z = \sum_{\lambda} P(\lambda)S(\lambda)\bar{z}(\lambda)\Delta\lambda, \tag{1.6}$$

where $\bar{x}(\lambda)$, $\bar{y}(\lambda)$, and $\bar{z}(\lambda)$ are the CIE color matching functions. So the only difference is the device-dependent color $D_{r,g,b}$ vs. the device-independent functions $\bar{x}, \bar{y}, \bar{z}$.

Many characterization techniques have been proposed. They largely fall into two categories: reflectance-based characterization, and characterization based on monochromatic light. Reflectance-based characterization usually requires a color target with known reflectances and a suitable sampling of the color space, such as a GretagMacbeth ColorChecker, of which a picture is taken. A direct mapping between the (raw) images RGB-values and the known XYZ values of the color target can be derived via linear regression. While these techniques are very easy to use, they are only valid for the current illumination condition as the illuminant $P(\lambda)$ is "baked in." The most common monochromator-based method uses a hollow white sphere, which is illuminated by a monochromator with an adjustable wavelength.

An image is taken for a number of wavelengths, which allows for a direct mapping between the device's color coordinates and CIEXYZ tristimulus values. While this is a time-consuming and expensive calibration method, it is vertically accurate.

CIEXYZ is the basis from which one can convert to many other common color spaces, such as sRGB. sRGB is notable because it has found widespread use, as it was designed for typical home viewing conditions and not darker environments that are used by professionals for color matching. It is a non-linear color space, with an overall gamma of about 2.2 but consisting of a linear plus a non-linear part.

HDR Imaging

High-dynamic range (HDR) imaging allows for the representation of a larger range of intensities than conventional images [Reinhard et al. 08, Reinhard et al. 10]. It is widely used by photographers and supported by software[1] to avoid saturated areas or under-exposed pixels. It is also used to acquire more precise illuminant information of the real world when modeling objects, or to guide image compositing for coherent common illumination when mixing real and virtual content.

Conventional camera sensors typically digitize luminance with 8 to 16 bits. Even when digitized with 16-bit accuracy natural scenes can still easily exceed the dynamic range of the sensor. There are many definitions of what is high-dynamic range. Some consider that non-linearly representing the range of luminance using 8 bits qualifies for HDR. Others consider HDR to be the full variation of the physical luminance of the real world that the human visual system is capable of adapting to, thus 10 orders of magnitude. Recently, a group of experts[2] came to the consensus that high-dynamic range should represent the perceptual range of intensities simultaneously perceivable by the human eye, thus 6 orders of magnitude, which can be stored on a 20-bit image.

There exist two main procedures to capture HDR content: by merging conventional camera images, or by providing enhanced hardware capability. In order to create an HDR image with a conventional camera, images are taken with different time exposures in order to capture different ranges of luminance. Combining these images requires two steps: radiometric calibration and merging values into HDR data. Radiometric calibration is necessary mostly if RAW sensor data are not available or very noisy. It consists of finding a linear color mapping from one image to another that are taken with different exposures. Merging values into HDR data consists of carefully selecting pixels from all images to form a coherent HDR image.

[1] e.g., Adobe Photoshop - http://www.adobe.com/fr/products/photoshop.html
[2] HDRi - COST Action IC1005 - http://www.ic1005-hdri.com/

Enhancing hardware capabilities corresponds to increasing the dynamic range of sensors. SpheronVR,[3] for example, provides cameras (photographic and video) with sensors capable pf covering 8 orders of magnitude. These cameras are not aimed at the general public, and some of them are at the stage of prototypes, limited by streaming and storage facilities. Other technologies involve using beam splitters [Aggarwal and Ahuja 04, Tocci et al. 11] to capture data at different intensities with a single camera and a single shot. Merging is done in a similar way as for sequential multi-exposure images. Finally, it is possible to adapt a mask in front of the sensing array with a pattern to reduce the incoming light to different degrees, and to produce spatially varying exposures [Nayar and Mitsunaga 00]. Beam splitter-based approaches as well as spatially varying exposure approaches provide the advantage that they can be directly applied to dynamic, time varying scenes since all images represent the same instant. These types of approaches, though, are limited in the captured range by their beam splitter capability and the spatially varying exposures respectively.

Radiometric Calibration Displays and cameras employ a response function to modify measured luminance to create pleasant overall colors when perceived by the human eye. For color image processing, radiometric calibration needs to be performed. In the case of high-dynamic range images, we need to find the inverse response function of the camera to linearize pixel color relations. Ideally, inter-image relation should lead to the radiometric relation for a 3D point that projects to the same image coordinates (x, y) of two images $\mathbf{I}^0$ taken at exposure time t_0 and $\mathbf{I}^1$ taken at exposure time t_1, linking the radiance E arriving at sensors and stored in images as RGB values:

$$E_{I_0}/t_0 = E_{I_1}/t_1 \tag{1.7}$$

RAW sensor information can be used directly with this equation to transform pixel color values to coherent radiance values in all images. However, depending on the camera, this is not always true, and even more when no access to RAW data is possible. There is a need to find the inverse camera function $g = f^{-1}$, with f non-linearly transforming the radiance values to color. Inverting the function is possible because values monotonically increase. Several methods have been proposed [Mann and Picard 95, Mitsunaga and Nayar 99, Grossberg and Nayar 04, Debevec and Malik 97]. They all fit a curve to selected values and therefore are approximative. However, this is generally sufficient, and remaining small errors can be compensated in the HDR reconstruction phase.

[3]https://www.spheron.com/home.html

HDR Reconstruction HDR Reconstruction is the process of merging values coming from different images into one coherent HDR value. The general equation for N images and the pixel colors $E_i(x, y)$ of each image i at coordinates (x, y) is:

$$E(x, y) = \frac{\sum_{i=0}^{N} \omega(I_{p_i}) \frac{g(E_i(x,y))}{t_i}}{\sum_{i=0}^{N} \omega(I_{p_i})} \tag{1.8}$$

The difficulty here is to chose the weights $\omega(I_{p_i})$ associated with the pixel I_{p_i} of image i. They are used to enhance or reduce the impact of pixel colors in the final HDR result [Granados et al. 10]. The weight function excludes under- and over-saturated pixels [Debevec and Malik 97] but can also be based on signal-to-noise ratio [Mitsunaga and Nayar 99].

This reconstruction approach assumes that images are perfectly aligned and that no movement occurred during sequence acquisition. If this is not the case, weights can also reflect the probability of a pixel to belong to the background [Khan et al. 06]. For motion registration or non-aligned cameras, more complicated operations need to be performed to register pixels before reconstruction can be achieved [Loscos and Jacobs 10, Bonnard et al. 13].

Multispectral Imaging

The quantum efficiency of silicon-based camera sensors is by itself a wavelength-dependent function. The Foveon sensor was able to detect color by measuring at three different penetration depths in the silicon. However, this concept has never been extended to more than three wavelength bands. The most common approaches for capturing more than three color channels are either to extend the Bayer pattern and include more colors, or to use a second optically aligned sensor with a Bayer pattern of different base colors.

If significantly more wavelength bands are required, there are basically two different approaches:

The first approach captures one wavelength band at a time using a filter wheel or a tunable filter. Tunable filters employ an electro-optic or acusto-optic effect to transmit only the selected wavelength band. The drawback of this filtering approach is that only a small fraction of the overall radiant power is captured in each band, resulting in a lengthy process to capture a multispectral image.

The second approach makes use of a prism or diffraction grating to split up the incoming light into its spectrum. Once spatially separated, the different wavelengths can be modulated individually and then recombined onto the sensor [Mohan et al. 08, Kim et al. 12]. The benefit is that the entire spectrum can be varied, although not necessarily in the same way for the entire image plane, rather than selecting only a single wavelength band. In order to produce a multi-channel image, this optical setup is often combined with compressed sensing approaches [Mohan et al. 08, Kim et al. 12].

1.6 Geometric Calibration

Applications such as 3D geometry reconstruction, view interpolation, and so on require understanding the mapping between 3D real scene points and image coordinates. The process of determining the actual value of the parameters that control that mapping is called geometric camera calibration. In this section, a common model for this mapping is presented, and the basic principles of lens distortion, intrinsic and extrinsic single-camera calibration are outlined. The simultaneous calibration of multiple cameras is explained in Section 2.3.

Camera Calibration Parameters

The geometric camera calibration parameters fall into four categories: sensor-related parameters, lens-related parameters, camera-lens assembly parameters, and extrinsic parameters.

Sensor-related paramaters include the *image width and height* in pixels, and the *pixel pitch*: the spacing of pixels in each row and between rows. They are usually known from camera specifications and region of interest settings.

Lens-related parameters do not depend on the camera the lens is mounted on, nor its position and orientation in space. They include the lens image formation model and lens distortion. Most lenses are rectilinear lenses, ideally mapping straight world lines to straight image lines. They are characterized by their *focal length*. Equidistant fish eye ($f\theta$) lenses can offer greater sharpness and less distortion for wide viewing angles. Also these lenses are characterized by their focal length f, which has however a different meaning than for rectilinear lenses. *Lens distortion* models quantify the deviation of a real lens from the ideal rectilinear or $f\theta$ model.

Camera-lens assembly parameters include the *principal point*, the *center of distortion* and *effective pixel aspect ratio and skew angle*. The principal point is the image coordinate of the intersection of the optical symmetry axis of a lens with the camera sensor plane. The center of distortion usually is equal to the principal point. The effective pixel aspect ratio and skew angle may deviate slightly from sensor specifications due to mechanical tolerances in lens and camera housing.

Table 1.1: Set of geometric camera calibration parameters.

symbol	parameter name	unit
w,h	pixel width and height	micrometers
$k_0, k_1, \ldots$	lens distortion coefficients	$1/cm^{\kappa}$
x_d	center of distortion	image coordinates
f	focal length	millimeters
x_c	principal point	image coordinates
a	effective aspect ratio	dimensionless
θ_{skew}	effective skew angle	degrees
α, β, γ	camera orientation Euler angles	degrees
C	optical center	world coordinates

The *position and orientation* of a camera in 3D real-world space are called the *extrinsic* parameters of the camera. Position is always relative to a particular choice of 3D real world coordinate system. The position that counts is the *optical center*: the point in 3D space where rays of light hiitting the lens would meet, if they were not bent to focus on the sensor. For humans, orientation is conventiently expressed by means of Euler angles (note there are 24 different interpretations of Euler angles [Schoemake 94]). In computations, quaternions, exponential maps, or a rotation matrix will usually be preferred.

Table 1.1 summarizes a typical set of geometric camera calibration parameters.

Mapping World Space Points to Image Coordinates

Mapping world space point X to image coordinates x basically takes four steps:

- mapping world space point X to camera space point X_c;
- application of the image formation model to map X_c to a location x_f on the lens focal plane, relative to the principal point;
- mapping lens focal plane position x_f to an ideal (undistorted) image coordinate $\bar{\mathrm{x}}$;
- applying the lens distortion model to obtain the observable (distorted) image coordinate x.

Lens distortion is part of image formation by the lens in physical reality. In visual computing, however, it is usually modelled as a correction to ideal image coordinates as described here.

Mapping image coordinates to world space rays takes the inverse of these steps, applied in reverse order.

The first step, is a simple translation taking the world origin to the cameras optical center C, and rotation $\mathbf{R}$ aligning the view to a canonical axis system, such as in OpenGL (view direction is negative Z, image right direction is X, image up direction is Y):

$$\mathrm{X}_c = \mathbf{M}\mathrm{X} \qquad \text{with} \qquad \mathbf{M} = \left[\mathbf{R}^\top | -\mathbf{R}^\top \mathrm{C}\right]. \tag{1.9}$$

The matrix $\mathbf{M}$ is called the camera *extrinsic matrix.*

For rectilinear lenses, the second step is a rescaling of X and Y by inverse depth $-1/Z$ and focal length f:

$$x_r = f\frac{X}{-Z} \quad y_r = f\frac{Y}{-Z} \quad r = f\tan\theta.$$

θ is the angle between the optical axis of the lens and the incident light ray direction. r is the distance in millimeters (if f is expressed in millimeters) of the light ray projection on the focus plane, relative to the principal point. For other lens models, other similar formulae apply (such as $r = f\theta$ for equidistant fish eye lenses). The minus sign is due to our coordinate system convention (OpenGL-style $Z < 0$ in front of the camera).

For a rectangular sensor pixel grid, and in absence of distortions causing pixel grid skew or aspect ratio abberations, the third step is a simple 2D scaling and translation from sensor plane position in millimeters relative to the principal point to pixel unit distance with respect to the top-left image corner or other chosen image coordinate origin. In general, it is a 2D shearing transform taking into account effective aspect ratio and skew angle.

For rectilinear lenses, steps two and three can be combined into a single matrix multiplication, yielding *homogeneous* undistorted image coordinates. These require *perspective division* of $\bar{x}$ and $\bar{y}$ by $\bar{z}$ in order to obtain affine image coordinates (Table 1.1):

$$\begin{bmatrix}\bar{x}\\ \bar{y}\\ \bar{z}\end{bmatrix} = \mathbf{K}\begin{bmatrix}X_c\\ Y_c\\ Z_c\end{bmatrix} \quad \text{with} \quad \mathbf{K} = \begin{bmatrix}\tilde{f} & -\tilde{f}\tilde{s} & -x_c\\ 0 & -\tilde{f}\tilde{a} & -y_c\\ 0 & 0 & -1\end{bmatrix} \tag{1.10}$$
$$\tilde{f} = f/w \quad , \quad \tilde{s} = -\tan\theta_{skew}. \quad , \quad \tilde{a} = a/\cos\theta_{skew}$$

This matrix $\mathbf{K}$ is named the *intrinsic camera matrix.* The minus signs in the definition of $\mathbf{K}$ are due to our coordinate system conventions (OpenGL style $Z < 0$ in front of the camera, Y pointing up, image y pointing down given image coordinate origin in the top-left corner).

For rectilinear lenses, the full mapping from homogeneous world coordinates to homogeneous undistorted image coordinates can be obtained as a single matrix-vector product:

$$\begin{bmatrix} \bar{x} \\ \bar{y} \\ \bar{z} \end{bmatrix} = \mathbf{P} \begin{bmatrix} X_w \\ Y_w \\ Z_w \\ 1 \end{bmatrix} \quad \text{with} \quad \mathbf{P} = \mathbf{KM} \tag{1.11}$$

The matrix $\mathbf{P}$ is called the *full camera matrix.*

A most common model for lens distortion, the fourth step, is the following [Brown 66, Slama 80, Heikkila and Silven 97, Zhang 00]:

$$\begin{aligned} \mathrm{x} &= \mathrm{x}_d + L(\bar{\mathrm{x}}') \quad , \quad L(\bar{\mathrm{x}}') = \bar{\mathrm{x}}' L_r(r) + L_t(\bar{\mathrm{x}}') \\ \bar{\mathrm{x}}' &= \bar{\mathrm{x}} - \mathrm{x}_d \quad , \quad r = \sqrt{\bar{x}'^2 + \bar{y}'^2} \\ L_r(r) &= k_0 + k_1 r^2 + k_2 r^4 + k_3 r^6 \\ L_t(x, y) &= (p_1 B(x, y) + p_2 D(x, y), p_2 C(x, y) + p_1 D(x, y)) \\ B(x, y) &= 3x^2 + y^2 \quad , \quad C(x, y) = x^2 + 3y^2 \quad , \quad D(x, y) = 2xy \end{aligned} \quad . \tag{1.12}$$

The model consists of a radial part L_r, modifying distance with respect to the *center of distortion* x_d, and a tangential part L_t.

Lens Distortion Calibration

Lens distortion is calibrated when a set of lens distortion parameter values has been found that warps a captured image into an image that shows straight world lines as straight image lines. The distortion parameters are the distortion coefficients k_i, as well as the center of distortion (x_d, y_d).

Auto-Calibration In order to calibrate lens distortion under uncontrolled circumstances, one or a few images of scenery exhibiting straight world lines suffices, such as windows or doors in an image of a building facade. Lens distortion parameters can be obtained by non-linear optimization, e.g., with the Levenberg–Marquardt algorithm. In each step of the optimization procedure, edge pixel locations in the input image(s) are warped using the (inverse) lens distortion model. The quality of the parameter set is evaluated by measuring to what extent the warped edge pixels form straight lines [Devernay and Faugeras 01]. A practical tool implementing a similar approach, is PTLens.[4]

Lens Distortion from Calibration Grids Often in stereo- or multi-view setups, lens distortion can be calibrated in controlled lab circumstances. Known patterns of features are filmed and analyzed. Often used patterns include

[4] http://epaperpress.com/ptlens/

planar checkerboard calibration patterns (using saddle-points) and rectangular grids of circular dots (using centroids).

In absense of lens distortion, the relation between the known 2D real world calibration grid feature positions and their image coordinates, is a planar perspective transform, also called a *2D homography* [Hartley and Zisserman 03, §2.3]. Distortion parameters can be estimated by iterative optimization algorithms that minimize the deviation of correspondences from a 2D homography. The distortion center can also be estimated using direct techniques [Hartley and Kang 05]. More direct estimation techniques, based on *lifted coordinates* are described in [Sturm et al. 11]. These techniques can be generalized to non-rectilinear lenses.

Intrinsic Calibration

Estimating the intrinsic camera calibration parameters is the determination of camera parameters determining the mapping between (ideal, undistorted) image coordinates and camera space ray directions. For rectilinear cameras, this mapping is governed by the intrinsic camera matrix K (Equation 1.10).

From camera specifications, pixel aspect ratio a and skew angle θ_{skew} are typically known to sufficient accuracy. Often, pixels are square. When a camera is equipped with a lens exhibiting lens distortion, the principal point x_c may be taken equal to the lens distortion center x_d calculated using above sketched methods. The main intrinsic parameters to be determined thus typically are the principal point $\mathrm{x}_c = (x_c, y_c)$ for a lens without significant distortion, and the focal length f.

Instrinsic camera parameters can be auto-calibrated from observations of orthogonal world lines and/or planes, or determined from 2D homographies relating a planar calibration grid with its image taken at different angles [Zhang 00].

These observations impose linear *constraints* on a particular symmetric 3×3 matrix $\omega = \left(\mathbf{K}\mathbf{K}^\top\right)^{-1}$, called the *image of the absolute conic (IAC)*. Consider, for instance, the vanishing points v_1 and v_2 of two (bundles of) orthogonal lines with direction vectors $\mathrm{D}_1 = (X_1, Y_1, Z_1, 0)$ and $\mathrm{D}_2 = (X_2, Y_2, Z_2, 0)$. Since the homogeneous component is zero, the relation between vanishing point v and affine direction vector $\mathrm{d} = (X, Y, Z)$, is:

$$\mathrm{v} = \mathbf{P}\mathrm{D} = \mathbf{K}\left[\mathbf{R}^\top | -\mathbf{R}^\top \mathrm{C}\right] \begin{bmatrix} X \\ Y \\ Z \\ 0 \end{bmatrix} = \mathbf{K}\mathbf{R}^\top \begin{bmatrix} X \\ Y \\ Z \end{bmatrix} \quad \Leftrightarrow \quad \mathrm{d} = \mathbf{R}\mathbf{K}^{-1}\mathrm{v}. \tag{1.13}$$

Since the direction vectors d_1 and d_2 are orthogonal,

$$0 = {d_1}^\top d_2 = {v_1}^\top \mathbf{K}^{-\top} \mathbf{R}^\top \mathbf{R} \mathbf{K}^{-1} v_2 = {v_1}^\top \mathbf{K}^{-\top} \mathbf{K}^{-1} v_2 = {v_1}^\top \omega v_2. \quad (1.14)$$

Such constraints on the IAC ω are stacked together into a homogenous linear system. This linear system is solved using singular value decomposition (SVD), with proper preconditioning. The thus found IAC is decomposed as $\omega = UU^\top$, U being an upper-triangular matrix, using Cholesky factorization. K is obtained as U^{-1}, and then decomposed into f, x_c, a and θ_{skew}, if required, and refined using Levenberg–Marquardt iterative optimization [Hartley and Zisserman 03, §8.6].

Extrinsic Calibration

Extrinsic calibration is the process of determining the location C and orientation $\mathbf{R}$ of a camera, or a set of cameras, with respect to a 3D world space coordinate system of choice.

The most straightforward way to find the location of fixed cameras is to measure them with a simple ruler or other distance measuring device. However, the exact location that matters is the optical center, which is the imaginary point in 3D space where the rays of light hitting the lens would meet if they were not bent to focus on the sensor. Its position relative to the camera body can be estimated typically only up to a few-centimeter precision.

Sometimes, location and/or orientation *tracker* devices, are used to measure camera positions and orientations. These devices can be based on mechanical, electrical, optical, magnetic, micro-electromechanical (MEM), electro-magnetic (EM, radio waves), or other principles [Danette Allen et al. 01]. GPS allows outdoor localization to an accuracy of about 1 meter, and update rate of one second typically. Compasses measure absolute orientation with respect to the earth magnetic field. For indoor use, optical tracking systems are often used, and regularly in combination with inertial tracking (with MEM devices). In all cases, it pays off to combine such measurements with visual tracking (Section 2.3).

Planar Calibration Grids The orientation $\mathbf{R}$ and location C of the camera, relative to a planar calibration grid, are easily obtained from a 2D homography $\mathbf{H}$, relating the grid with its image, and the intrinsic matrix $\mathbf{K}$.

Assume the calibration grid is in the world XY-plane ($Z = 0$). Let p_1, p_2, p_3, p_4 denote the columns of the full camera matrix $\mathbf{P}$. Image points x are related with their corresponding calibration grid points

$\mathrm{X} = (X, Y, 0, 1)$, as follows:

$$\mathrm{x} = \mathbf{P}\mathrm{X} = \begin{bmatrix} \mathrm{p}_1 & \mathrm{p}_2 & \mathrm{p}_3 & \mathrm{p}_4 \end{bmatrix} \begin{bmatrix} X \\ Y \\ 0 \\ 1 \end{bmatrix} = \begin{bmatrix} \mathrm{p}_1 & \mathrm{p}_2 & \mathrm{p}_4 \end{bmatrix} \begin{bmatrix} X \\ Y \\ 1 \end{bmatrix}$$

Therefore, $\mathbf{H} \doteq \begin{bmatrix} \mathrm{p}_1 & \mathrm{p}_2 & \mathrm{p}_4 \end{bmatrix}$, and since $\mathbf{P} = \mathbf{K}\begin{bmatrix} \mathbf{R}^\top | -\mathbf{R}^\top \mathrm{C} \end{bmatrix}$:

$$\mathbf{H} \doteq \mathbf{K}\begin{bmatrix} \mathrm{r}_1 & \mathrm{r}_2 & \mathrm{t} \end{bmatrix}. \tag{1.15}$$

Apart from sign and normalization, the first two columns r_1 and r_2 of $\mathbf{K}^{-1}\mathbf{H}$ provide the first two rows of $\mathbf{R}$. The third row is the cross product $\mathrm{r}_1 \times \mathrm{r}_2$. The optical center follows as $\mathrm{C} = -\mathbf{R}\mathrm{t}$. The fourfold ambiguity is resolved by testing each possible solution against the actual data.

This method is used in the popular camera calibration approach by Zhang [Zhang 00] and implemented in the camera calibration toolbox of Bouguet,[5] which is available in MATLAB and OpenCV.

3D Ground Control Points There is no straightforward way to estimate pose (position and orientation) of a camera relative to a set of world space points with known coordinates from their image projections. The full camera matrix P, however, can be estimated from 3D-2D correspondences as follows. When intrinsics are known, pose then can be obtained after full matrix estimation, as $\mathbf{M} = \mathbf{K}^{-1}\mathbf{P}$.

$$\mathrm{x} \doteq \mathbf{P}\mathrm{X} \iff \begin{cases} kx = \mathrm{p}^1\mathrm{X} \\ ky = \mathrm{p}^2\mathrm{X} \\ kw = \mathrm{p}^3\mathrm{X} \end{cases} \implies \begin{cases} \left(w\mathrm{p}^1 - x\mathrm{p}^3\right)\mathrm{X} = 0 \\ \left(w\mathrm{p}^2 - y\mathrm{p}^3\right)\mathrm{X} = 0 \end{cases}$$

k makes the scale ambiguity in $\mathrm{x} \doteq \mathbf{P}\mathrm{X}$ explicit. p^i denotes the i-th row of $\mathbf{P}$. Cross-multiplication of the first two equations with the third, yields the right-most form.

Equation pairs resulting from each given correspondence are stacked together into a homogeneous linear system. The solution is as usual obtained from SVD of the system matrix, with proper preconditioning, as the right-singular vector corresponding with the smallest singular value, and refined using Levenberg–Marquardt iterative optimization [Hartley and Zisserman 03, §7]. This is the basis of the often used calibration method of Tsai [Tsai 87], which among other things, also iterates the above sketched approach with lens distortion optimization.

In practice, the POSIT algorithm [DeMenthon and Davis 95] allows more efficient and robust pose estimation, when other camera parameters

[5] http://www.vision.caltech.edu/bouguetj/calib_doc/

are not needed. POSIT follows an iterative approach. It starts by estimating pose assuming an orthogonal projection model with scaling, rather than a full perspective model. The pose, obtained by making this assumption, is refined in a few, fast, iterations until good agreement is found with the actual observations. A linear algorithm for pose estimation was proposed in [Quan and Lan 99].

Example: Calibration of a Trifocal Camera Rig

Figure 1.3(top-left) depicts a camera rig consisting of a broadcast camera and lens with two auxiliary machine vision style cameras on its sides.[6] The auxiliary cameras are synchronized with the main camera, and provide additional views from which left-right stereo pairs can be generated in post-production, allowing control of convergence angle and baseline distance in post. These parameters need to be chosen differently, depending on the size of the screen the outcome is viewed on (ranging from mobile phone types of displays, over TV screens, to cinema). The lenses are wide-angle lenses, exhibiting distortion. The cameras were calibrated independently using the principles outlined in this chapter, as follows.

The camera rig is placed in front of a TV flat screen, on which a sequence of black-and-white patterns is displayed. The patterns consist of circular dots. The sequence of patterns reminds one of binary patterns used in structured light scanning, and allows to identify each dot in each camera. The dot centers form thousands of high-quality "labeled" rather than "anonymous" correspondences between the three captured views, also in cases in which patterns are not fully in view. Figure 1.3(bottom-left) shows the analysed pattern for the broadcast camera. Note how lens distortion is significant, even on the high-end broadcast camera lens used.

The dots form an equidistant grid in physical 3D space, and thus define a metric calibration world plane. Lens distortion is calibrated by first estimating the distortion center using a direct approach [Hartley and Kang 05]. Radial distortion parameters are obtained by iterative fitting of a homography to the image-to-calibration-grid correspondences minimizing deviation. Figure 1.3(bottom-right) shows that after lens calibration fitted and observed image coordinates correspond very well. Two observations of the calibration grid, from different angles (by moving the TV screen), were used.

The principal point was taken equal to the center of distortion, and pixel pitch and aspect ratio were taken from sensor specifications. Focal length was estimated using the IAC approach based on two observations of the metric plane defined by the calibration grid. In principle, a single

[6] http://www.20203dmedia.eu

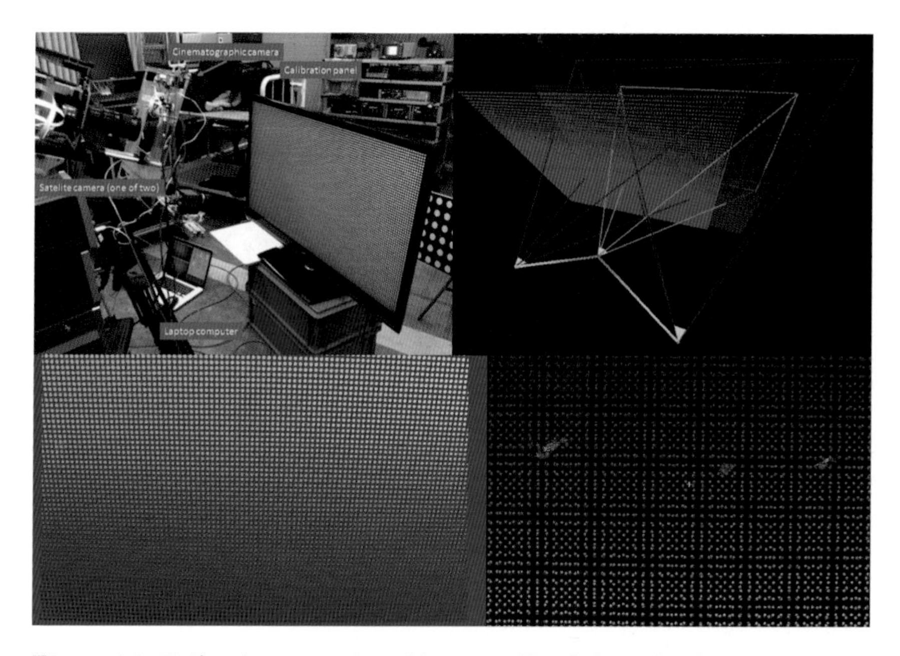

Figure 1.3: Trifocal camera rig calibration. Top-left: trifocal-camera set-up in front of computer-driven TV flat screen. Bottom-left: decoded calibration pattern on the central (broadcast) camera. Red corresponds with world-X, and green with world-Y coordinates. Note how lens distortion is significant, even on the high-quality broadcast lens used. Bottom-right: after lens distortion calibration, the fitted and observed image coordinates coincide very well. Top-left: calculated camera positions and orientations, and the relative position of the calibration grid, which was observed from two different angles.

observation would have done. The focal length of the auxiliary cameras was within 1% of lens specification.

Finally, the position and orientation of the cameras, and the calibration grid (viewed from two different angles), were calculated from the homographies and intrinsic matrices. The result, after refinement by iterative optimization, is shown in Figure 1.3 (top-right).

1.7 Summary

Going from uncalibrated, raw camera images to fully calibrated, meaningful, measured data is an essential prerequisite for many subsequent image analysis tasks and visual computing applications. It must be noted, however, that the traditional image calibration pipeline described here, and in fact most computer vision algorithms, make implicit assumptions about the image acquisition process that hold only approximately for real-world photos and videos: For example, computer vision algorithms frequently assume that the *image exposure time* is indefinitely short. Physically, of course, any photo or video frame has been exposed for a finite period of time, potentially giving rise to motion blur [Sellent et al. 11]. Similarly, the *rolling shutter effect* of CMOS sensors and *streaking* of CCD sensors is frequently ignored. *Depth-of-field* caused by the, necessarily finite, lens aperture is also regularly not accounted for, nor is the fact that the widely used sRGB color signal constitutes a non-linear color model. Research into how to process images taken with commodity cameras to obtain physically meaningful measurement data has only just begun.

2

Stereo and Multi-View Video

Laurent Lucas, Céline Loscos, Philippe Bekaert, and Adrian Hilton

2.1 Introduction

The problem of multi-view acquisition relates to the capture of synchronized video data representing different viewpoints of a single scene. In contrast to video surveillance systems that aim to cover a large area with multiple cameras, the purpose here is to cover a single, often fairly restricted, physical space from multiple perpectives and to use the footage for 3D scene reconstruction to facilitate free-viewpoint rendering.

Depending on the final application, the number, layout, and settings of cameras can vary considerably. The most common configurations include lateral or directional camera arrays vs. global or omnidirectional multi-view cameras. In the first case, these devices provide multiple views, e.g., 2-views for binocular systems [Dubois 01, Peinsipp-Byma et al. 09] from close-together viewpoints, placed on the same side of a scene. They produce media adapted to stereoscopic displays. With regularly spaced cameras, autostereoscopic displays can be driven. In contrast, in omnidirectional systems multiple cameras are deployed around the target space. They are mainly designed for performance capture and free viewpoint applications. Finally, wide-baseline setups of multiple synchronized video cameras facilitate video-based, free-viewpoint rendering [Magnor 05].

Besides these purely video-based solutions, hybrid systems adding depth sensors to video footage are also currently being explored (Chapter 4). So far and regardless of the chosen capture device, all systems share the need to synchronize and calibrate (often even geometric and/or colorimetric corrections) information captured by different cameras, either online or as a post-process.

2.2 Multi-Video Capture Geometry

Configurations Overview

The generation of planar multi-stereoscopic views requires capturing with a synchronized multi-view camera [Perlin et al. 00, Dodgson 02]. Because such systems are intended to follow the natural binocular depth cue, it is not surprising that their design shows a striking similarity to the human eye. For example, the interaxial distance between the optical centers of the left- and right-eye camera lenses is usually chosen in relation to the interocular distance [Hill 53]. Furthermore, similar to the convergence capability of the eyes, it must be possible to adapt a camera to a desired convergence condition or zero parallax setting, i.e., to choose the 3D scene part that is going to be reproduced exactly on the display screen. This can be achieved by different camera configurations:

- "Toed-in" setup: with this setup the cameras are rotated inwards so that their optical axes converge at a single point. Since images have different trapezoid distortions, it is necessary to apply a systematic correction to enable the perception of 3D [Son et al. 07].
- "Parallel" setup: with this approach the cameras are parallel and the optical axes of the cameras converge at infinity. Without any post-processing shifting of the images to correct the zero parallax distance, objects at infinity will be cast at the surface of the display, and all other images will be cast in front of the display [Yamanoue 06]. Nothing appears behind the screen surface (no positive screen parallax).
- "Decentered parallel" setup: with this setup the cameras are again parallel. However, the images are shifted either in post-production, or in the camera by shifting two imaging sensors behind the lenses [Dodgson 02]. The shift of the images has the effect of changing the zero parallax distance (sometimes called the convergence distance).

Geometric principles described below are based on the latest model of camera. A more detailed description can be found in [Prévoteau et al. 10].

Viewing Geometry

The analysis of the characteristics of viewing geometry relies on a global reference frame defined in relation to the display device r centered in $\mathbf{o_v}$ (Figure 2.1(a)). The 3D display system mixes $n \times m$ images within its ROI (region of interest) with the dimensions L (width) and H (height). Each of these images (denoted by $\mathbf{I} = (i_1, i_2) \in \mathbb{N}_n \times \mathbb{N}_m$) is presumed to be

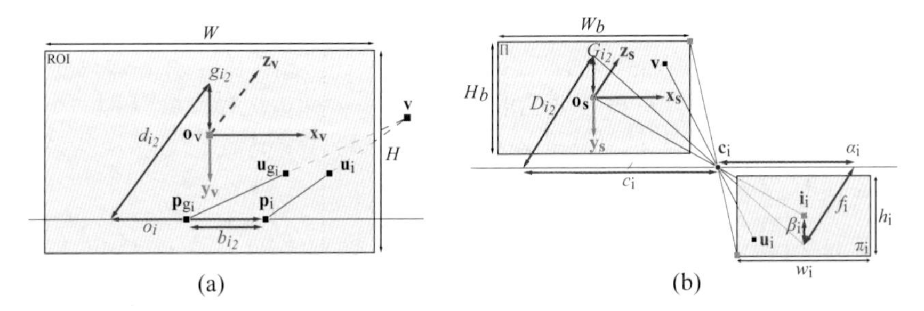

Figure 2.1: Characterization of geometry of: (a) the 3D multiscopic display using co-localized planar mixing; (b) the decentered parallel shooting setup.

correctly visible, at least from the preferred selected position $\mathbf{p}_i$. These positions are arranged on m lines parallel to the ROI lines situated at a distance of d_{i_2} from the system's region of interest.

Preferential positions are placed on these lines to ensure that the viewer, whose binocular gap is b_{i_2}, with the eyes parallel to the lines on the display, will have his right eye in $\mathbf{p}_i$ and his left eye in $\mathbf{p}_{g_i}$. The right eye in $\mathbf{p}_i$ will see image number $\mathbf{I}$ while the left eye $\mathbf{p}_{g_i}$ will see image number $\mathbf{I}_{g_i}$, knowing that $\mathbf{I}_{g_i} = \mathbf{I} - (q_{i2}, 0)$ where q_{i_2} represents the gap between image numbers composing coherent stereoscopic couples which are visible with a binocular gap of b_{i_2} at a distance of d_{i_2}. Combining the preferential positions of both the left and right eyes yields $\mathbf{p}_i = \mathbf{p}_{g_i} + b_{i_2}\mathbf{x}_v$ and $o_i = o_{g_i} + b_{i_2}$. The preferential position lines are placed on the vertical axis by g_{i_2} which represents the drop, i.e., the vertical gap between line i_2 of preferential positions and the center $\mathbf{o}_v$ of the ROI. When $m = 1$, the device does not create any vertical separation and any drop is acceptable a priori.

Supposing that the pixels $\mathbf{u}_i$ and $\mathbf{u}_{g_i}$ are stereoscopic homologues in the images $\mathbf{I}$ and $\mathbf{I}_{g_i}$, their perception by the right and left eye in $\mathbf{p}_i$ and $\mathbf{p}_{g_i}$, leads the viewer's brain to perceive the point $\mathbf{v}$ in 3D by stereopsis.

Shooting Geometry

The analysis of shooting geometry relies on the global shooting reference frame R, which is centered at the desired point of convergence $\mathbf{o}_s$ (which is also the center of the shared base Π in the scene) and is directed so that the first referential vectors are co-linear to the axes of the shared base Π in the scene and are therefore co-linear with the axes in the capture areas. In addition, the first axis is presumed to be parallel to the lines in the capture areas, and the second axis is parallel to the columns in these areas. The size of the shared base Π has the dimensions W_b and H_b. This reference frame defines the position and direction of all the projection pyramids representing the capture areas by specifying the direction of observation $\mathbf{z}_s$ and the m alignment lines of the optical centers. In line with these principles, the $n \times m$ shooting pyramids are specified by:

- optical axes in the direction $\mathbf{z_s}$;
- the optical centers $\mathbf{c_i}$ aligned on one or several (m) straight lines parallel to the lines in the shared base and therefore the direction $\mathbf{x_s}$;
- rectangular capture areas $\pi_\mathbf{i}$.

The capture areas must be orthogonal to $\mathbf{z}_s$ and therefore parallel to each other and the shared base Π as well as the straight lines holding the optical centers $\mathbf{c}_i$ (which are defined by their distances to $\mathbf{o}_s$, D_{i_2} in relation to $\mathbf{z}_s$, G_{i_2} in relation to $\mathbf{y}_s$, and $c_\mathbf{i}$ in relation to $\mathbf{x}_s$). These capture areas are placed at distances of $f_\mathbf{i}$ in relation to $\mathbf{z}_s$, $\beta_\mathbf{i}$ in relation to $\mathbf{y}_s$, and $\alpha_\mathbf{i}$ in relation to $\mathbf{x}_s$ from their respective optical centers $\mathbf{c}_i$. Their dimensions are w_i and h_i. They are decentered in relation to their respective optical axes in the points $\mathbf{i}_i$ such that the straight lines $(\mathbf{i}_i\mathbf{c}_i)$ which define the target axes intersecting at the fixed point of convergence $\mathbf{o}_s$. The centers $\mathbf{c}_i$ and $\mathbf{c}_{g_i}$ must be on the same center line and spaced from B_i in relation to $\mathbf{x}_s$ $(\mathbf{c}_i = \mathbf{c}_{g_i} + B_i\mathbf{x}_s$ and $c_i = c_{g_i} + B_i)$.

This kind of shooting configuration provides a depth perception on a multiscopic system with co-localized planar mixing with the possibility of a protruding as well as hollow image effect. However, this does not ensure that the perceived scene will not be distorted in relation to the original scene. The absence of distortion implies that the viewing pyramids are perfect homologues of the shooting pyramids (structures formed from Π with $\mathbf{c}_i$ as origins), i.e., they have exactly the same opening and deviation angles in both horizontal and vertical directions. Any deviation to this shooting and viewing pyramids homology involves a potentially complex distortion of the 3D image perceived in relation to the captured scene.

Distortion Analysis

According to the previous analyses of viewing and shooting geometries, it is possible to connect the coordinates $(\mathbf{x}_s, \mathbf{y}_s, \mathbf{z}_s)$, in the reference frame R, from the point $\mathbf{v}$ in the scene captured by the previously identified cameras with the coordinates $(x_1, y_\mathbf{i}, z_i)$ in the reference frame r of its homologue v_i perceived by an observer of the display device, placed in a preferential position (the right eye in $\mathbf{o}_i$).

The relation between the 3D coordinates of the scene point and those of its images perceived by a viewer may be characterized under homogeneous coordinates by:

$$a_i \begin{bmatrix} x_i \\ y_i \\ z_i \\ 1 \end{bmatrix} = \begin{bmatrix} k_{i_2} \left| \begin{matrix} \mu_i & & \gamma_i \\ & \rho\mu_i & \delta_i \\ & & 1 \end{matrix} \right| & \begin{matrix} 0 \\ 0 \\ 0 \end{matrix} \\ \begin{matrix} 0 & 0 & \frac{k_{i_2}(\epsilon_i - 1)}{d_{i_2}} \end{matrix} & \epsilon_i \end{bmatrix} * \begin{bmatrix} X \\ Y \\ Z \\ 1 \end{bmatrix}$$

The above equation can be seen as the analytic distortion model for observer position i which matches the stereoscopic transformation matrix given in [Jones et al. 01]. As such this model clearly exhibits the whole set of distortions to be expected in any multiscopic 3D experience, whatever the number of views implied or the very nature of these images (real or virtual). It shows, too, that these distortions are somehow independent from one another and may vary for each observer position i.

The above model exhibits some new parameters quantifying independent distortion effects. Those parameters may be analytically expressed from geometrical parameters of both shooting and rendering multiscopic devices. Their relations to geometrical parameters in and impact on distortion effects are as follows:

- $k_{i_2} = d_{i_2}/D_{i_2}$, control the global enlarging factor,
- $\epsilon_i = (b_{i_2}/B_i) \times (W_b/W)$, control the potential non-linear distortion which transforms a cube into a pyramid trunk according to the global scale factor $a_i = \epsilon_i + k_{i_2}(\epsilon_i - 1)\frac{Z}{d_{i_2}}$ possibly varying along Z,
- $\mu_i = b_{i_2}/(k_{i_2}B_i)$, control width over depth relative enlarging rate(s), or horizontal/depth anamorphose factor,
- $\rho = (W_b/H_b) \times (H/W)$, control height over width relative enlarging rate(s), or vertical/horizontal anamorphose factor,
- $\gamma_i = (c_i b_{i_2} - e_i B_i)/(d_{i_2}B_i)$, control the horizontal "shear" rate of the perceived depth effect,
- $\delta_i = (p_{i_2}B_i - G_{i_2}b_{i_2}\rho)/(d_{i_2}B_i)$, control(s) the vertical "shear" rate(s) of the perceived depth effect.

The previously detailed analysis of this model and its further inversion offers a multiscopic shooting layout design scheme acting from freely chosen distortion effects and for any specified multiscopic rendering device. Moreover, this model makes it possible to quantify those distortions for any couple of shooting and viewing settings by simple calculus based upon their geometric parameters.

2.3 Calibration and Synchronization of Cameras

In order to be able to relate imagery from multiple cameras with each other with the goal of 3D stereo viewing, view interpolation, 3D reconstruction, or other purposes, the cameras need to be synchronized and calibrated, both geometrically and photometrically.

A multi-camera system is geometrically calibrated as soon as all component cameras are calibrated. In principle, the methods explained in Section 1.6 suffice to calibrate also a multi-camera setup. However, correspondences between images of the different cameras, taken at the same time instance, can and shall be exploited for faster and better result. In this section, it is explained how image correspondences can be exploited for intrinsic and extrinsic calibration. Practical recommendations are given. The section concludes with suggestions concerning colorimetric calibration and synchronization.

Correspondences between images are typically provided automatically by feature point detectors and feature point description matching, such as SIFT. Feature detection, description, and matching are explained at an introductory level in text books such as [Szeliski 11], or, more in-depth in [Tuytelaers and Mikolajczyk 07].

Intrinsic Multi-Camera Auto-Calibration

In Section 1.6, the key to intrinsic camera calibration is to exploit certain scene features, such as orthogonality of 3D world lines and/or planes as observed in an image. These observations lead to constraints on a particular symmetric 3×3 matrix $\omega = \left(\mathbf{K}\mathbf{K}^{\top}\right)^{-1}$, called the *image of the absolute conic (IAC)*. The intrinsic camera parameters are obtained by decomposing ω using Cholesky factorization and Eq.(1.10).

The images of the absolute conic, by different cameras, at different positions and in different orientations, are related with each other. Exploiting these relationships allows to calculate the intrinsic (and extrinsic) parameters of the cameras in certain cases without the need to capture images of a calibration target, and even under completely uncontrolled circumstances.

It was explained in Section 1.6 that the vanishing points v_1 and v_2 in the image by a camera with intrinsic matrix $\mathbf{K}$, of world space points at infinity (direction vectors) D_1 and D_2, are related by $\mathrm{v}_2^{\top}\omega\mathrm{v}_1 = 0$, with $\omega = \mathbf{K}^{-\top}\mathbf{K}^{-1}$ the image of the absolute conic.

Now consider a second camera, with potentially different intrinsic matrix $\mathbf{K}'$, orientation $\mathbf{R}'$ and location C'. The vanishing points v_1' and v_2' in the image by this second camera, of the same world points at infinity D_1 and D_2, satisfies $\mathrm{v}_2'^{\top}\omega'\mathrm{v}_1' = 0$. $\omega' = \mathbf{K}'^{-\top}\mathbf{K}'^{-1}$ is the IAC of the second camera.

The vanishing points by the two cameras are related by a homography $\mathbf{H}_\infty$, called the *infinity homography* of the second camera with respect to the first:

$$\mathrm{v}' = \mathbf{K}'\mathbf{R}'^\top \mathrm{d} = \mathbf{H}_\infty \mathrm{v} \quad \text{with} \quad \mathbf{H}_\infty = \mathbf{K}'\mathbf{R}'^\top \mathbf{R}\mathbf{K}^{-1}. \tag{2.1}$$

Note that camera location doesn't matter in imaging points at infinity.

The infinity homography transfers ${\mathrm{v}'_2}^\top \omega' \mathrm{v}'_1 = 0$ to a new relation between the vanishing points in the first view:

$${\mathrm{v}'_2}^\top \omega' \mathrm{v}'_1 = 0 \quad \Longrightarrow \quad \mathrm{v}_2^\top \mathbf{H}_\infty^\top \omega' \mathbf{H}_\infty \mathrm{v}_1 = 0$$

Since this relation holds simultaneously with $\mathrm{v}_2^\top \omega \mathrm{v}_1 = 0$, for any pair of vanishing points corresponding to perpendicular world space directions, $\mathbf{H}_\infty^\top \omega' \mathbf{H}_\infty$ must model the same conic as ω:

$$\omega \doteq \mathbf{H}_\infty^\top \omega' \mathbf{H}_\infty. \tag{2.2}$$

Each component of ω is linearly related to any component of ω' by above equation, up to a common scale factor. These relationships allow to transfer constraints on either one to the other.

The challenge, in practice, is to automatically determine the infinity homography $\mathbf{H}_\infty$ from image point feature correspondences in the presence of parallax, instead of from vanishing points. This is the basis of a multitude of auto-calibration methods, for rotating cameras, planar camera motion, turntable motion, moving stereo rigs, video sequences captured with a handheld video camera with zooming lens, and so on. [Hartley and Zisserman 03, §19].

Extrinsic Multi-Camera Calibration: Epipolar and Trifocal Constraints

Extrinsic calibration consists of locating camera optical centers and estimating the camera orientation. Methods for single-camera extrinsic (and full) camera calibration have been presented in Section 1.6. The key to extrinsic multi-camera calibration is to exploit epipolar and trifocal geometry.

Consider two cameras, seeing a world space point X at normalized image coordinates (camera space ray directions) $\hat{\mathrm{x}}$ respectively $\hat{\mathrm{x}}'$. Let the first camera define the world coordinate system (optical center at the origin 0, orientation matrix $\mathbf{I}$ is unity). Assume the second camera is at position C and in orientation $\mathbf{R}$, relative to the first. Non-coincident camera centers 0 and C form together with a non-collinear world space point X a plane, named an *epipolar plane*. The cross product of $\mathrm{X} - \mathrm{C} \doteq \mathbf{R}\hat{\mathrm{x}}'$ and $\mathrm{X} - 0 \doteq \hat{\mathrm{x}}$ is perpendicular to the *epipolar axis* $\mathrm{C} - 0$, so

$$(\mathbf{R}\hat{\mathrm{x}}' \times \hat{\mathrm{x}}) \cdot \mathrm{C} = 0 \quad \Longleftrightarrow$$

$$\hat{\mathrm{x}}^\top \mathbf{E}^\top \hat{\mathrm{x}}' = 0 \quad \text{with} \quad \mathbf{E}^\top = [\mathrm{C}]_\times \mathbf{R} \quad \text{with} \quad [\mathrm{C}]_\times = \begin{bmatrix} 0 & -C_Z & C_Y \\ C_Z & 0 & -C_X \\ -C_Y & C_X & 0 \end{bmatrix} \tag{2.3}$$

The matrix $\mathbf{E}^\top$ is called the *essential matrix* of the first camera with respect to the second. The matrix $\mathbf{E}$, with $\hat{\mathrm{x}}'^\top \mathbf{E}\hat{\mathrm{x}} = 0$ for all corresponding $\hat{\mathrm{x}}$ and $\hat{\mathrm{x}}'$, is the essential matrix of the second with respect to the first. The essential matrix encodes the relative position and orientation of two cameras of any kind, with central projection model.

For rectilinear cameras, normalized image coordinates and image coordinates are related as $\hat{\mathrm{x}} = \mathbf{K}^{-1}\mathrm{x}$, and $\hat{\mathrm{x}}' = \mathbf{K}'^{-1}\mathrm{x}'$, so that:

$$\mathrm{x}'^\top \mathbf{F}\mathrm{x} = 0 \quad \text{with} \quad \mathbf{F} = \mathbf{K}'^{-\top}\mathbf{E}\mathbf{K}^{-1}. \tag{2.4}$$

The matrix $\mathbf{F}$ is called the *fundamental matrix* of the second camera with respect to the first one. This relation states the well-known fact that the image x' of X in the second camera, must lie along a line $\mathrm{x}'^\top \mathrm{l}' = 0$: the intersection line of the epipolar plane of X with the second image. The lines' parameters are $\mathrm{l}' = \mathbf{F}\mathrm{x}$. Such lines are called *epipolar lines.* The same is true on the first image: $\mathrm{l}^\top \mathrm{x} = 0$, with $\mathrm{l} = \mathbf{F}^\top \mathrm{x}'$. All epipolar lines meet at a single point in an image, named an *epipole.* The epipole is the projection of the others camera position in the first camera's view. It shows the direction of travel of a moving camera.

The fundamental matrix encodes the complete projective geometry of two rectilinear cameras. Similar relations exist for non-rectilinear cameras, but epipolar lines become curves as dictated by the lens image formation model.

Given corresponding normalized image coordinates $\hat{\mathrm{x}}$ and $\hat{\mathrm{x}}'$, Eq.(2.3) is a homogeneous linear equation in the unknown coefficients of the essential matrix. The same is true for image coordinate correspondences and the fundamental matrix (Eq.(2.4)). Each correspondence yields such an equation. The matrix can be computed by stacking these equations together into a homogeneous linear system, which is solved using SVD with proper preconditioning and singularity constraint enforcement. False correspondences reported by automatic image feature detection and matching algorithms are filtered by means of a procedure named *random sampling consensus (RANSAC).* The result is iteratively refined using the Levenberg–Marquardt non-linear optimization algorithm, and decomposed into relative camera position and orientation. These procedures are explained in detail in, for instance, [Hartley and Zisserman 03, §10 and 11].

It is important to realize that calibration, and 3D reconstruction, from epipolar geometry alone is fundamentally ambiguous by an arbitrary 3D homography. This can be understood as follows: if a camera matrix $\mathbf{A}$ images a space point X into x and $\mathbf{B}$ into x$'$, Then $\mathbf{AH}$ and $\mathbf{BH}$ image $\mathbf{H}^{-1}$X at exactly the same image locations x and x$'$. The fundamental matrices of any such pair of camera matrices are identical. Removing the projective ambiguity is possible only if cameras are also intrinsically calibrated, using for instance intrinsic multi-camera auto-calibration as outlined above.

Similar relations exist between three camera views, both in terms of image coordinates, and in terms of camera space ray directions (normalized image coordinates)

$$\sum_{i,j,k=1,2,3} \mathcal{T}_{ijk} \mathrm{x}_i \mathrm{x}'_j \mathrm{x}''_k = 0. \tag{2.5}$$

The $3\times3\times3$ table of numbers $\mathcal{T}$ is called a *tri-focal tensor*. There are 27 such relations for each triple-point correspondence, of which 4 are independent. Trifocal tensor computation is more involved than essential or fundamental matrix computation [Hartley and Zisserman 03, §16], but the resulting constraints are stronger: the locations of the image coordinates need to match exactly. With two-view constraints, image coordinates still have a 1D degree of freedom of moving along their epipolar line. For this reason, a stronger filtering of image correspondences results, as well as a more stable and accurate calibration and 3D reconstruction.

In practice, camera calibration can be obtained as a side result of 3D reconstruction software packages such as bundler[1] or VisualSFM.[2] These packages implement the principles outlined here, and refine calibration together with 3D position of image feature correspondences in one big nonlinear optimization procedure called *bundle adjustment*.

Multi-Camera Matrix Factorization

A final noteworthy calibration principle is multi-camera matrix factorization. Consider a set of n world space points X_i, $i = 1 \dots n$, which are all visible in a set of m cameras with full camera matrices $\mathbf{P}^j$, $j = 1 \dots m$. The image of the i-th point by the j-th camera is $\mathrm{x}_i^j = \mathbf{P}^j \mathrm{X}_i$. Making the scale factors explicit, the matrix of image points can be written as:

$$\mathbf{W} = \begin{bmatrix} k_1^1 \mathrm{x}_1^1 & k_2^1 \mathrm{x}_2^1 & \cdots & k_n^1 \mathrm{x}_n^1 \\ k_1^2 \mathrm{x}_1^2 & k_2^2 \mathrm{x}_2^2 & \cdots & k_n^2 \mathrm{x}_n^2 \\ \vdots & \vdots & & \vdots \\ k_1^m \mathrm{x}_1^m & k_2^m \mathrm{x}_2^m & \cdots & k_n^m \mathrm{x}_n^m \end{bmatrix} = \begin{bmatrix} \mathbf{P}^1 \\ \mathbf{P}^2 \\ \vdots \\ \mathbf{P}^m \end{bmatrix} \left[\mathrm{X}_1 | \mathrm{X}_2 | \cdots | \mathrm{X}_n\right].$$

[1] http://www.cs.cornell.edu/~snavely/bundler/
[2] http://ccwu.me/vsfm/

W is the product of a $3m \times 4$ matrix and a $4 \times n$ matrix, so it has rank 4. The $3m \times n$ matrix **W** can be decomposed in such product by means of SVD. The key to make this idea work in practice is in calculating the unknown scale factors on the fly from epipolar constraints [Sturm and Triggs 96].

Svoboda et al. used this as the basis for building a popular, complete and fully automatic multi-camera calibration MATLAB package [Svoboda et al. 05]. Metric calibration is obtained by exploiting multi-camera intrinsic constraints as explained in Section 2.3. Multi-image point correspondences may be generated by waving a laser pointer.

Example: Calibration of a Multi-Camera Studio for Human Motion Capture

Figure 2.2 illustrates the calibration of a multi-camera capture studio at the British Broadcasting Corporation BBC.[3] The studio contains twelve to fourteen broadcast TV cameras, capturing human motion from viewpoints all around. The cameras are genlocked, guaranteeing that they capture at exactly the same rate. The phase at which they sample is adjusted to be equal by observing a set of pulsed LEDs.

Broadcast cameras are photometrically calibrated by a white and black balance. For that they are pointed to a white sheet, and their "white balance" function activated. Next, their iris is closed completely, and the "black balance" function activated. This is a standard line-up procedure in TV broadcasting and provides pretty good results especially if cameras are of the same type. One potential problem specific to systems with cameras viewing a scene from a wide range of angles is that if the lighting is not uniform, and the white sheet used is not totally matte, then it might look slightly different when viewed from cameras in very different positions due to light reflections. In this case, the sheet may need to be moved or re-angled when white-balancing different cameras.

The cameras are calibrated geometrically using LED pole and multi-camera calibration tools developed by BBC,[4] based on bundle adjustment. Image correspondences are established by identifying images of a moving wand with pulsed LEDs mounted on a pole through the working space. The LEDs at the end have a different color, so observed LEDs are still suited for analysis even if the whole wand is not in view. This is particularly useful for more tightly zoomed-in cameras, like the bottom right example showing an actor's face in Figure 2.2. The top-right image in Figure 2.2 shows the LEDs observed by a camera. The bottom-right image in Figure 2.2 shows calibrated camera positions and orientations, as well as the position and orientation of the wand at analyzed instances.

---

[3] http://react-project.eu

[4] Developed by Julien Pansiot and Oliver Grau.

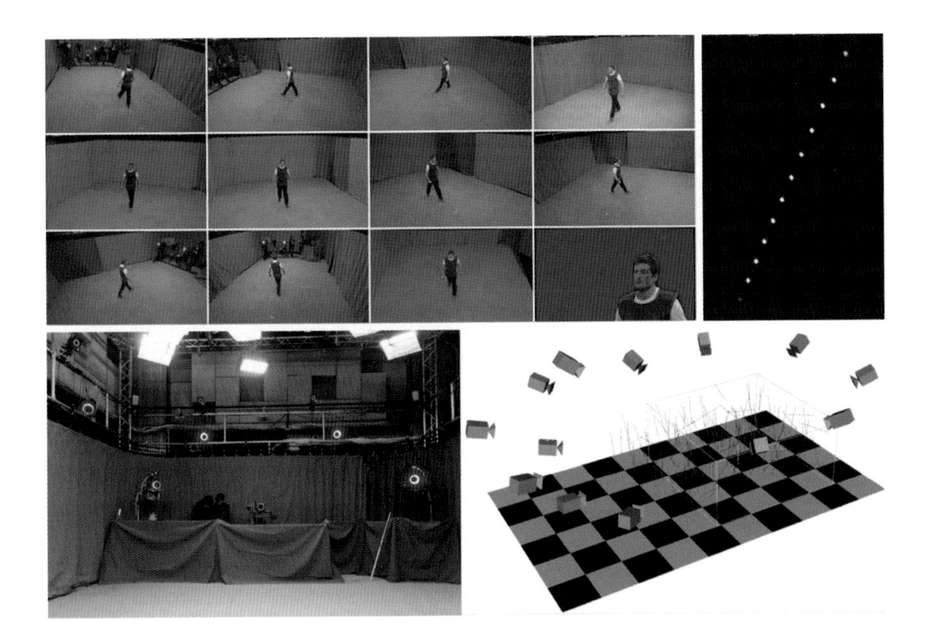

Figure 2.2: Multi-camera human capture studio calibration at BBC: top-left: views provided by the cameras of the setup; bottom-left: a view of the studio; top-right: camera view of the calibration LED wand; bottom-right: visualization of calibrated camera and wand positions and orientations.

Calibration in Practice

It pays off to spend effort in obtaining accurate calibration, to sub-pixel accuracy for 3D reconstruction and similar applications. Unfortunately, this is non-trivial even in controlled environments, and off-the-shelf software and approaches aiming for fully automatic calibration will often fail to provide the required accuracy if used as a "black box."

One problem is that several calibration parameters, such as position and orientation, or lens focal length and distortion parameters, are pseudo-dependent: small variations in different parameters cause almost identical effects and errors in one parameter but may be countered by errors in others. It may be tricky to decouple them when fitting everything together. Calibration procedures that proceed in steps, each addressing a set of independent parameters, can avoid this issue.

Many practitioners calibrate cameras using a mix-and-match approach: they combine the techniques (and software) presented in this text in ad-hoc ways, often in part using their own implementations which are adapted to own-designed calibration grids and wands, camera setups, and so on. For instance, BBC have developed a wand with several pulsed LEDs on a line, as a more accurate alternative to laser pointer waving in Svobodas approach. In the example on page 20, a TV flat screen is used for producing sequences of calibration patterns producing "labeled" rather than "anonymous" calibration features.

It is beneficial to exploit all knowledge one can have about the setup. For instance, focal length may be known from fixed focal lens specifications; relative camera positions can be measured to a few centimeters' precision using a simple ruler in a fixed setup, or pan-tilt angles tracked with a MEMS sensor.

As usual in data modeling problems such as camera calibration, in addition to computed values, an indication of reliability also needs to be provided in the form of covariance estimates or confidence limits [Press et al. 07, §14].

Expressing camera calibration parameter values in human-understandable units helps to assess the plausibility of the result. If lens focal length, expressed in millimeters, is way off the lens indicator reading, for instance, or positions strongly disagree with what a ruler measures, there is most probably something seriously wrong with the calibration. The obtained results may explain world and image coordinates well for the particular calibration dataset, but there is no guarantee that the mapping will also be right for whatever other scenery.

Photometric Calibration

Many algorithms for processing multi-camera imagery assume color constancy, that is: an object seen by different cameras is observed with the same color. Obviously, this will only be the case for diffuse ("Lambertian") reflective objects, but even in the presence of non-diffuse objects, it is beneficial to make sure that colors from different cameras match each other as closely as possible.

This will be the case if the cameras are individually color calibrated and characterized as explained in Section 1.5. In practice, this means that a correct ICC profile is available and being used for each camera, and cameras are operating with equal settings, in particular with respect to white balancing.

If using color-calibrated cameras, or if cameras in a multi-camera setup are all of the same make and model, and operate with equal settings, good color consistency is often achieved using very simple means: a per-camera gain, compensating for slightly different lens apertures mainly.

In general, it pays off to color calibrate multi-camera setups using a Gretag–MacBeth color chart before actual shooting, even for calibrated cameras. The color chart allows to white balance the recording for the actual lighting being used, and calibrate consistently under that lighting.

For machine vision cameras, red and blue gain factors are set such that the neutral color patches of the chart appear as grayish as possible. Subsequently, the camera response on the neutral patches is inspected and a luminance curve calculated, linearizing response. Finally, a per-camera RGB color matrix is computed by non-linear optimization that equalizes the cameras response to the non-neutral color patches of the chart.

For small baseline setups, color matching by equalizing the cumulative histogram in the red, green, and blue channels is a fast and practical approach to ensuring color consistency between the cameras.

Even with the most sophisticated color equalization algorithms, however, there is no guarantee that all surfaces being captured will exhibit equal colors. Most materials in reality are not perfectly Lambertian, and colors will be different depending on illumination and view angle. Lens flares, due to interreflections of incident light inside camera lenses, is another major cause of color inconsistencies and local loss of contrast in practice. This is particularly noticeable with sunlight in outdoor situations, but also with studio spot lights.

Camera Synchronization

No matter how well cameras are calibrated for a static scene, geometrically and photometrically, without synchronization no sensible results will be obtained for moving scenes or moving cameras.

With the broadcast style of cameras, a genlock device will ensure that captured frames are transferred at the same pace for all cameras. This does not mean that cameras take pictures at the very same moment: there will in general still be capture phase differences between cameras. Broadcast cameras usually allow control of that phase. The phase can be calibrated by observing an old-style CRT monitor, or a set of fast flashing LEDs.

The machine vision type of cameras typically have an external trigger input, that allows control of the very moment that the camera shutter is opened. Driving it can be as simple as connecting these GPIO (general purpose IO) inputs to the lanes of a parallel port computer output, and sending byte codes to that parallel port. Or the GPIO exposure output signal of a master camera is hard wired to the external trigger input pins on the other cameras to slave cameras to the master, which is left running free at a desired frame rate.

This ensures that frames are captured simultaneously; however, different models of cameras will in general deliver frames with different delays.

Further delays are possible due to driver software and operating system issues on the capturing computer(s). In order to keep corresponding frames together, frames shall be tagged on the cameras by a time stamp or frame count in some way.

Broadcast cameras allow to embed SMPTE time code stamps with captured frames. SMPTE time code is received from a time code master clock.

Machine vision cameras allow to embed a camera clock and frame count. Camera clock drift will after a while cause different time stamps even for simultaneously captured frames. There may be one frame time difference as soon as after half an hour. Camera time stamps therefore must be synchronized by correlating with computer time or another external time source. Cameras need to be started appropriately to ensure consistent time stamp and frame count origin. Machine vision camera SDK documentation typically explains how to do this. A more powerful way to ensure consistent tagging of simultaneously captured frames from machine vision cameras is to feed additional IO signals and embed their state as frame metadata.

With consumer cameras, such as handycams, action cameras, and DSLR cameras, none of this is possible. Figuring out what frames are approximately captured at the same time is possible by inspecting the audio signal. Even with very cheap cameras, sound and image are usually recording synchronously to a sufficient extent. The audio signal will in general also allow to measure the time difference between different cameras shuttering for cameras of the same make and same firmware version. If audio and video are captured independently of each other, a clapper board and/or flash light or other sudden and fast movement or moving object will help to identify corresponding frames from different cameras. However, always keep in mind that, even when consumer cameras are started exactly at the same moment, using for instance their LANC input, there is no guarantee that their shutters will open and close simultaneously.

2.4 3D Cameras in Practice

A broad range of cameras is being used in 3D production. In case of binocular systems, capturing requires generally the use of two cameras connected by a rigid or articulated mechanical device known as a stereoscopic rig. Their optical design can be subdivided into two groups, multi-lens *vs.* single-lens systems (Figure 2.3). In both groups there is an even further distinction between single sensor and multiple sensor subclasses.

Single lens with single sensor setups follow Wheatstone's principle except that the camera is fixed to a slide bar. One image is taken at one point, the camera is moved along the bar to a fixed separation, then the second

Figure 2.3: Examples of 3D cameras and other devices: (a) STEREOTEC™side-by-side live rig; (b) 3DTV™Octocam camera array; (c) Panasonic™Lumix G 3D lens; (d) Panasonic™HDC-Z10000 stereoscopic twin-lens camcorder.

image is taken. Such a design allows for clear alignment and matching is much easier. However, they operate only for stationary scenes.

Conversely, single lens with multi-sensor setups involve problems of synchronization between sensors. Another drawback is cross talk between images since the resolution of the images is less than half of the sensor resolution. They also require adding components that need to be aligned and rely on minimal lens aberrations on the boundaries of the lens surface.

Multi-lens with single sensor setups combine views in a single sensor (Figure 2.3(c)). Merging all of the sensors makes it easier for matching and synchronization even if it does make it harder for the system to be versatile.

Finally, multi-lens with multi-sensor setups are probably the most common and most studied category of stereo cameras as they largely consist of two or more separate cameras in various orientations (Figure 2.3(a), (b) and (d)). Thus, cameras can be mounted at 90 degrees, looking through a half-mirror angled at 45 degrees or more simply, side by side.

2.5 Summary

Calibrating the setup of multiple cameras is a necessary prerequisite if image content is later to be inter-related across cameras and jointly processed. It can constitute a tedious and time-consuming procedure, especially when recording large scene volumes, e.g., outdoors. If small baselines suffice, e.g., for stereoscopic capture, rigidly mounted stereo or multi-view camera rigs can be used that are either already pre-calibrated or need to be calibrated only once.

3
Omni-Directional Video

Peter Eisert and Philippe Bekaert

3.1 Introduction

Omni-directional capture and display of scenes have a long tradition. Already in the 18th century, the English painter Robert Barker created large accessible panoramas showing a real scene from a particular viewing position, providing an immersive feeling of being there. The step from static scenes to omni-directional capturing of dynamic scenes was showcased at the World Fair of 1900 in Paris using 10 mechanical film cameras and projectors. During the last century, several commercial panoramic video systems followed, e.g., Disney's Circle Vision 360 or IMAX movie theaters. The advent of high quality digital cameras and projectors simplified the handling and processing of omni-directional video, i.e., the synchronization, calibration, warping, and blending, and made it widely applicable. Nowadays, there exists a large number of different high-resolution, multi-projection systems (Figure 3.1), that are used for immersive omni-directional cinema or presentation of panoramic multimedia events [Fehn et al. 07].

Besides the pure omni-directional reproduction of real-world scenes for entertainment purposes, many new applications have emerged. Today, omni-directional video displays are being used in simulators for professional training [Foote et al. 04], in modern digital planetariums for edutainment purposes [Lantz 07], as digital backlot for the cost-efficient representation of background scenes in movie productions [Schreer et al. 13] (Chapter 23), or they serve as an intuitive interface for large video collections [Tompkin et al. 13]. Many of these applications require synchronized high-resolution, omni-directional capturing of real-world scenes.

The simplest way of capturing omni-directional video is to use a single camera with a fisheye lens or a catadioptric system to image the scene hemispherically onto a single image sensor [Nayar 97, Krishnan and Nayar 08]. Although synchronization and stitching problems are avoided, acquisition resolution of such systems limits their use to mainly surveillance purposes. Instead, the focus here is on multi-camera systems that capture sub-parts of a scene in parallel. The multiple video streams are then stitched together again, resulting in a high-resolution panorama. The main

Figure 3.1: TimeLab: Panoramic 3D cinema with 7K resolution.

challenges in stitching and seamless blending of the individual parts consist of geometric and photometric corrections which are caused by non-exact focal point alignment of all cameras, differences in color response, view dependent scene appearance, and optical effects like lens flare. In addition, omni-directional scene capture typically requires high-dynamic range (HDR) technology (Section 1.5), because light sources as well as regions in shadow must be simultaneously imaged [Eden et al. 06].

In order to minimize geometrical corrections, the focal points of all cameras need to be as close together as possible. If the size of the cameras is small, a star-like configuration can be realized where all cameras face outward (Section 3.2). For larger cameras, or to avoid disparity correction, mirrors can be used that allow, at least theoretically, to place all focal points into the same position (Section 3.3). The creation of stereoscopic panoramic video, finally, faces the contradicting problem of providing large disparities for stereo viewing while requiring small disparities for stitching (Section 3.4).

3.2 Mirror-Free Panoramic Capture

In order to capture parallax-free omni-directional video with multiple cameras, ideally the cameras must be set up in such a way that their optical centers coincide. Unfortunately, this point typically lies somewhere within the lens objective, near the front of the lens for wide-angle lenses or more toward the back for lenses with long focal length. Consequently, it is physically impossible to arrange several cameras such that their optical centers coincide.

The only practical way to ensure parallax-free omni-directional video capture is by using a single camera lens, or using multiple cameras with mirrors that are arranged in such a way as to effectively mirror the optical centers of all cameras into one virtual focal point (Section 3.3). On the downside, mirrors cause a multi-camera rig to be larger, heavier, and more delicate than a bunch of closely spaced cameras looking outward to capture a complete view of the world around. Moreover, it is not possible to realize a full 360×180 degrees omni-directional video capture rig this way without having the cameras or part of the mirrors being visible in the image itself.

Mirror-less camera rigs and tools: A common practice in high-end media and movie production today is to capture using a ring-shaped rig of digital still-image or movie cameras equipped with wide angle lenses, and to stitch the images together using commercial video stitching software like kolor[1] or open-source panorama tools like PTGUI.[2] Although by far the most panoramic and 360 video camera rigs being used are still custom-built,[3] some omni-directional cameras are also commercially available, e.g., the Point Grey Ladybug cameras,[4] or rig mounts for the popular GoPro action camera (Figure 3.2).[5]

Basic stitching: A convenient, real-time approach to stitch together multiple images into a single panorama is by virtually projecting the camera imagery onto the surface of a cylinder or, more accurately, a sphere. The sphere is viewed from its center. The virtual projection loci correspond to the relative positions of the cameras in the rig. Orientation, field of view, lens distortion of the virtual projectors must all match that of the real cameras. The virtual projection can be implemented conveniently using projective texture mapping, e.g., using OpenGL. The same shaders allow to perform feathering with alpha values based on the distance to the edge of the camera images raised to a power allowing to control the sharpness of the blending. This produces perfect stitching at a single distance to the spectator only. Everything closer or further away will appear doubled to some extent. However, the distance can be controlled at will. This basic stitching approach already yields useful results in many practical cases, given judicious camera placement and parameter control (sharpness and distance).

Calibration: Before stitching can be performed, the relative orientation and position of the cameras in the rig as well as lens properties

[1] www.kolor.com
[2] www.ptgui.com
[3] http://www.radiantimages.com/blog/9-red-hot-chilli-peppers
[4] www.ptgrey.com
[5] www.360heros.com, www.freedom360.us

Figure 3.2: Mirror-less panoramic and omni-directional multi-camera rigs: 360-hero rig with GoPro cameras (top left), Point Grey Ladybug (top second from left), and other custom-built rigs (Hasselt University, Belgium).

(i.e., focal length, principal point, lens distortion; cf. Section 1.6) must be precisely calibrated. In principle, generic camera calibration methods and procedures can be used (Section 2.3). For improved results, additional a priori knowledge can be exploited. For example, the relative orientation and position of all cameras is known from rig design. Fine-mechanical constructions like camera rigs are typically realized to a precision of one tenth of a millimeter in position and one tenth of a degree in orientation. Also, typically fixed focal distance (prime) lenses are used in omni-directional and panoramic camera rigs whose focal length match lens specifications to one tenth of a millimeter. If the result of the calibration procedure disagrees by more than the afore-mentioned tolerances from the mechanical rig design or lens specifications, most probably the calibration procedure is overfitting or suffers from pseudo-dependencies between parameters, such as orientation and principal point or focal length and lens distortion. The most important unknown parameters in calibration are the location of the optical center within the camera-lens system, the principal point (image coordinate at which the optical axis of the lens intersects the camera sensor),

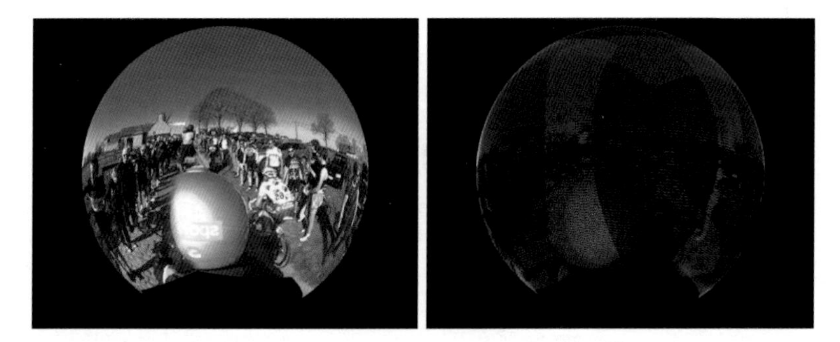

Figure 3.3: Basic stitching: using projective texturing, camera images are mapped onto the surface of a sphere. The virtual spectator is located at the center. The right image shows the projections as wireframes to visualize the composite of the stitched result at the left. This simple method produces exact results at one fixed scene distance only. Everything closer or farther away shows ghosting artifacts, e.g., the motorcyclist helmet.

and lens distortion parameters. Lens distortion can be accurately measured by observing straight lines or rectangular grid patterns (Section 1.6). The lenses' principal points can usually be determined accurately by shooting a distant scenery in which the distance between camera centers no longer causes noticeable parallax. Once geometric relationships are established, photometric properties like color reproduction and lens vignetting must be measured for each camera and adjusted (Section 2.3).

Advanced stitching: A good starting reference for studying warping and stitching for panoramic and omni-directional photography and video is [Szeliski 06]. It describes the basic theory as well as a full range of methods on how to align and register component imagery and to compose them properly. Ideally, a stitching algorithm transforms the set of captured images in such a way that the resulting composite is identical to a panorama image captured with an imagery camera rig with exactly coincident optical centers. To correct for local parallax in small-baseline camera rigs, one promising approach is to build upon stereo- and multiple view interpolation [Scharstein and Szeliski 02, Rogmans et al. 09]. By interpolating overlapping camera images novel views from coincident locations can be synthesized, resolving any parallax issues. In theory, the depth estimates resulting from such algorithms may also be used to generate stereoscopic views and allow for 3D reconstruction, just as with conventional stereo camera rigs (Section 17.3). In order to provide high effective image resolution, however, practical mirror-less panoramic and omni-directional multi-camera rigs feature only little overlap between camera views. Still,

Figure 3.4: As long as camera placement and algorithm parameters (sharpness and distance) are carefully selected, basic stitching already provides acceptable results in many practical cases, even in the presence of close-by scenery and large depth variations.

view interpolation helps to reduce parallax artifacts in the overlapping area [Szeliski 96, Kang et al. 04, Qi and Cooperstock 08, Adam et al. 09]. In practice, a remaining obstacle is the rather unpredictable and ungraceful failure characteristics of current view interpolation algorithms. There is steady progress in this area, however (Section 17.2), and it may become the preferred approach to omni-directional video capture in the long run.

Yet another approach consists of carving images along irregular seams [Ozawa et al. 12]. While this works well for still panoramas, optimal seam carving often results in objectionable flickering when applied to video.

3.3 Mirror-Based Panoramic Capture

For star-like camera arrangements as described in the previous section, the focal points of all cameras lie on a circle whose minimal radius is determined by the physical size of the cameras and lenses used. This inevitably leads to parallax effects that make stitching and blending of the overlapping areas difficult, especially for objects close to the cameras. In principle, these problems can be avoided by using a pyramidal mirror system [Tan et al. 04, Hua et al. 07, Weissig et al. 12] (Figure 3.5). By correctly placing

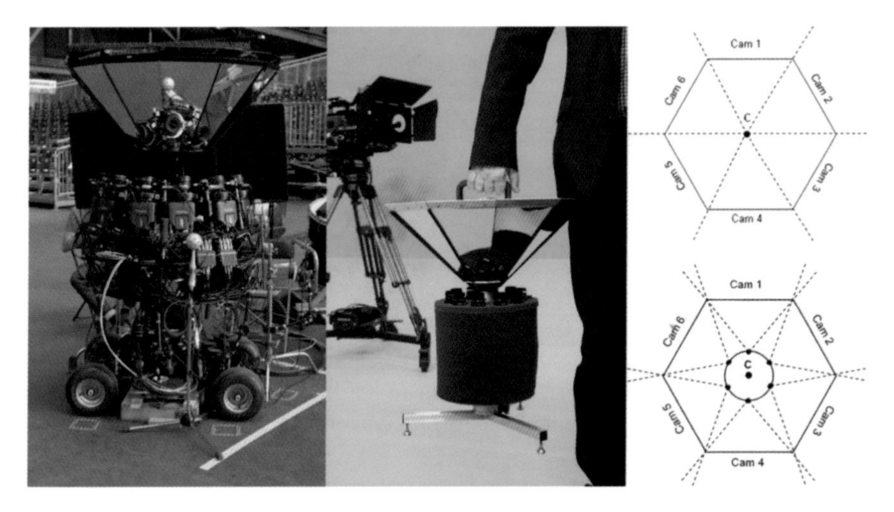

Figure 3.5: Prototype of a mirror-based omni-directional cameras for 3D panorama acquisition [Weissig et al. 12].

the cameras under a rig of differently oriented planar mirrors, the optical centers of all cameras can be made to virtually coincide at the same 3D position, independent of camera size. This allows for the use of high-end cameras, e.g., six ARRI Alexa digital cinema cameras in the OmniCam prototype (Figure 3.5 left). The system can record video panoramas at a resolution of 6K x 2K pixels and at 16 bits of dynamic range. A less bulky and mobile version consists of ten micro HD cameras capturing 360° panoramas and a vertical viewing range of 60° at 10K x 2K pixels (Figure 3.5, middle). For applications that require larger viewing angles, two mirror pyramids can be placed on top of each other [Tan et al. 04, Hua et al. 07]. By tilting the mirrors less than 45°, the focal points of upper and lower cameras can still be brought into alignment, at the price of not fully exploiting full sensor area anymore due to trapezoidal distortion.

While the mirrors allow for perfect alignment of all cameras' optical centers, the resulting individual images do not feature any overlapping area anymore which makes seamless stitching and blending of the individual tiles almost impossible. Sufficient overlap of adjacent images can be created by shifting the focal points of the cameras slightly outward (Figure 3.5 right). This off-center placement enables the adjustment of image overlap at the expense of introducing small image disparities that, however, can be corrected for as described in Section 3.2. The OmniCam prototype system (Figure 3.5), for example, employs a radial shift of 5 mm that results in an image overlap of about 10 pixels. It allows for parallax-free stitching of scenes with a depth range from about two meters to infinity.

Figure 3.6: Crosstalk at mirror borders can lead to ghosting artifacts. Small blinds between mirror segments help prevent crosstalk.

One practical problem of the pyramidal mirror setup is the introduction of small blurring artifacts at the image borders between mirror segments that become visible at high resolution. Along the mirror edges, due to the finite aperture of the camera lenses a camera captures reflections from both mirror segments (Figure 3.6). By integrating light over two mirror planes, object points from different directions may be projected onto the same pixel resulting in blurring or ghosting. This effect can be reduced by closing the aperture such that the mirror is in focus, but this would significantly reduce the light falling onto the sensor and make capturing in dark environments difficult. Instead, small black blinds can be added between the mirror segments to avoid crosstalk (Figure 3.6). The blinds mask the neighboring mirror segments from the view of each camera such that each camera integrates only over one mirror tile and the black blind, leading merely to some darkening of edge pixels. Since this darkening occurs in the overlapping area where the final image is additively blended from two segments, it remains unnoticeable if photometric calibration and correction is properly performed. With all processing steps diligently applied, the resulting panorama is indistinguishable from that obtained with a single-chip panorama camera (Figure 3.7).

3.4 Stereoscopic Panoramic Capture

In contrast to the 2D case, capturing stereoscopic panoramic video is much more challenging due to the conflicting requirements of little or no disparity for seamless stitching and large disparities for stereoscopic viewing experience (Chapter 18). A star-like camera arrangement does not work well for stereoscopic panorama capture (Figure 3.8 left). The unavoidable baseline

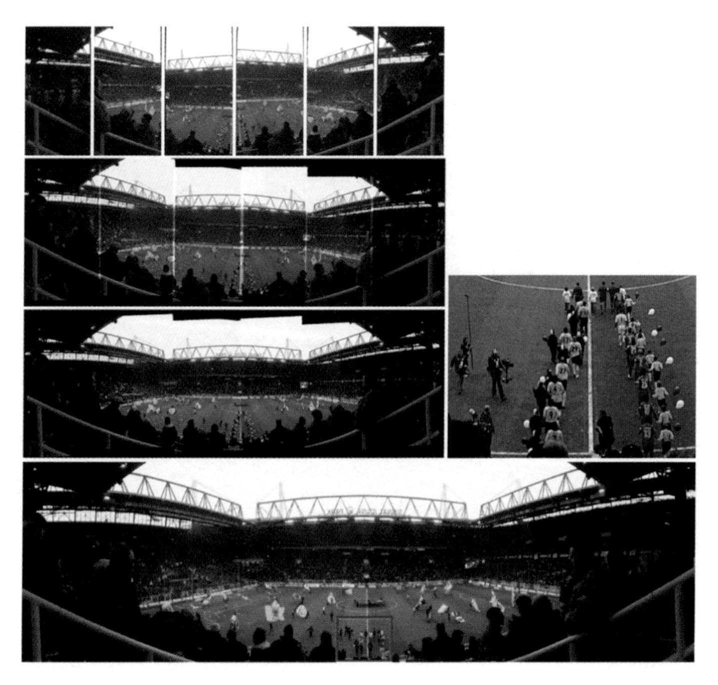

Figure 3.7: Subsequent processing steps necessary to create a panorama image using a multi-camera capture rig. From top to bottom: original camera views; geometric correction; photometric correction and blending; final cut-out of panorama with additional close-up views illustrating the high resolution [Weissig et al. 12].

B between adjoining cameras for the left and right view is smaller than the baseline S between neighboring camera pairs. This renders stitching very difficult and is likely to introduce annoyingly visible disparity artifacts between camera pairs. The problem becomes less severe when using a larger number of smaller cameras, but the unfavorable discrepancy between baselines B and S persists.

One common method to overcome this problem is to eccentrically rotate two slit cameras (or one regular area camera) and to concatenate image columns over time to stitch a panorama. Since the virtual camera positions are extremely close together, no disparity problems occur during stitching. This is, for example, exploited in omni-stereo or concentric mosaic capturing [Peleg et al. 99, Shum and He 99, Peleg et al. 01] and has been successfully employed in the creation of high-resolution stereo panoramas [Richardt et al. 13]. The disadvantage of the rotating camera setup is its inherent limitation to static scenes only. This problem has been

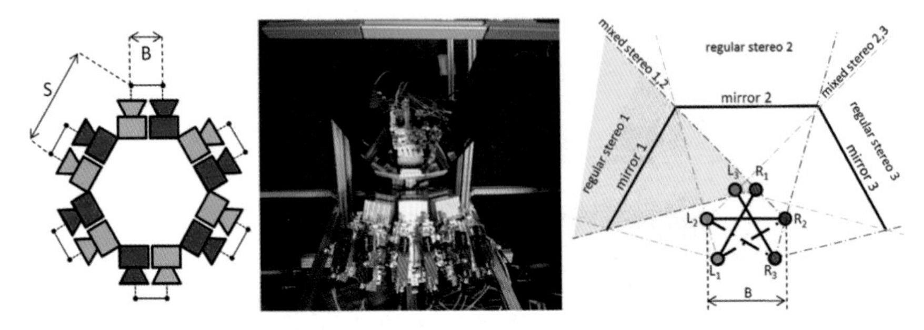

Figure 3.8: Left: Illustration of the conflict between desired stereo baseline B and unwanted parallax S between neighboring camera views for the mirror-free design. Middle: 3D omni-directional camera with 3 camera pairs for 180° capture. Right: Geometric design of a stereoscopic omni-directional camera.

addressed by [Belbachir et al. 12] who developed a high-speed rotating camera that can capture stereo panoramic video at 10 fps and 1800 pixels for 360° cylindrical panoramas. While the approach can, in principle, be extended to even higher frame rates and resolution, the resulting very short shutter times require extremely bright scenes.

Instead of sequentially scanning the scene, Peleg et al. used a sophisticated system of spiral lenses and parabolic beam splitters in order to simultaneously capture two panoramas [Peleg et al. 00]. A system relying on a less complex optical system is proposed in [Schreer et al. 12]. The 3D OmniCam shown in Figure 3.8 is based on a pyramidal mirror system with 3 camera pairs capturing 180° (or 6 pairs for full 360°). Similar to the 2D case, the mirrors position the focal points of all cameras along a small circle. The advantage of this approach compared to the star-like setup is that disparity for stereoscopic viewing comes from cameras on the circle opposite to each other while stitching is performed between the much closer neighboring (virtual) camera views (Figure 3.8 right). $L1$ and $R1$ are the virtual focal points for a pair of cameras looking through mirror segment 1. This design makes sure that distance $\overline{L_1R_1}$ is larger than $\overline{L_1L_2}$. The real cameras of baseline $B = 6$ cm are toed-in such that their optical axes cross on the mirror surface. This ensures maximal overlap of left and right views on a mirror segment. Still, in the sector indicated "mixed stereo 1 2" only one camera (L_1) captures the scene via mirror segment 1. For this region, the corresponding image data for the right view must be taken from R_2 looking through segment 2. As illustrated in Figure 3.9 (left), the final left and right panoramas are stitched together such that there are regions that originate from left and right cameras of the same mirror segment, or, for

Figure 3.9: Left: Illustration of the different regular and mixed stereo segments. Right: Simulation of left and right panorama stitching using a geometrical setup [Schreer et al. 12].

the mixed stereo case from neighboring mirror segments. Similar to the 2D setup (Section 3.3), the cameras' focal points are shifted slightly outward such that the baselines do not intersect the mirror center. This again leads to an overlap in viewing angles such that seamless stitching across mirror segments can be accomplished. The shift must be adjusted such that the regular baseline $\overline{L_1R_1}$ is the same as the baseline $\overline{L_1R_2}$ for a mixed stereo pair. With this mirror setup, the conflicting requirements on disparity for stitching and viewing can be elegantly solved, enabling the capture of high-quality stereoscopic omni-directional video [Schreer et al. 12].

3.5 Summary

The advent of small digital cameras has enabled the development of cost-efficient high-resolution, omni-directional video capture devices. High-quality video panoramas can reproduce arbitrary views of a scene from a fixed vantage point. Omni-directional video has many interesting applications like background representations in movie productions or in format-agnostic productions for different displays and devices. 2D omni-directional video is relatively straightforward and can be realized compactly by a star-like camera arrangement. In this setup, due to the physical size of the cameras parallax effects cannot be avoided and have to be corrected for by appropriate stitching algorithms. For very high quality results, bigger cameras and lenses are needed, increasing the disparity problem. A mirror rig allows for virtually placing all focal points at the same position, enabling, in principle, parallax-free stitching. In practical scenarios, however, focal points must be radially shifted somewhat to create overlapping image regions between mirror segments. Stereoscopic omni-directional video is much more difficult to acquire due to the conflicting requirements of little disparity for seamless stitching while needing large disparities to enable stereoscopic viewing experience (Chapter 18). For static scenes, a rotating camera can sequentially scan the environment. For stereoscopic panoramic video capture, more sophisticated capture setups are necessary. One possibility is again to use a mirror rig to place all focal points on a ring around the center, leading to large baselines of opposite camera pairs for stereo impression and small baselines between neighboring cameras for stitching. This enables capturing convincing high-resolution, stereoscopic, omni-directional video while avoiding parallax artifacts.

4
Range Imaging

Andreas Kolb and Fabrizio Pece

4.1 Introduction

A vast number of applications benefit from or even require geometric information acquired from real environments, such as cultural heritage, virtual and augmented environments, human machine interaction, safety in an industrial or automotive context, to name just a few.

The notion of range imaging subsumes contact-free techniques for acquiring per-pixel distance information with respect to a current pose using sensors comprising a 2D array-like arrangement of sensor elements (*pixels*). Even though a single acquired frame formally results in a 2.5D information, a large number of applications use a series of these 2.5D datasets in order to recover a complete 3D model; therefore the data delivered by range sensing systems is usually denoted as 3D data.

From the technical perspective, there are two basic principles for range imaging, namely time-of-flight (ToF) and triangulation-based methods. The ToF technology is based on measuring the time that light emitted by an illumination unit requires to travel to an object and back to the sensor array. This is used in LIDAR (light detection and ranging) scanners for high-precision, point-by-point distance measurements. In the last decade, this principle has found realization in microelectronic devices, i.e. chips, resulting in new range-sensing devices, the so-called *ToF cameras*. Triangulation-based methods, on the other hand, utilize the disparity effect, i.e., an object is displaced in the image plane as the observing 2D-camera is moving in (approximately) lateral direction. As the amount of displacement relates to object distance as well as to camera displacement (*baseline*), the distance can be reconstructed in case the baseline is known. Triangulation-based systems can be realized as *passive devices* resulting in the *stereo vision* or *multi-view vision* approaches of Chapter 8. Light-field cameras can also be used for passive triangulation (Chapter 5). *Active triangulation-based systems*, on the other hand, realize the triangulation principle using (at least) one active illumination device instead of passive 2D cameras. The illumination unit, e.g., a projector or a laser sheet, is used to "code" the spatial directions from the position of the active device in such a way that the observing 2D camera can identify these directions in the image plane. Again, the two directions from the active and the passive device together with the baseline form a triangle in space from which the distance to the observed object point can be deduced. Specific instantiations of active triangulation methods are structured light scanners which use a beamer to code the spatial directions by one or several patterns, and laser triangulation which uses a sweeping laser sheet to allow for a row-by-row acquisition of the object's geometry.

The recent trend in range imaging is clearly toward very fast and low-cost off-the-shelf 3D imaging acquisition systems which are able to acquire range images in video frame rate, i.e., with 30 FPS or more. The two currently most prominent representatives are discussed in this chapter: *ToF-cameras* and the Kinect^{TM}sensor, issued by Microsoft Corp. in conjunction with the XBox 360 game console. Recently Microsoft issued a new Kinect^{TM}version which is based on ToF imaging technology. At the time of writing Microsoft is preparing to issue a new Kinect 2^{TM}version which will use ToF imaging technology.

4.2 Structured Light Cameras—Kinect^{TM}

Even though the principle of structured light (SL) range sensing is comparatively old, the launch of the Microsoft Kinect^{TM}in 2010 as interaction device for the XBox 360 clearly demonstrates the maturity of the underlying principle.

Technical Foundations SL cameras, such as the first version of the Microsoft Kinect^{TM}, typically employ an IR laser projector combined with a monochrome CMOS camera which captures depth variations by analyzing the way a projected pattern deforms when striking an object surface in the scene. This is the same principle of structured light 3D scanners. Historically, structured light scanners operated at low frame-rate. Recently though, Liu et al. have made a major speed breakthrough in 3D laser scanning introducing a technique that can reach data processing performance of 120 Hz [Liu et al. 10a]. By utilizing the binary defocusing technique, Zhang and Huangeven have shown that even higher working frame-rate, well over 500 Hz, can be achieved without loss of precision [Zhang et al. 10b].

While the general principle to realize SL systems is based on analyzing the deformation of a projected IR light pattern, different approaches to achieve this have been investigated in the literature. Zhang and Huang propose a real-time SL system which runs on specialized hardware [Zhang and Huang 04]. The authors present a real-time scanner that uses digital

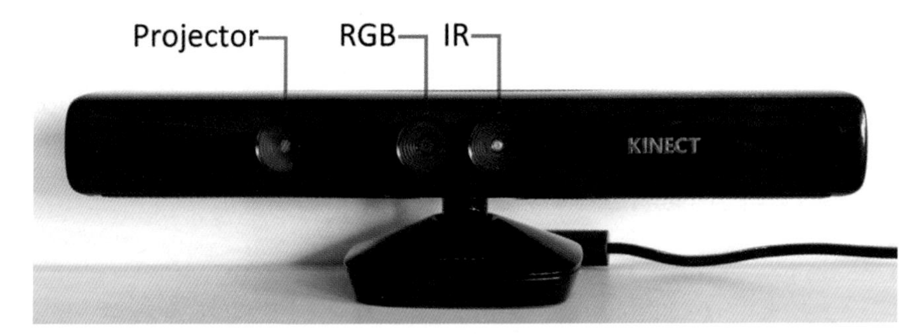

Figure 4.1: Sensor placement within a Kinect™sensor. The baseline is of approximately 7.5cm.

fringe projection and phase-shifting techniques to capture, reconstruct, and render high-density details of dynamically deformable objects such as facial expressions. The proposed system is capable of working at high rates, up to 40 Hz. Scharstein and Szeliski are able to capture high-resolution depth maps of complex scenes by combining multiple SL projectors [Scharstein and Szeliski 03]. [Zhang et al. 02] propose a variation of classic structured light algorithms that uses a pattern of stripes of alternating colors to match observed edges in the scene and to recreate 3D shapes; similarly, [Fechteler et al. 07] propose a fast and high-resolution 3D scanner that recreates textured 3D shapes from just two images. Hall-Holt and Rusinkiewicz introduce a SL method to obtain real-time structured light range scanning based on a new set of illumination patterns [Hall-Holt and Rusinkiewicz 01]. Such patterns are based on coding the boundaries between projected stripes. The stripe boundary codes allow range scanning of moving objects at 60 Hz with 100 μm accuracy over a 10 cm working volume. The system uses a standard video camera and digital light processing projector (DLP) to produce dense range images.

Kinect™ technology is based on the classic SL approach. The unit comprises two cameras, one RGB and one IR, and one laser-based IR projector (Figure 4.1). The IR camera and the IR projector form a stereo pair with a baseline of approximately 7.5 cm. The IR projector sends out a fixed pattern of light and dark speckles. The pattern is generated from a set of diffraction gratings that are designed to lessen the effect of the zero-order propagation, i.e., to avoid a centered bright dot [Zalevsky et al. 07].

Depth calculation is performed by triangulating the known pattern emitted by the projector, that is stored on the unit. For each new frame, depth is estimated at each pixel p_i by sliding a correlation window on the recorded IR frame. The window is typically small (9×7 or 9×9 pixels). It is used to compare the recorded pattern at p_i with the corresponding stored pattern. The best match gives an offset from the known depth, in terms of pixels, also known as disparity. The device performs an interpolation of the best match to get sub-pixel accuracy to $\frac{1}{8}$ of a pixel. Given the known depth of the stored pattern and the disparity value, an estimated depth for each pixel is calculated by triangulation.

Challenges Working with the Kinect^{TM}sensor is relative easily as plenty of drivers and resources are provided for a variety of operating systems. However, challenges arise when trying to calibrate the internal sensors to obtain depth-plus-color information (i.e., RGBD pixels), when the systematic depth error needs to be calibrated, or when multiple units have to operate together.

Camera Calibration The IR and RGB cameras are separated by a small baseline of approximately 2.5 cm horizontally. Hence, if color and depth information are needed for each pixel, calibration of one sensor to the other has to be performed. As the IR camera can be considered a variation of a gray-scale camera, the classic checkerboard method (Section 1.6) can be employed for extrinsic calibration. Such calibration routines assume that the intrinsic parameters of the two sensors are known; if this is not the case, intrinsic calibration can be performed by using a zero-distortion model for the IR camera, and a distortion and de-centering model for the RGB camera. Typical values for the translations between the two cameras will be close to zero along Y and Z axis, and approximately 2.5 cm on the x axis. The rotation between the sensors is minimal, with values smaller than 1 degree.

The above solution, while based on an established calibration method for which a variety of tools are readily available [Bouguet 08, Bradski 00], is unfortunately prone to error as it relies on accurate matches of image features across both cameras. Accurate matches between RGB and IR cameras, however, is surprisingly difficult to establish. Therefore, other calibration routines have been designed for the Kinect^{TM}. [Herrera C. et al. 12] present an algorithm that simultaneously calibrates two RGB cameras and a depth camera as well as the relative pose between them. The method is designed to calibrate a Kinect^{TM} unit to an external RGB image, and it proves to be more accurate and reliable than the classic checkerboard method. The algorithm requires only a planar surface to be imaged from various poses. The calibration does not use depth discontinuities in the depth image which makes it flexible and robust to noise. In addition to camera calibration, the authors presented a new depth distortion model for the

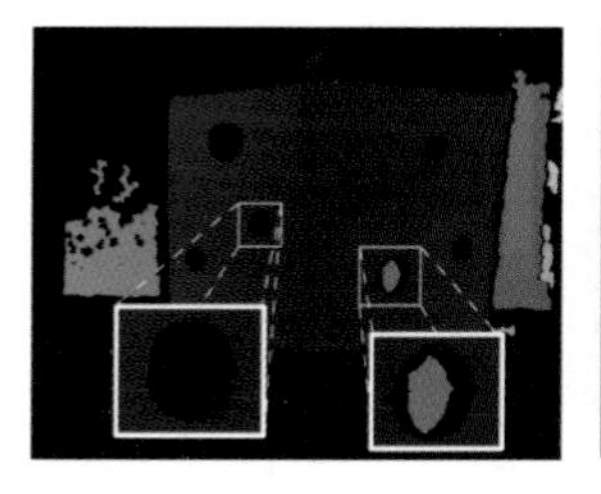 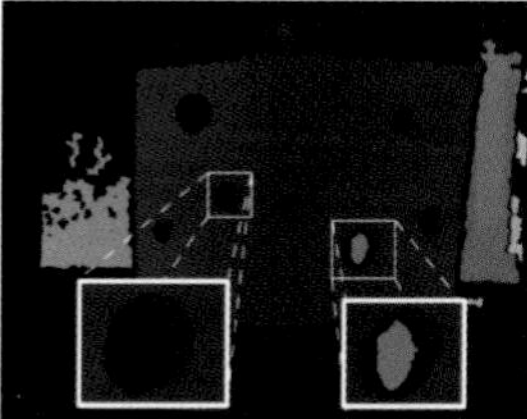 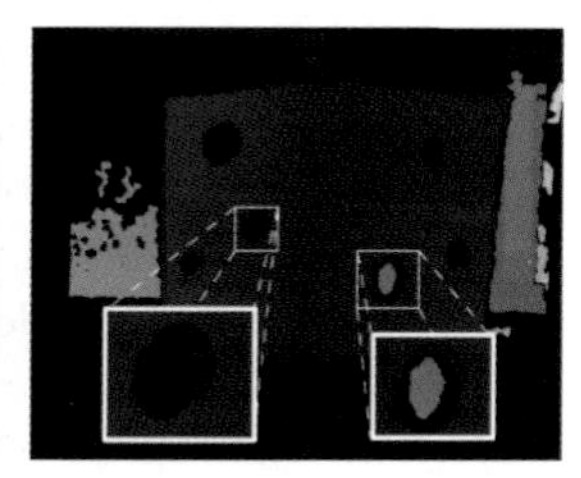

Figure 4.2: Comparison of two calibration solutions' accuracy. An object with holes is acquired by a Kinect^{TM}sensor, and the RGB image is overlaid on the depth-map after calibrating the two cameras. The [Raposo et al. 13] method yields accurate results with small misalignment (left). The [Herrera C. et al. 12] solution introduces significant misalignment without distortion correction (center). The misalignment becomes even more severe if distortion correction is applied (right).

depth sensor which can be used to calibrate the unit systematic error. [Raposo et al. 13] modify the above work to employ only 6–10 disparity-image pairs of a planar checkerboard pattern, dramatically speeding up the calibration time and requiring only a few input images. Figure 4.2 shows the comparison of the two methods.

A different approach is introduced by [Zhang and Zhang 11]. The authors present a maximum likelihood solution for the joint depth and color calibration based on two principles. First, co-planarity of the checkerboard points in the depth image is enforced. Second, additional point correspondences between the depth and color images are manually specified or automatically established to help improve calibration accuracy. Finally, a variation on intrinsic calibration is presented by [Teichman et al. 13]. The author presents a new generic approach to the calibration of depth sensor intrinsics based on simultaneous localization and mapping (SLAM) [Smith and Cheeseman 86]. In particular, no specialized hardware, calibration target, or hand measurement is required, making the calibration routine completely unsupervised. Compared to the supervised calibration approach, the proposed technique requires only a few minutes of data input from unmodified, natural environments. This is particularly advantageous in situations where humans cannot be involved, e.g., to hold checkerboards in front of the camera, such as for a robot that needs to calibrate automatically while in the field.

Systematic Error Correction Similar to most range cameras, the Kinect^{TM} suffers from systematic error in depth estimation. Khoshelham and Elberink analyzed the accuracy and resolution of the sensor depth data, showing that the systematic error is generally smaller than 3 cm, but that it increases on the periphery of the sensor or when depth measurements are collected further away from the unit [Khoshelham and Elberink 12]. Interestingly, the error seems also to be stronger when depth measurements are collected close to the camera sensor [Smisek et al. 11]. There are several approaches to handle the systematic error, including the one presented in [Smisek et al. 11]. [Herrera C. et al. 12] proposed a distortion model to correct the systematic unit error. A different approach is introduced by [Yamazoe et al. 12]. The general principle beyond their calibration routine is that, as the SL principle is based on both emitter and receiver, the intrinsic parameters of both the IR camera and projector should be taken into account. Hence the authors present a depth correction model that is based on joint estimation of depth-camera and projector intrinsic parameters, achieved by showing only a planar board to the depth sensor.

Multiple Units Integration Combining several Kinect™cameras is non-trivial due to potential interference problems given by multiple IR patterns projected into the scene. To combat this, one can carefully align multiple units to avoid IR overlaps, but this requires a tedious manual calibration. A more general solution, based on constant shake of the units, has been recently presented by [Butler et al. 12]. The authors propose to associate to each unit a motor with an offset weight. The motor shakes the camera, and consequently also the IR projector and camera are moved. As the shaking is constant for both sensors, the depth estimation algorithm still works reliably for the single unit. However, from the viewpoint of another Kinect™, the pattern of the other projector moves around and interferes with its own pattern only for a small amount of time. This results in reduced interference between both cameras (Figure 4.3). A different solution to mitigate interference errors is introduced by [Berger et al. 11]. The authors apply a set of fast rotating disks to multiple Kinect™ units, effectively creating a time division multiple access (TDMA). Each disk contains a gap large enough to allow a laser beam to pass through it. Hence, each unit's laser diode is blocked by the disk, except for the time when the gap is allowing the laser to project its pattern into the scene. Each Kinect ™is equipped with such a disk rotating at the same speed but with a different phase, ensuring that only one laser projects the pattern into the scene at any given time.

4.3 Time-of-Flight Cameras

Technical Foundations In general, there are two main approaches which can be employed in order to realize a ToF camera system without

Figure 4.3: Depth-map acquired by a single unit placed within a multi-Kinect setup. Left: significant error in depth due to cross-talk, including depth values being hallucinated. Right: the method in [Butler et al. 12] applied.

requiring expensive coherent light illuminations: continuous wave intensity modulation [Xu et al. 98, Hostica et al. 06, Oggier et al. 05] or pulse-based optical shutter approaches [Iddan and Yahav 01, Yahav et al. 07].

Continuous Wave (CW) Intensity Modulation is the most common approach used in ToF cameras. The general idea is to actively illuminate the scene under observation using near infrared (NIR) intensity-modulated light with a modulation frequency f_m (Figure 4.4). Due to the distance between the camera and the object (sensor and illumination are assumed to be at the same location), and the finite speed of light c, a frequency shift

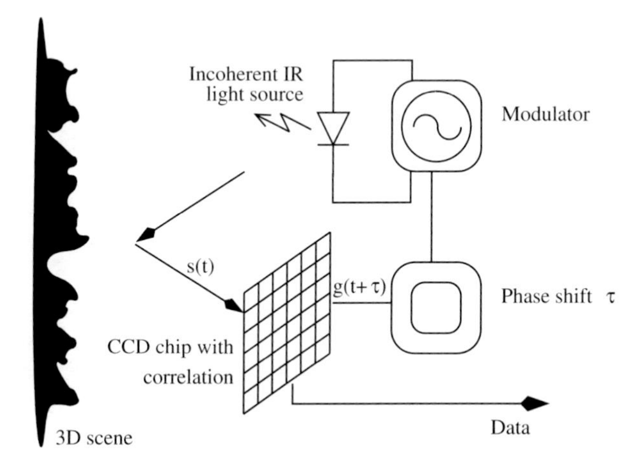

Figure 4.4: The ToF phase-measurement principle.

ϕ is caused in the optical signal which is detected for each sensor pixel by mixing. This information can be easily transformed into the sensor-object distance as the light has to travel the distance twice, i.e.,

$$d = \frac{c\phi}{4\pi f_m}.$$

More technically, the incident optical signal s on each pixel, reflected by the scene, is correlated with the reference generator signal g, possibly with an internal phase offset τ. This appoach, which is also called *mixing*, yields the correlation function which is sampled in each pixel

$$C(\tau) = s \otimes g = \lim_{T\to\infty} \int_{-T/2}^{T/2} s(t) \cdot g(t+\tau)\, dt.$$

For simple sinusoidal signals $g(t) = \cos(2\pi f_m t)$ the optical signal yields $s(t) = b + a\cos(2\pi f_m t + \phi)$, where a and b are the amplitude and the correlation bias of the incident optical signal, respectively. Some basic trigonometric calculus yields $C(\tau) = \frac{a}{2}\cos(f_m\tau + \phi) + b$.

The demodulation of the correlation function C, i.e., the computation of the phase shift ϕ, is done using several samples of the correlation function C obtained by four sequential phase images with different internal phase offset τ: $A_i = C(i \cdot \frac{\pi}{2})$, $i = 0, \ldots, 3$:

$$\begin{aligned} \phi &= \operatorname{arctan2}(A_3 - A_1, A_0 - A_2), \quad I = \tfrac{A_0+A_1+A_2+A_3}{4}, \\ a &= \tfrac{\sqrt{(A_3-A_1)^2+(A_0-A_2)^2}}{2}, \end{aligned} \tag{4.1}$$

where $\operatorname{arctan2}(y, x)$ is the angle between the positive x-axis and the point given by the coordinates (x, y).

The new Kinect 2^{TM} uses a more complex modulation signal, consisting of various frequencies and different signal shapes. It still applies the basic principle of mixing the optical and the reference signal in order to detect the phase shift.

Figure 4.5 shows recent CW ToF-camera models from different manufacturers. The resolution of these cameras is 120×160 pixel for the CamBoard pico XS and 512×424 pixel for the new $\text{Kinect}^{\mathrm{TM}}$.

Pulse-Based Optical Shutter The pulse-based optical shutter approach is an alternative ToF principle based on the indirect measurement of the time of flight using a fast electronic shutter technique [Iddan and Yahav 01, Yahav et al. 07]. As pulse-based devices are not further discussed, this paragraph only sketches the basic concept of this sensor. The illumination unit sends short NIR light pulses $[t_{\text{start}}, t_{\text{end}}]$ which represent a depth

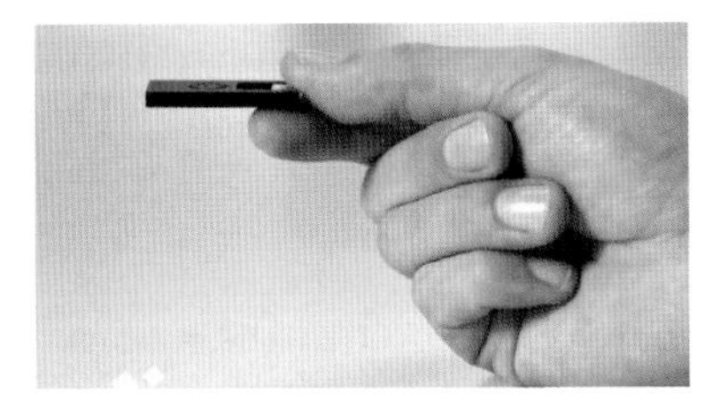

Figure 4.5: Current ToF cameras: The CamBoard XS (left) from pmdtechnologies and the new Kinect^{TM} 2 from Microsoft (right).

range of interest ("light wall," Figure 4.6). The optical signal is reflected by the scene objects leading to a "distorted" light wall, resembling the objects' shapes. An electronic shutter integrated with the standard CCD chip cuts the front (or rear) portion of the optical signal at the gating time $t_{\text{gate}} = t_{\text{start}} + \Delta_t$; Δ_t resembles the depth offset from the camera to the depth region of interest. The resulting intensity I_{front} is proportional to the distance of the corresponding object's surface. The object's reflectivity and the light attenuation are compensated by using the relation to the completely reflected pulse intensity without gating: $I_{\text{front}}/I_{\text{total}}$. Larger depth ranges may be acquired using several exposures with varying gating parameters [Gvili et al. 03].

Challenges of Time-of-Flight Cameras ToF cameras are active imaging systems that use standard optics to focus the reflected light onto the chip area. Therefore, the typical optical effects like shifted optical centers and lateral distortion need to be corrected for, which can be done using classical intrinsic camera calibration techniques (Section 1.6). However, even though some of the newest ToF cameras achieve almost nearly VGA resolution, most cameras have a resolution in the QQVGA-range or

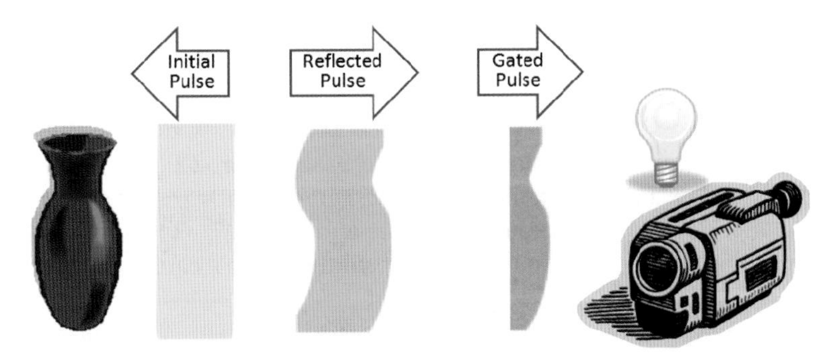

Figure 4.6: The pulse-based optical shutter principle.

even below. This resolution is rather small in comparison to standard RGB- or grayscale-cameras. Thus, for these cameras at or below QQVGA resolution, standard calibration techniques have to be applied with care [Lindner and Kolb 06].

As with any other camera system, ToF cameras can suffer from oversaturation in case of too long exposure times in relation to the ambient background light and the objects' distance and/or reflectivity. Some camera vendors provide a suppression of background intensity on the chip, thus allowing also for outdoor applications. Object areas with extremely low reflectivity or objects far from the sensor lead to a low incident optical signal to the ToF camera, resulting in a bad signal-to-noise-ratio (SNR). Compared to dense stereo, laser scanners and similar approaches, the depth measurement quality is still in the range of several millimeters up to a few centimeters for real-world scenes; thus more accurate reconstructions require spatial data fusion (Chapter 9). Similar to active sensing, the parallel use of several cameras may lead to interference problems, i.e., the active illumination of one camera influences the result of another camera. This kind of interference can be circumvented by using different modulation frequencies.

The ToF principle itself possesses several sensor-specific challenges, which are discussed in detail in the following.

Systematic Distance Error Practically, the generated signal s is not sinusoidal, thus applying the phase reconstruction formulas in Eq.(4.1) yields a systematic error, also called "wiggling" (Figure 4.7 top left). The mean of the systematic error is typically in the order of ± 5 cm after any bias in the distance error has been removed. Furthermore, the systematic error depends on the exposure time of the camera which can be considered as a constant offset with respect to a systematic error at a reference exposure time.

There are several standard approaches to handle the systematic error. First, one has to acquire reference data ("ground truth"). This can be done using track lines [Steitz and Pannekamp 05, Lindner and Kolb 06] or a robot in order to locate the camera in a global reference frame with respect to a known plane [Fuchs and May 08]. Both approaches require rather cost-intensive equipment. Alternatively, vision-based approaches are applied to estimate the extrinsic parameters of the camera with respect to a reference plane, e.g., a checkerboard [Lindner and Kolb 07]. These approaches, however, can only be applied to "high resolution" ToF cameras, additional cameras, or multiple data acquisitions. Correction schemes simply model the depth deviation using a look-up-table [Kahlmann et al. 07] or function fitting, e.g., using b-splines [Lindner and Kolb 06].

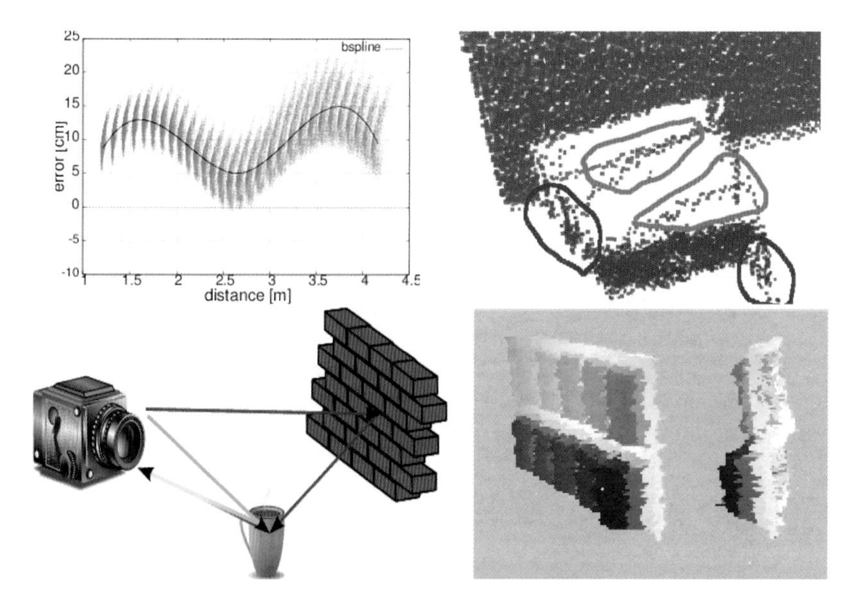

Figure 4.7: Error sources of ToF cameras. Top left: Systematic (wiggling) error for all pixels (gray) and fitted mean deviation (black). Top right: Motion artifacts (red) and flying pixels (green) for a horizontally moving planar object in front of a wall. Bottom left: Schematic illustration of multipath effects due to reflections in the scene. Bottom right: Acquiring a planar grayscale checkerboard reveals the intensity related distance error.

The systematic error is usually handled in the vendor's device driver. Alternatively, there are open calibration tools like the one from the University of Kiel.[1]

Depth Inhomogeneity At object boundaries, a pixel may observe inhomogeneous depth values. In this case, the mixing process results in a superimposed signal caused by light reflected from different depths, leading to wrong distance values ("flying pixels," Figure 4.7(top right)). There are simple methods relying on geometric models that give good results in identifying flying pixel, e.g., by estimating the depth variance which is extremely high for flying pixel [Sabov and Krüger 10]. Alternatively, more complex approaches such as splitting of pixels, respectively depth image upscaling can be used in combination with gradient methods [Lindner et al. 08] or with additional information from high resolution 2D cameras [Guomundsson and Sveinsson 11].

[1] www.mip.informatik.uni-kiel.de/tiki-index.php?page=Calibration

Multi-Path Effects and Intensity-Related Distance Error *Multi-path effects* relate to an error source common to active measurement systems: The active light may not only travel the direct path from the illumination unit via the object's surface to the detector, but it may additionally travel *indirect paths*, i.e., being scattered by highly reflective objects in the scene or within the lens systems or the housing of the camera itself (Figure 4.7(bottom left)). Within a camera pixel, these multiple responses of the active light are superimposed leading to an altered signal not resembling any meaningful distance information anymore.

One simple approach to compensate for multiple reflections is proposed by [Falie and Buzuloiu 08]. Here the assumption is that the indirect effects are of rather low spatial frequency compared to the direct effects. The authors use differential phase images from neighboring pixels, thus trying to compensate the low-frequency indirect component. Unfortunately, this approach does not work very robustly. Taking only a single pixel into account, it can be shown that any correction scheme requires several modulation frequencies in order to identify superimposed signals in a single pixel. Assuming a perfectly sinusoidal signal, [Dorrington et al. 11] present an analytic formulation for the signal superposition resulting in a highly non-linear optimization scheme which exhibits unstable behavior for specific constellations.

There is another visually obvious artifact which is currently assumed to be a specific form of a multi-path effect, i.e., the *intensity-related distance error*. As expected, darker object regions exhibit a worse SNR resulting in a stronger variation of the depth measurement. The intensity-related error manifests itself as an additional non-zero biased distance offset depending on the amount of incident active light, e.g., resulting from a variation of the objects reflectivity (Figure 4.7(bottom right)).

The specific intensity-related error has mainly been tackled using phenomenological approaches. [Lindner et al. 10] explicitly measure the depth derivation caused by the intensity variation and correct this behavior using a b-spline function fitting.

Dynamic Errors One key assumption of ToF cameras is that each pixel observes a single object point during the acquisition of all phase images. This assumption is violated in case of moving objects or moving cameras, resulting in *motion artifacts*. In real scenes, motion may alter either the observed depth and/or the reflectivity observed by a pixel during the acquisition. Processing the acquired phase images while ignoring the motion present in the acquisition leads to erroneous distance values at object boundaries (Figure 4.7(top right)).

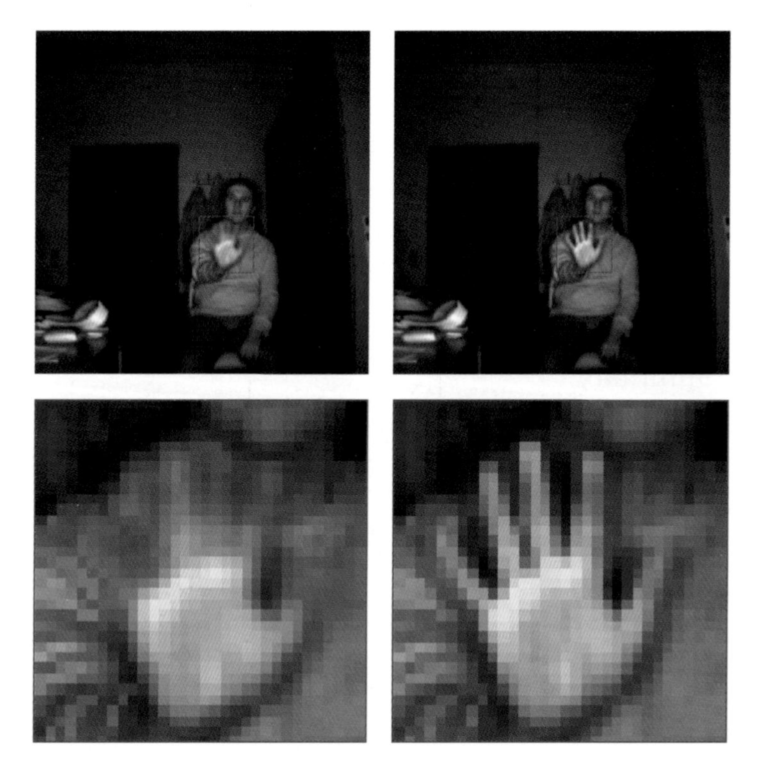

Figure 4.8: Motion compensation: Direct distance computation without compensation (left), and corrected distance (right).

[Schmidt and Jahne 11] detect motion artifacts using temporal gradients of the phase images. In case the gradient in one of the phase images exceeds a certain threshold, motion is detected in this pixel. Correction is performed using extrapolated information from prior frames.

Since motion artifacts result from in-plane motion between subsequent (phase) images, one alternative approach is to use optical flow methods in order to align the individual phase images. [Lindner and Kolb 09] apply a fast optical flow algorithm proposed by [Zach et al. 07] in order to align the three phase images A_1, A_2, A_3 to the first phase image A_0. Optical flow algorithms rely on a brightness consistency assumption which the phase images do not obey, as their "intensity" is the result of a mixing process. The brightness consistency constraint can be fulfilled if the full intensity values for each phase image are available (only few ToF cameras have this option and applying it usually reduces the camera's frame rate due to bandwidth limitations). Additional intensity calibration is applied in order to correct for different gain behavior of pixels and for inhomogeneous illumation patterns. The optical flow approach is very expensive, resulting in low overall frame rates. [Lefloch et al. 13] propose an optimized approach using only two optical flow computations per depth image (Figure 4.8). A faster approach is to use block-matching techniques applied only to pixels where motion has been detected. They use the direct intensity variation of the phase images which is zero if no motion occurs [Högg et al. 13]. Assuming a linear motion between the phase images and applying a brute force search in the pixel's neighborhood, the flow can be efficiently corrected.

4.4 Summary

Range sensing is one of the longest researched technical challenges in computer vision and photogrammetry. Very recent developments make full-view range depth information available at video frame rates. Fast triangulation-based techniques and time-of-flight cameras offer cheap and easy access to range data at 30 Hz or more. Very current examples are the Kinect™, and time-of-flight cameras with a focus on continuous-wave approaches.

As mentioned above, the main limitations of current range sensing cameras lie in the limited resolution and the relatively high noise level. The low lateral resolution leads to unwanted effects at object boundaries, e.g., flying or masked-out pixels. As for any active range measuring system, objects with specular or low-reflectivity surfaces cause additional problems, leading to erroneous measurements for both Kinect™and ToF cameras. For the multi-exposure ToF cameras, temporal errors may occur for fast moving objects and/or cameras.

Despite all the limitations, the development over the last five years is stunning. All the mentioned limitations have been pushed back significantly and it can be expected that there will be further improvements in the upcoming years. These expectations are also backed by the recent release of the second-generation Kinect™cameras which employ a ToF sensor for depth sensing.

5
Plenoptic Cameras

Bastian Goldlücke, Oliver Klehm, Sven Wanner, and Elmar Eisemann

5.1 Introduction

The light field, as defined by Gershun in 1936 [Gershun 36] describes the radiance traveling in every direction through every point in space. Mathematically, it can be described by a 5D function which is called the *plenoptic function*, in more generality sometimes given with the two additional dimensions time and wavelength. Outside a scene, in the absence of occluders, however, light intensity does not change while traveling along a ray. Thus, the light field of a scene can be parameterized over a surrounding surface; light intensity is attributed to every ray passing through the surface into any direction. This yields the common definition of the light field as a 4D function. In contrast, a single pinhole view of the scene only captures the rays passing through the center of projection, corresponding to a single 2D cut through the light field.

Fortunately, camera sensors have made tremendous progress and nowadays offer extremely high resolutions. For many visual-computing applications, however, spatial resolution is already more than sufficient, while robustness of the results is what really matters. Computational photography explores methods to use the extra resolution in different ways. In particular, it is possible to capture several views of a scene from slightly different directions on a single sensor and thus offer single-shot 4D light field capture. Technically, this capture can be realized by a so-called plenoptic camera, which uses an array of microlenses mounted in front of the sensor [Ng 06]. This type of camera offers interesting opportunities for the design of visual computing algorithms, and it has been predicted that it will play an important role in the consumer market of the future [Levoy 06].

The dense sampling of the light field with view points lying close together may also offer new insights and opportunities to perform 3D reconstruction. Light fields have thus attracted quite a lot of interest in the computer vision community. In particular, there are indications that *small changes* in view point, are important for visual understanding. For example, it has been shown that even minuscule changes at occlusion boundaries from view point shifts give a powerful perceptional cue for depth [Rucci 08].

5.2 4D Light Field Acquisition

Considering the special case that the light field is recorded on a planar surface, the 4D light field in this sense can be viewed as an intensity function that not only depends on the 2D position on the imaging plane, but also on the 2D incident direction. Many ways to record light fields have been proposed and can be classified into three main categories [Wetzstein et al. 11]. *Multi-sensor capture* solves the problem essentially on the hardware level. One can assemble multiple (video) cameras into a single array, with the cameras lying on a common 2D plane [Wilburn et al. 05]. This solution is quite expensive and requires careful geometric and photometric calibration of the sensors [Vaish et al. 04], as well as considerable effort to process and store the huge amount of data streamed by the array in real-time. However, with temporal synchronization of the camera triggers, one can also apply camera arrays to the recording of dynamic scenes [Wilburn et al. 05]. Furthermore, they allow some interesting applications due to their wide baseline.

In contrast, with *time-sequential imaging* one is limited to static scenes, but only a single sensor is required. Different view points of the scenes are captured consecutively by moving the camera [Levoy and Hanrahan 96, Gortler et al. 96], rotating a planar mirror [Ihrke et al. 08], or programmable aperture, where only parts of the aperture are opened for each shot, allowing to re-assemble the light field from several such images by computational means [Liang et al. 08]. Besides cost considerations, an advantage of the sensor being the same for all views is that calibration is simplified (Chapter 1).

Finally, a technology which recently has become available in commercial cameras is *single-shot multiplexing* where a 4D light field is captured with a single sensor in a single shot, which also makes it possible to record videos. In all cases, one faces a trade-off between resolution in the image ("spatial") and view point ("angular") domain. In plenoptic cameras [Ng 06, Bishop and Favaro 12, Georgiev et al. 11, Perwass and Wietzke 12], spatial multiplexing is realized in a straightforward manner by placing a lenslet array in front of the sensor, which allows to capture several views at the same time. Other techniques include coded aperture imaging [Lanman et al. 08] or, more exotically, a single image of an array of mirrors can be used to create many virtual view points [Manakov et al. 13].

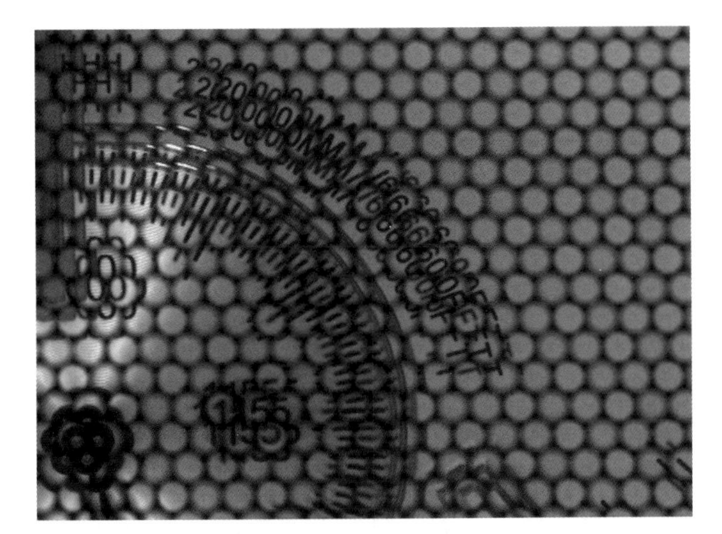

Figure 5.1: Detail of a raw image captured by a plenoptic 2.0 camera by Raytrix. Objects closer to the camera are visible in more microlens images. The camera makes use of different types of microlenses to increase depth of field, which can be distinguished in this image by comparing the sharpness of the projections.

Of the light field acquisition techniques above, plenoptic cameras are gaining increasing interest in the vision community since they are now commercially available as affordable consumer hardware.

5.3 Plenoptic Cameras

While normal 2D cameras only record irradiance from different directions at a single view point in space, plenoptic cameras capture the complete 4D light field on the sensor plane. The idea originates in the early 20th century. First described using a grid of pinholes inside a camera by Ives in 1903 [Ives 03], Lippmann proposed the use of microlenses in front of the image plane in 1908 [Lippmann 08]. Several improvements to the design have been proposed. For example, cameras manufactured by the company Raytrix employ multiple types of microlenses to accomplish a larger depth of field (Figure 5.1).

At the time of writing, plenoptic cameras are commercially available from two manufacturers. The Lytro camera is based on the "plenoptic 1.0" design and targeted at the consumer market, while the Raytrix camera is based on the "plenoptic 2.0" design and targeted at industrial applications. This is reflected in both price as well as the bundled software.

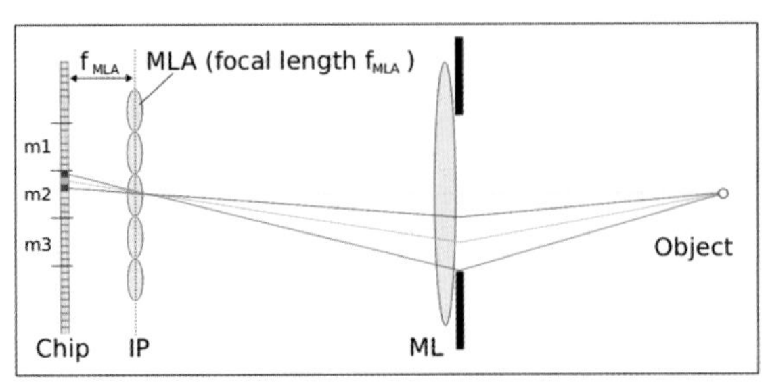

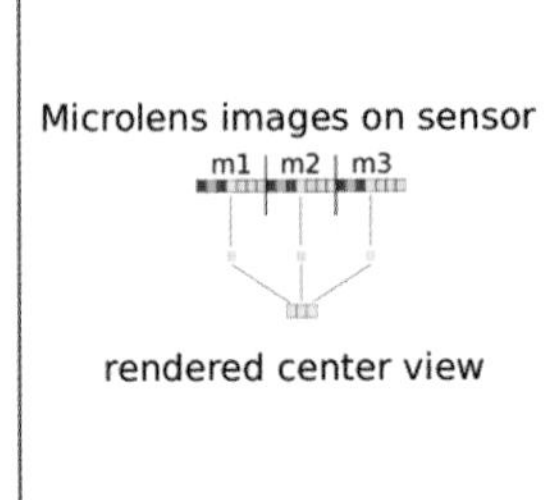

Figure 5.2: *Left:* one-dimensional sketch of a plenoptic camera 1.0 setup. Light rays emitted by the object are focused by the main lens (ML). The microlens array (MLA) is placed at the image plane (IP) of the main lens and thus separates the rays according to their direction. *Right:* a single view point, indexed by (s, t), here the center view, is extracted by collecting the corresponding pixels of each micro image m_i.

The plenoptic camera 1.0 (Lytro camera) design is based on a usual camera with a digital sensor, main optics, and an aperture. In addition, a microlens array is placed in the focal plane of the main lens exactly at the focal length f_{MLA} from the sensor (Figure 5.2). This way, instead of integrating the focused light of the main lens on a single sensor element, the microlenses split the incoming light cone according to the direction of the incoming rays and map them onto the sensor area behind the corresponding microlens. In particular, one has direct access to the radiance $L(u, v, s, t)$ of each ray of the light field by choosing the micro-image of the microlens corresponding to spatial position (s, t) and pixel corresponding to direction (u, v) of the underlying micro-image. The size of each microlens is determined by the aperture or f-number of the main optics. If the microlenses are too small compared to the main aperture, the images of adjacent microlenses overlap. Conversely, sensor area is wasted if the microlenses are too large. Since light passing the main aperture also has to pass a microlens before being focused on a pixel, what actually happens is that the camera integrates over a small 4D volume in light field space. The calibration of unfocused lenslet-based plenoptic cameras like the ones commercially available from Lytro is discussed in [Dansereau et al. 13].

The main disadvantage of the 1.0 design is the poor spatial resolution of the rendered views, which is equal to the number of microlenses. By slightly changing the optical setup, one can increase the spatial resolution dramatically. As another way to compactly record 4D light fields, the focused plenoptic camera has been developed, often called the plenoptic camera 2.0 (Raytrix camera) [Lumsdaine and Georgiev 09, Perwass and Wietzke 12].

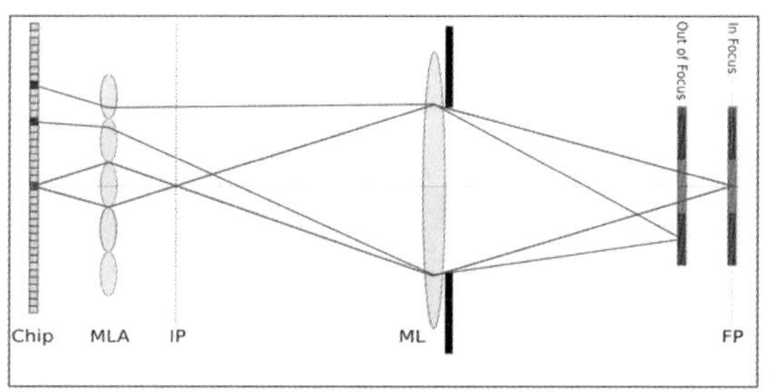

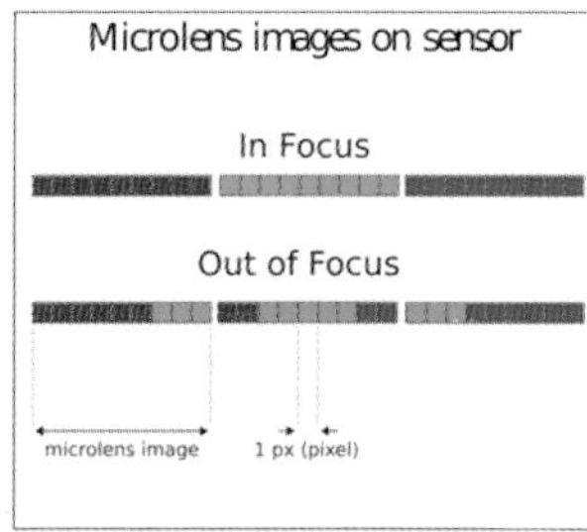

Figure 5.3: *Left:* one-dimensional sketch of a plenoptic camera 2.0 setup. Light rays emitted by the object are focused by the main lens (ML) onto the image plane (IP). The microlens array (MLA) is placed so that the microlenses are focused onto the image plane of the main lens, mapping fractions of the virtual image onto the sensor. Green rays originate from an object in focus of the main lens (FP), blue rays from an object away from the principal plane of the main lens. *Right:* resulting micro-images of an object in and out of focus.

The main difference in the optical setup between the cameras is the relative position of the microlens array. The microlenses are no longer placed at the principal plane of the main lens, but are now focused onto the image plane of the main lens. In effect, each microlens now acts as a single pinhole camera, observing a small part of the virtual image inside the camera. This small part is then mapped with high spatial resolution onto the sensor. The scene points have to lie in a valid region between the principal plane of the main lens and the image sensor. Scene features behind the principal plane cannot be resolved.

Scene points that are not in focus of the main lens but within this valid region are imaged multiple times over several neighboring microlenses, thus encoding the angular information over several micro-images (Figure 5.3 [Lumsdaine and Georgiev 09, Perwass and Wietzke 12]). Angular information is encoded while at the same time preserving high resolution. Due to multiple imaging of scene features, however, rendered images from this camera have a lower resolution than the inherent sensor resolution promises. The light field is encoded in a complicated way, and it is necessary to perform an initial depth estimate at least for each microlens in order to decode the sensor information into the standard 4D light field data structure [Wanner et al. 11]. External and internal calibration of plenoptic 2.0 cameras has been investigated in [Johannsen et al. 13].

Figure 5.4: One way to visualize a 4D light field is to think of it as a collection of images of a scene, where the focal points of the cameras lie in a 2D plane. The rich structure becomes visible when one stacks all images along a line of viewpoints on top of each other and considers a cut through this stack (denoted by the green border). The 2D image one obtains in the plane of the cut is called an *epipolar plane image (EPI).*

5.4 4D Light Field Structure and Depth Reconstruction

Since a 4D light field can be understood as a dense collection of multiple views, off-the-shelf correspondence search techniques can be applied to infer 3D structure (Chapter 8). Due to the rich information content in the light field data, however, also specialized methods can be developed, which work more efficiently and robustly.

One line of research follows the philosophy of the earliest works on the analysis of epipolar volumes [Bolles et al. 87], and rely on the fact that 3D scene points project to lines in the epipolar-plane images. The reason is that a linear camera motion leads to a linear change in projected coordinates (Figure 5.4). These lines can be more robustly detected than point correspondences which have been exploited in several previous works [Bolles et al. 87, Berent and Dragotti 06, Criminisi et al. 05]. A recent advanced method aims at accurate detection of object boundaries and is embedded in a fine-to-coarse approach, delivering excellent results on very high-resolution light fields [Kim et al. 13].

In the same spirit, an efficient and accurate approach, which is however limited to only small disparity values and thus has a limited depth range, computes a direct estimate of the local orientation of the pattern [Wanner and Goldlücke 14] (Figure 5.5). Here, orientation estimation is performed using an eigenvector analysis of the first-order structure tensor of the EPI. This approach can be extended to detect multiple overlaid

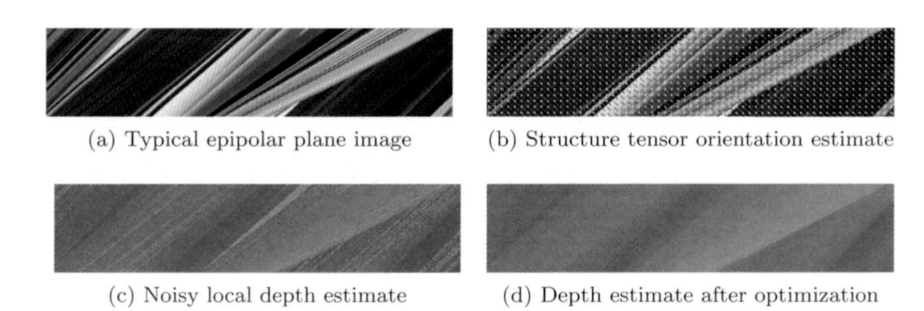

(a) Typical epipolar plane image

(b) Structure tensor orientation estimate

(c) Noisy local depth estimate

(d) Depth estimate after optimization

Figure 5.5: Depth estimation on an epipolar plane image (a). Standard 2D pattern analysis using the structure tensor yields a robust orientation estimate (b), whose slope encodes the (still noisy) depth map for the EPI (c). Global optimization techniques result in a consistent estimate across all views (d).

patterns to efficiently reconstruct reflections or transparent objects [Wanner and Goldlücke 13]. Since local depth estimates from any source (including, e.g., stereo matching) are usually noisy, global smoothing schemes can be employed to improve the result. By careful construction of the regularizers and constraints, one can obtain consistent estimates over the complete light field which respect occlusion ordering across all views [Goldlücke and Wanner 13, Wanner and Goldlücke 14].

Other 3D reconstruction methods specific to light fields exist, including focus stacks in combination with depth-from-focus methods [Nayar and Nakagawa 94, Perez and Luke 09]. Multiple methods that make use of depth maps to warp individual light field views to densify the light field from a sparse set of views have been proposed (Section 17.2).

5.5 Spatial and Angular Super-Resolution

Since plenoptic cameras trade off sensor resolution for the acquisition of multiple view points, it is not surprising that super-resolution techniques are one focus of research in light-field analysis. Such methods have been investigated using priors regarding statistics of natural images [Bishop and Favaro 12] as well as modified imaging hardware [Lumsdaine and Georgiev 09].

In the classical Bayesian approach, an image formation model is set up to obtain the known input images from the desired super-resolved target image. In particular, when one transforms the target image into the image domain of an input image and performs a downsampling operation (usually

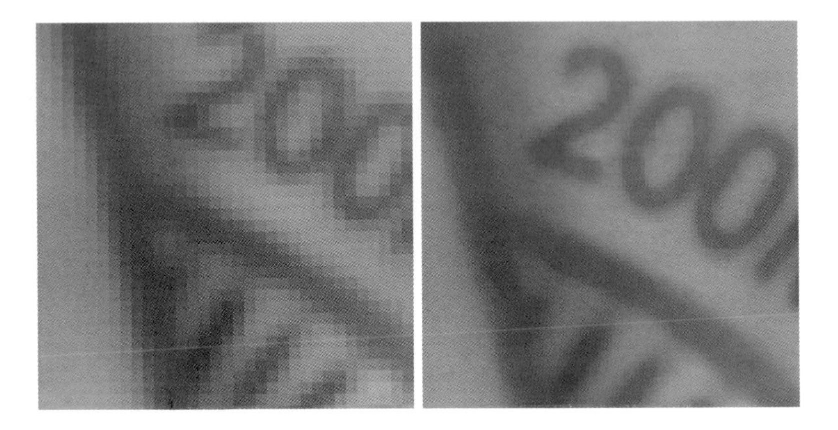

Figure 5.6: By solving a single inverse problem, one can create super-resolved novel views from a 4D light field captured with a Raytrix plenoptic camera [Wanner and Goldlücke 12]. Above are close-ups of one of the 7×7 input views (left) and the result from the super-resolution algorithm (right).

modeled via a blur kernel), one should obtain an exact copy of the input image. In practice, however, this property will not be satisfied exactly due to sensor noise or sampling errors. Thus, the set of equations is enforced as a soft constraint in a minimization framework, where the desired super-resolved image appears as the minimizer of some energy functional [Bishop and Favaro 12, Wanner and Goldlücke 12].

In particular, some frameworks also allow to generate views in new locations, thus solving an image-based rendering task in the same step [Wanner and Goldlücke 12]. In some recent work, a Bayesian framework was explored which also models uncertainties in the depth estimates and which is able to mathematically derive many of the heuristics commonly used in image-based rendering [Pujades et al. 14]. The topic of image-based rendering is explored in detail in Chapter 17.

5.6 Refocusing and Other Applications

In this section, methods are presented that allow to simulate the intrinsics of a usual camera by relying on a light field as input. The two additional dimensions of a 4D light field compared to a conventional 2D image (quantities: radiance $[\mathrm{W\,m^{-2}\,sr^{-1}}]$ vs. irradiance $[\mathrm{W\,m^{-2}}]$) make it possible to produce effects such as changing the aperture or refocusing (adjusting the focal plane) even *after* a photo has been taken.

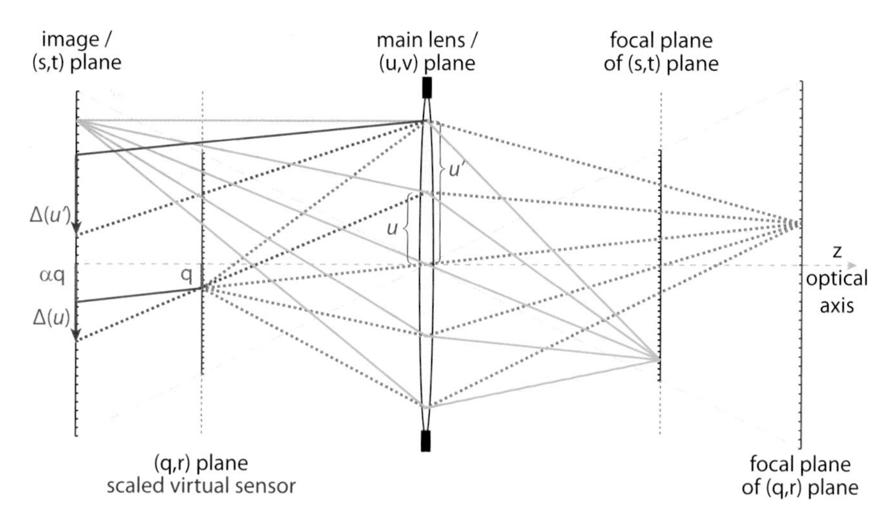

Figure 5.7: Refocusing example: a new virtual sensor at plane (q, r) is introduced, causing a different focal plane. The mapping from coordinates (q, r) in the local space of the scaled sensor to coordinates (s, t) in the space of the original image plane reduces to a constant translation $\Delta(u, v)$.

For many of these effects, a depth image has to be computed first, which can be directly derived from the light field (Section 5.4). In particular, the plenoptic camera 2.0 requires a reasonable depth estimate to reconstruct any meaningful image from the captured light field. This depth reconstruction is possible because the light field stores partially redundant information. More precisely, objects in a scene tend to have similar appearance under slightly different viewing angles. While the redundancy can be directly used for compression of light field data [Levoy and Hanrahan 96], it has recently been exploited for the reconstruction of a light field from very sparse data [Marwah et al. 13].

The basis of the following examples is to sample or integrate the 4D light field to synthesize a new 2D image. The classical (u, v, s, t) parameterization [Levoy and Hanrahan 96] of a light field uses two distinctive planes that are aligned with the optical axis z: $[u, v]^T$ denotes the coordinates on the plane at the main lens (ML) z_{UV} and $[s, t]^T$ the coordinates on the focal plane of the main lens. As points on the focal plane uniquely map to points on the image plane (IP), $[s, t]^T$ also denotes the coordinates on the image plane at z_{ST}.

The light field can be used to fetch radiance for a new plane (q, r) at distance z_{QR}, parallel to the (s, t) plane. The mapping to the original coordinates is simple as it only requires to find the intersection of the ray,

Figure 5.8: Example of a refocusing sequence. Left to right: the focal plane is moved from front to back, shifting the focus from the buddha statue to the pirate. The scene was captured with the kaleidoscope camera add-on [Manakov et al. 13] with an 50mm f/1.4 main lens. While this light-field camera add-on only captures nine directions, these views are sufficient to estimate depth and perform view interpolation, allowing for smooth out-of-focus blur.

originating at $[q, r, z_{\text{QR}}]^T$ with direction $[u, v, z_{\text{UV}}]^T - [q, r, z_{\text{QR}}]^T$ with the (s, t) plane at z_{ST}. The $[s, t]^T$ coordinates of the intersection point can be determined in two steps: first, a scaling α of $[q, r]^T$ depending on the positions of the (s, t), (q, r), and (u, v) planes is computed: $\alpha = \frac{z_{\text{UV}} - z_{\text{ST}}}{z_{\text{UV}} - z_{\text{QR}}}$. Second, a translation by $\Delta(u, v) = -\beta \cdot [u, v]^T$ with $\beta = \frac{z_{\text{QR}} - z_{\text{ST}}}{z_{\text{UV}} - z_{\text{QR}}}$ yields the final coordinates in the (s, t) plane: $[s, t]^T = \alpha \cdot [q, r]^T + \Delta(u, v)$ (Figure 5.7).

While a pinhole camera could, in theory, have an infinitesimal aperture, such a camera would not produce any image, because no light would be detected. Hence, cameras rely on a larger aperture and use a lens to refocus the rays. All points on a so-called focal plane project to exactly one location on the sensor; outside the focal plane, points can project to several locations. Adjusting the focal plane right is a major challenge in photography. Imagining light rays leaving a camera, all rays from a given pixel will meet on the focal plane. Traversing these light rays in the opposite direction, all rays will be integrated at the given sensor pixel. With 4D light fields, it is possible to perform this integration in a post-capture process (Figure 5.8).

A usual camera with the sensor at the image plane (IP) is simulated by integrating over all directions, hence, the (u, v) plane. Roughly, for a plenoptic camera 1.0, all pixels under a microlens are summed up as: $L(s, t) := \sum_u \sum_v L(u, v, s, t)$. The focal plane depends on the distance of the IP to the ML. Assuming the thin lens model, the original focal plane is at a distance $d_{\text{org}} = (1/f - 1/(z_{\text{UV}} - z_{\text{ST}}))^{-1}$ from the main lens, where f is the focal length of the main lens. A virtual move of the image plane to a (q, r)-plane at z_{QR} causes the focal plane to change. Precisely, the new focal plane will be located at a distance $d_{\text{refocus}} = (1/f - 1/(z_{\text{UV}} - z_{\text{QR}}))^{-1}$. To evaluate the result with the new focus plane, from a point on (q, r) all rays toward (u, v) are integrated, whereby (u, v, q, r) is mapped to (u, v, s, t) coordinates by the ray/plane intersection method described above.

This approach can be rendered more efficiently by splatting individual views (each indexed by their (u, v) coordinates (Figure 5.2 right)). Entire scaled views indexed by (u, v) can be accumulated on the sensor: $L(q, r) = \sum_u \sum_v L(\alpha q + \Delta(u), \alpha r + \Delta(v), u, v)$ with $\Delta(u), \Delta(v)$ denoting the first respectively second component of Δ.

The main challenge of refocusing is that it requires a very high number of different view points in order to achieve a smooth out-of-focus blur. For a large blur kernel, banding or ghosting artifacts can remain visible. As none of the existing plenoptic cameras provides a sufficiently high number of view points, it is often essential to perform view interpolation (Sections 5.5 and 17.2).

In photography, *Bokeh* defines the rendering of out-of-focus areas by a camera lens. For small and very bright out-of-focus lights, this effect can be strong and is used as a stylization method. The shape of the Bokeh is indirectly defined by the shape of the lens aperture. As the lens aperture in a standard camera cannot be changed, photographers often attach an additional aperture with reduced size in front of the lens. The attachment simply blocks incoming light from certain directions. With the 4D light field, it is very simple to simulate such a behavior: $L'(u, v, s, t) = b(u, v)L(u, v, s, t)$ with b being a function mimicking the aperture shape. In order to control the aperture, incident light rays are thus scaled by a weighing factor (usually a binary mask). Additionally, it is possible to make this influence depend on the wavelength.

As refocusing practically requires interpolation in the (u, v) domain to generate additional views (Sections 5.4 and 5.5), the same pipeline can also perform extrapolation. Extending the available directional domain corresponds to photography with a larger aperture, which allows for very narrow depth-of-fields. Manakov et al. [Manakov et al. 13] report the simulation of a lens with an aperture of up to $f/0.7$ from a single snapshot light-field.

In practice, pixels of a light-field camera do not correspond to exact rays. Instead, each pixel records the incident irradiance within a small cone of directions. Each view point that relates to the microlens corresponds to an image taken with a lens of small aperture (Figure 5.2 right). Consequently, these views also exhibit depth-of-field, and any refocusing operation is limited to the depth-of-field range imposed by these optics. Similarly, the captured light field might not be sufficient to deal with large apertures as some light rays necessary for the border pixels might be missing.

In a 4D light field, when keeping (s, t) constant, varying the (u, v) parameters results in a view of the scene (Figure 5.2 right), which roughly corresponds to a capture of the scene with a pinhole camera centered at (u, v). Changing the (u, v) parameters causes a "lens-walk" and offsets the

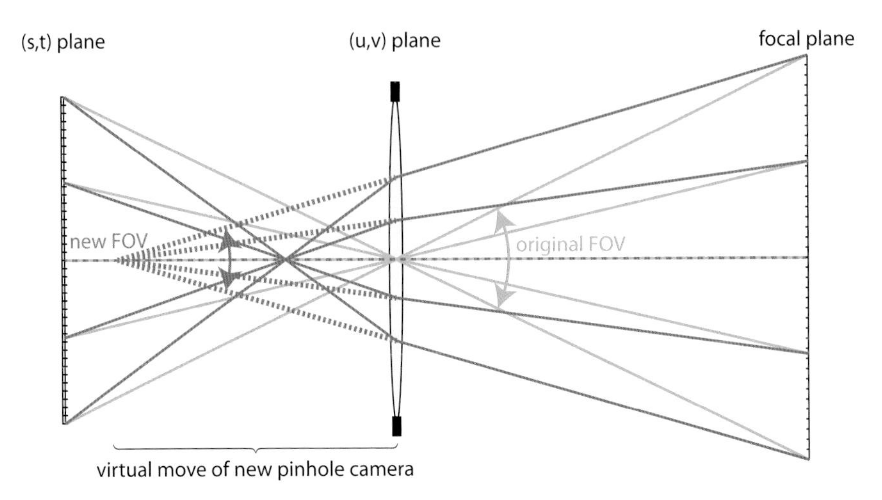

Figure 5.9: Dolly zoom: the light field allows to pick a (u, v) per (s, t). While the center view (green) corresponds to a pinhole camera with center of projection at the origin of the (u, v) plane, it is possible to simulate a moving pinhole camera with different field-of-view. Here, objects in front of the focal plane shrink while objects behind the focal plane grow in projected size (and are potentially cut off).

corresponding image. This 2D effect relates to the Ken Burns-effect using a 2D pan and zoom, but with a light field this walk can also be extended to 3D by varying the (s, t) coordinates. Hereby, a parallax effect is induced due to the different view points. A trivial extension is the generation of stereo images by sampling two (u, v) views. More details on multi-view-stereo are described in Section 8.3.

A computationally more involved effect is the "dolly zoom" or "Hitchcock zoom," where a change of the field-of-view (FOV) and camera motion along the viewing direction are synchronized, while focusing on an object in the scene. It causes out-of-focus regions to warp due to the changing FOV while the focal plane position and its imaged extent remain the same. Typically, this effect is used to shrink/grow the background, while keeping the object in focus at the same scale for a dramatic effect. To achieve this result, the image is rendered by: $L(s, t) = L(s, t, \gamma s, \gamma t)$ with γ defining the strength and direction of the effect. Here, a single ray sample is taken from each view (Figure 5.9).

While refocusing processes thousands of views for high-quality rendering, the dolly zoom requires a single view per pixel. In turn, the computational complexity stems from the fact that it requires dense directional information. In practice, angular interpolation is strictly needed. While splatting of entire views, as in the refocusing application, is not possible, some computational simplifications can be made. The effect is most efficiently implemented by coupling the ray selection $(s, t, \gamma s, \gamma t)$ and directional interpolation in a GPU shader.

5.7 Summary

With the advent of consumer-grade plenoptic cameras, light field imaging has become comparatively cheap. Acquisition of a 4D light field is now as simple as taking a picture with a standard digital camera. Consequently, in addition to the traditional light field applications in computational photography and image-based rendering, a lot of research interest has been geared recently to leverage light fields for computer vision challenges like non-Lambertian 3D reconstruction.

Related to this chapter, Chapter 8 deals with 3D reconstruction from light field correspondence estimation, while Chapter 17 covers image-based rendering in more detail.

6
Illumination and Light Transport

Martin Fuchs and Hendrik P.A. Lensch

6.1 Introduction

Correctly modeling scene illumination is crucial for many applications in real-world visual computing. Many applications require painstaking control and repeatable conditions—for instance, whenever illumination response is used to infer scene geometry, such as in photometric stereo or shape-from-shading. Other applications can abstract from illumination, once it is precisely known. Some methods enable editing the effect of incident illumination on a scene, a technique known as *relighting*, in a way which takes into account both direct reflections, especially diffuse color and specular highlights, as well as more complex indirect effects such as interreflections, subsurface scattering or caustics.

This chapter is concerned with the last group of techniques. It deals with the problem of digitally recording illumination and its effects on scene appearance. Recently, Heide et al. and Velten et al. have proposed sophisticated optical implementations which are able to temporally resolve light transport and create visualizations of light propagating through a scene [Heide et al. 13, Velten et al. 13]. In contrast, this chapter addresses the cumulative effect of illumination on a scene, especially the connection between incoming and observed light as mediated by global light transport.

Exhaustive recording of light transport will likely remain intractable for the foreseeable future. Accordingly, the design of acquisition setups is intrinsically intertwined with choosing an underlying, possibly simplifying, light transport model; as a result, this chapter discusses global modeling of light transport before addressing practical issues of recording.

6.2 Modeling Illumination

Illumination can be exhaustively expressed as a spectral radiance distribution of a 4D incident light field (Chapter 5). Many visual computing applications, however, make the simplifying assumption that illumination is *distant* from the scene being recorded, that is, it is so far away that its variation over the scene surface points is negligible, and hence, in essence, it changes only with incident direction.

In this case, measured illumination can be stored in a 2D data structure, the *environment map* [Blinn and Newell 76]. Environment maps may be parameterized in a variety of ways, the choice of which enables multiple trade-offs regarding *storage efficiency* (parameterizations which fill a single or at least a small number of rectangular texture without wasting space are advantageous here), *rendering efficiency* (parameterizations which can be efficiently sampled, and/or permit a rasterization pipeline to compute the environment map are useful), and finally *low distortion* regarding straight lines and variation of the ratio between solid angle and mapped area. The latter needs to be considered especially when an environment map is used for image-based illumination in rendering, as distortions in solid angle need to be numerically compensated when integrating over the incident illumination. Unfortunately, these trade-offs can generally not be optimized simultaneously.

Aside from distant illumination, other modeling assumptions may also enable storage as a flat image: in cases where illumination can be expressed as a projective mapping between a plane and a ray distribution—as would be the case for digital projectors, as long as the pinhole model is applicable—the parameterization as coordinates in this plane arises naturally as a reasonable choice.

6.3 Measuring Illumination

It is possible to record an environment map by using a camera with a very wide field-of-view (*"fish-eye lens"*), or perform panoramic stitching of several images of a single camera, which is rotated by its optical center (Chapter 3). For recording environment maps special camera rigs, such as the PointGrey Ladybug or the Google Street View cameras have been developed. Spheron offers high-resolution HDR cameras with a rotating line sensor. A more common, cheaper approach, instead, is to record an HDR image of a *light probe*, an object of known geometry and reflectance, and infer the incident illumination from its appearance.

The most common light probe design consists of a ball shape with a mirroring surface material [Debevec 98] (Figure 6.1). This serves two purposes: the rotationally symmetric geometry of a sphere makes registration, that is, the mapping between camera pixels and observed surface geometry, easier, as merely the silhouette of the sphere needs to be detected in a recorded picture. The mirror surface enables a simple environment map

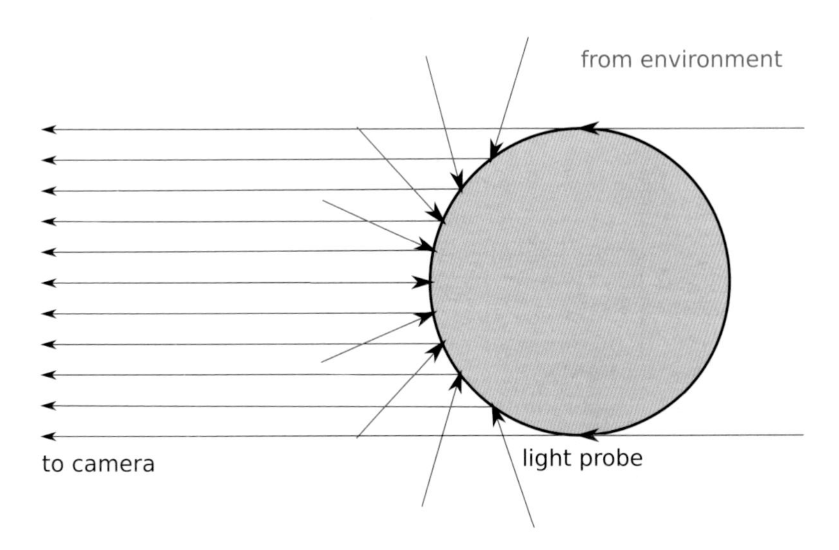

Figure 6.1: A spherical light probe ideally provides a $360° \times 180°$ view of the environment to an orthographic camera. As this illustration of light ray distribution shows, equidistant sampling on the camera creates parameterization problems at the silhouette of the sphere, reducing the usability of grazing-angle observations.

recovery scheme: any ray casting implementation may be used to trace rays through the camera pixels, reflect them at the surface of the estimated sphere, and deposit the image information in the pixel value as radiance measurement in the output environment map. There, the found values need to be interpolated to fill eventual gaps.

In an ideal recording scenario with an orthographic camera, a single image of the sphere reveals a full $360° \times 180°$ environment map. In practice, even though a long focal length can be used to approximate an orthographic projection, a blend between two recordings taken from opposite directions is preferable: environment map directions behind the light probe are observed in pixels close to the sphere silhouette where mis-registration of the sphere position has the largest effect and where the sampling density between orthogonal directions is most unevenly distributed. Additional recording positions can be useful in order to avoid the central pixels which show the recording camera being reflected on the sphere.

Even putting aside specifically crafted optical implements, several options for procuring a mirror ball are available, the least expensive being a holiday ornament such as a Christmas bauble. However, while the manufacturing techniques result in a high-quality surface finish, the material shape is typically so uneven that a precise measurement of the geometry in combination with elaborate registration of the sphere orientation may be required. More precise geometry may be found in ball bearing balls, the surface of which, however, has little resistance to scratching and which are only widely available in limited diameters up to a few centimeters.

Fuchs et al. use a black snooker ball as a light probe, which has both a smooth surface, sharp highlights, and is comparably resistant mechanically [Fuchs et al. 05]. However, due to its non-metallic material, the pixel values cannot be used directly to reconstruct an environment map.

More precise HDR estimates can be obtained by combining a mirror, a dark specular, and a diffuse sphere [Stumpfel et al. 04] to recover the environment map, very bright light sources, and the correct total irradiance. For outdoor scenarios, a low-parameter sky and sun illumination model can be indirectly recovered from a single image without any specific light probe [Lalonde et al. 09].

6.4 Modeling Light Transport

In contrast to just capturing the incident illumination in the form of an environment map, the light transport in an arbitrary scene is a complex, non-local phenomenon which for opaque surfaces in free space is governed by the *rendering equation*

$$L(\mathbf{x}, \boldsymbol{\omega}_o) = L_e(\mathbf{x}, \boldsymbol{\omega}_o) + \int_\Omega f_r(\boldsymbol{\omega}_o, \mathbf{x}, \boldsymbol{\omega}_i) L(\mathbf{x}, \boldsymbol{\omega}_i) \cos\theta \, d\boldsymbol{\omega}. \tag{6.1}$$

The radiance $L(\mathbf{x}, \boldsymbol{\omega}_o)$ leaving point $\mathbf{x}$ in direction $\boldsymbol{\omega}_o$ combines the self emission L_e and the light L incident from all possible directions Ω that is scattered at $\mathbf{x}$ toward $\boldsymbol{\omega}_o$ by the *bidirectional reflectance distribution function (BRDF)* f_r. Here, the light transport can become quite complex as the light reflected at one scene point might indirectly illuminate other scene points due to interreflections, causing so-called global effects. In scenes with participating media and transparent or translucent materials even more complex interaction will occur.

Reasoning about light transport becomes significantly easier when writing the rendering equation in operator notation

$$\mathbf{L} = \mathbf{L}_e + \mathbf{KL}. \tag{6.2}$$

The process of scattering and reflection typically acts linearly on the light field $\mathbf{L}$ and the transformation of the light field due to single scattering events can be modeled by the linear operator $\mathbf{K}$, effectively representing the convolution of the incident light field with the BRDF. Global light transport thus is the solution of Eq.(6.2):

$$\mathbf{L} = (\mathbf{I} - \mathbf{K})^{-1}\,\mathbf{L}_e \tag{6.3}$$

When measuring the light transport in a scene one typically can only measure a subset $\mathbf{C} \subseteq \mathbf{L}$ of the entire light field, typically a single image or a collection of images. Denoting the present illumination $\mathbf{L}_e$ by $\mathbf{L}$ and the transport operator $\mathbf{T} = (\mathbf{I} - \mathbf{K})^{-1}$, most measurement approaches are governed by a simple linear equation:

$$\mathbf{C} = \mathbf{TL} \tag{6.4}$$

Provided some controlled illumination $\mathbf{L}$ and the measurements $\mathbf{C}$, the task is to identify the operator $\mathbf{T}$. Knowing $\mathbf{T}$ allows for synthesizing novel images $\mathbf{C}$ under arbitrary illumination, a process called *relighting*.

The transport operator $\mathbf{T}$ often is also referred to as a *reflectance field* R [Debevec et al. 00] that maps the incident light field to the reflected light field incorporating all global light transport effects. Written as an integration over all incoming directions $\boldsymbol{\omega}_i \in \Omega$ and scene points $\mathbf{x}_i \in S$,

$$C(\mathbf{x}_o, \boldsymbol{\omega}_o) = \int_\Omega \int_S R(\mathbf{x}_i, \boldsymbol{\omega}_i, \mathbf{x}_o, \boldsymbol{\omega}_o) L(\mathbf{x}_i, \boldsymbol{\omega}_i)\, ds\, d\boldsymbol{\omega} \tag{6.5}$$

Insights on the particular structure of $\mathbf{T}$ can be found in [Veach 97, Garg et al. 06, Ramamoorthi and Hanrahan 01c, Seitz et al. 05]: Veach [Veach 97] analyzes the structures that appear in the context of rendering synthetic scenes. Garg et al. [Garg et al. 06] show that direct reflections cause a sparse manifold in $\mathbf{T}$ while global effects tend to introduce partially dense interaction in $\mathbf{T}$ albeit typically with rather low-rank structure. Ramamoorthi and Hanrahan [Ramamoorthi and Hanrahan 01c] analyze the frequency response of light transport from first principles and derive resulting bandwidth constraints when recovering reflectance or illumination. The invertibility of the operator given partial information has been investigated by Seitz et al. [Seitz et al. 05].

One important property of the light transport operator is that it incorporates Helmholtz reciprocity: in the same optical medium, the reflectance observed along a path will be the same as if observed along the same path just in the opposite direction. Thus, with the right setup one measurement can provide information from both sides. This can be used for stabilizing or accelerating the measurement process in the form of dual photography approaches [Sen et al. 05, Garg et al. 06, O'Toole et al. 12].

Techniques for both relighting or light transport measurement can be categorized based on the way the transport operator $\mathbf{T}$ is represented or identified.

By discretizing the light field or camera image as well as the incident illumination, the transport operator maps from illumination to pixel basis functions. For example, the complete reflectance field between a one megapixel camera and the illumination from a one megapixel environment map already yields a matrix $\mathbf{T}$ of $1,000,000 \times 1,000,000$ entries. As the size of $\mathbf{T}$ quickly becomes impractical, many variants have been proposed.

Some approaches directly sample $\mathbf{T}$ as the impulse response to a discrete set of illumination stimuli. Most often, point light sources (or projector rays) are turned on in sequence each time capturing one image (or light field) corresponding to one column of the transport matrix ([Debevec et al. 00, Goesele et al. 04, Fuchs et al. 07]). Relighting then is achieved by vector-matrix multiplication, i.e., each input image is scaled by the desired light source color for that particular direction and the relit image is obtained by summing up all scaled images.

Modeling the illumination in a lower-dimensional space with correspondingly fewer basis functions, e.g., spherical harmonics, the number of required measurements can be drastically reduced [Ghosh et al. 10, Tunwattanapong et al. 13] by measuring the system response to each basis function.

Adaptive schemes try to identify regions in the reflectance function that require sampling with higher resolution compared to others. For example, when the observed reflectance at a novel sampling location can already be explained by interpolation of the coarser samples, the local resolution is already sufficient [Fuchs et al. 07]. Closely related to adaptive schemes is the representation of $\mathbf{T}$ by hierarchical bases. Approaches use, for example, wavelets [Peers and Dutré 03, Sen et al. 05, Peers et al. 09] or hierarchical, low-rank approximations such as H-matrices [Hackbusch 99, Garg et al. 06].

Further simplifications are possible by constraining the space of possible transport matrices. In the simplest form, one focuses on direct reflections only. By identifying only the peak in each row one can derive a simple scheme to project an image onto an arbitrary surface. Inverting the direct component, one can efficiently reduce the influence of the spatially varying reflectance of the surface [Wetzstein et al. 07]. In environment matting [Zongker et al. 99, Matusik et al. 02], the contribution to each camera pixel is limited to a Gaussian distribution around the peak transport coefficient, allowing to approximate effects such as blurred refraction or reflection, but not to truly express global light transport.

Related to the problem of measuring global transport is the measurement of local appearance where the appearance of a specific surface or 3D object is to be captured. Spatially varying reflectance then is parameterized by the surface and the appearance for each surface point modeled by some surface reflection model. Weyrich et al. provide a concise overview of these techniques [Weyrich et al. 08].

Note that the equations given are expressed for total quantities of radiance but can just as well be interpreted as referring to wavelength bands of spectral radiance or color channels. Wavelength-changing effects can be incorporated by an additional dimension to integrate over and expressing distributions over incoming and scattered wavelength.

6.5 Measuring Light Transport

The problem of actually acquiring light transport is closely related to the chosen representation of both the reflectance field and the incident illumination.

Choice of Illumination

Early research focused on point light sources providing highly controllable illumination. The positions of the light sources here are either fixed, as in a light stage [Debevec et al. 00], manually or adaptively positioned [Masselus et al. 02, Lensch et al. 03, Fuchs et al. 07] or treated as unknowns to be recovered in the reflectance estimation process. In order to achieve denser sampling of the incident hemisphere, monitors may be used as light sources. They provide addressable resolution in the millions of pixels, but, as a consequence, a direct sampling approach is practically intractable. Instead, hierarchical schemes are employed [Zongker et al. 99, Peers and Dutré 05, Sen et al. 05].

The use of projectors [Masselus et al. 03, Goesele et al. 04, Sen et al. 05, Garg et al. 06] allows for capturing both directional dependency as well as localized spatial scattering phenomena, but at the same time it increases measurement complexity as 4D illuminations are used.

Reconstruction Algorithms

As the cost of brute-force sampling of all light sources quickly becomes too expensive, some algorithms explore the structure of the light transport, reconstructing the constrained operator from fewer measurements rather than measuring it directly.

Using varying natural illumination conditions and capturing the actual incident illumination with a light probe, Matusik et al. reconstruct the reflectance field for outdoor relighting by least squares minimization [Matusik et al. 04]. Fuchs et al. capture indoor scenes under varying illumination, using the incident environment map implicitly as an illumination basis [Fuchs et al. 05]. For relighting, a Bayesian framework projects the intended illumination into the basis of captured illuminations, producing a set of weights that are used to blend the input images directly.

Based on adaptive, hierarchically refined illumination patterns, Garg et al. recover a block-wise rank-1 approximation of the light transport based in H-matrices [Garg et al. 06]. Blocks and the corresponding illumination patterns get refined as long as the rank-1 approximation does not hold.

General low-rank approximations of the overall light transport can be recovered using the Kernel–Nyström method by successively exploring a higher-rank reconstruction with each additional measurement [Wang et al. 09]. O'Toole and Kutulakos directly measure the eigenvectors of the light transport operator by computing the next illumination pattern using the Krylov subspace method [O'Toole and Kutulakos 10]. By illuminating with the next illumination vector, the scene itself carries out the necessary vector-matrix multiplication.

Using a fixed set of wavelet noise illumination patterns, Peers et al. successively refine a wavelet representing the reflectance field such that the observations can be explained in a least squares sense [Peers and Dutré 05]. This reconstruction problem is similar to compressive sensing where a fixed set of noise patterns are employed to recover a sparse wavelet representation by minimizing the L_1 error [Peers et al. 09]. Sen et al. show that the same set of fixed illumination patterns can be used to recover a sparse representation independent of the chosen basis [Sen and Darabi 09].

In some applications, it is not necessary to recover the complete transport operator. Instead, qualitatively different parts of the light transport are identified. An example is the separation of direct from global light transport effects [Nayar et al. 06]. Providing high-frequency illumination patterns, direct reflections are immediately affected while the global component tends to produce constant response due to smoothing. Similarly, both effects can be separated using time-resolved transient imaging as global effects propagate over longer, more complex paths [Wu et al. 12b]. High-frequency patterns in the angular domain can be employed to separate specular from diffuse reflectance [Lamond et al. 09].

6.6 Measuring Light Transport—Practical Issues

It is good practice to design the measurement process around the scene types to be recorded. One important aspect is invariance: taking many pictures takes significant time, and this may rule out some methods or setups altogether. Human faces, for instance, can be kept motionless for at most a few seconds at a time, and even then, fine structures such as hair may be susceptible to air flow—accordingly, the entire measurement process may not take time in excess of a few seconds, ruling out moving a light source and enforcing a setup with statically configured, quickly switchable light sources [Wenger et al. 05]. But even seemingly inanimate objects may vary their appearance surprisingly quickly; freshly cut flowers, for instance, may change their shape over a matter of minutes.

Measurement Setup: Illumination

In active approaches, the design of actively controlled illumination plays an important role. Needless to say that external influences are to be minimized, including daylight or other uncontrolled light leakage into the lab in which the recording takes place. Black cloth has been used to great effect [Goesele et al. 00]; it serves a dual purpose of not only isolating the measurement space from the outside, its low albedo also reduces the influence of bounce light, increasing the signal-to-noise ratio. Evenly spread cloth is often preferable to folds: while folds permit less bounce light on average, the bounce light shows visible directional variation, and the folds are revealed in mirror-like reflections, while uniform black cloth may be visually hidden under the camera noise floor. Rough materials are preferable to satin-like appearances as they distribute bounce light more evenly over the outgoing directions.

Both for active and passive approaches, the light source needs to be calibrated: either by recording the incident illumination [Matusik et al. 04] while the primary measurement takes place, or as a separate calibration step. In an active setup, the latter is usually preferable in order to prevent the calibration of the light source from interfering with the measurement itself. For an active illumination measurement pipeline, there is also a choice: one can either observe the light distribution (position and shape of the light source) [Goesele et al. 03] or use technical means to create it (for instance, applying robotics to create a precisely moving light source, or displaying a repeatable pattern on a computer monitor).

Due to the linearity of light transport, it is relatively easy to correct for uneven brightness between illumination sources in software: it is sufficient to multiply a linearized input image to match an exact value. Correcting a deviant position, however, is nearly impossible without knowing scene geometry and reflectance in advance (in which case there is no need to record it anyway). Spectral differences cover a middle ground: for photographic applications, (matrix) color correction may result in satisfactory results; for precise measurements of spectral behavior, the illuminants need to be carefully pre-selected to match precisely.

Depending on the specific illuminants used in recording, variations of light source brightness over the course of the measurement play a role as well: mercury-vapor lamps need a few minutes to heat up, then stay stable for some time, but may subtly fade in brightness over matters of hours. Light-emitting diode (LED) light sources are stable for a longer time and are instantly ready—many white LEDs rely on phosphors, though, to create a spectrally broad illumination, which may require heat-up times in the order of fractions of a second before the emitted spectrum is stable.

Besides stability, overall brightness of light sources is an important design choice as there is a trade-off on light source brightness and camera exposure time: in principle, less brightness can be compensated for with longer exposure time. The effect of some noise types, however, such as dark current noise (Section 1.3), increases with exposure time. On the other hand, brighter light sources increase the ambient temperature in a usually enclosed space, thus contribute to thermal noise, and may negatively impact the scene to be recorded.

As components for home entertainment systems, digital projectors have recently become relatively inexpensive, and increasingly attractive to use for light transport acquisition purposes. Yet, considered as measurement light sources, they come with challenges of their own. Professional systems still have advantages when it comes to synchronization; at the low end, even seemingly unnecessary problems may be encountered, such as the native output resolution not being available to the input video interfaces. For all technologies, contrast is a serious limitation. Some manufacturers achieve high contrast between successive frames with global shutters. However, for measurement purposes, local contrast plays a bigger role and remains limited: while individual pixels can be reliably switched on, the brightness of a single pixel cannot offset the black level of two million others in a HD projector; while the equivalent of dark frame subtraction—take a picture with the projector instructed to show a black frame—helps in theory, the minute increase in brightness of switching on a single pixel over a uniform, but large frame can rarely be observed with an off-the-shelf digital camera due to its limited in-frame dynamic range (Section 1.3).

Geometrically, projectors can approximately generate light field slices for rays that originate from a single point, the center of projection. As they are designed to illuminate a planar surface with high efficiency, they usually come with large apertures and hence only achieve shallow depth of field, limiting the working volume. Laser projectors, both sweeping line and modulating plane designs, usually do not have this restriction, but come with problems of their own: for one, laser light is coherent and interference effects produce speckle on the camera sensor which may dominate the effects to be measured.

In addition, laser projectors share a problem with RGB LED projectors: the color is mixed from very pure primaries. This makes for excellent color reproduction in projection as perceived by a human observer. However, for measurement purposes, broader spectra deliver better color rendition.

Comparing liquid crystal (LCD) and digital mirror (DLP) technologies reveals differences in how intermediate brightness levels are created. LCDs can maintain constant intermediate levels, but need minimal time to vary brightness and usually have less local contrast than their DLP counterparts. DLPs, on the other hand, switch practically instantly and have better contrast, but dither between minimal and maximal brightness over space and time. Hence, precise synchronization between projector and camera is an even more important issue. As a minimal measure, the camera exposure time should be chosen as an integer multiple of the projector's frame time. This is especially important for DLP projectors which use a spinning color wheel to create colors, as integer-frame exposure time helps to counter the "rainbow" color artifacts. For multi-projector setups, precise synchronization is important, as well.

Measurement Setup: Supporting Equipment

Beyond illumination devices, other components of the measurement setup need to be picked carefully in order to ensure that light direction and shape are the only variants. This implies high requirements regarding the dependability of the equipment. Tripod mounts must be stable: even if they have a stable "feel," slight movements, which may only amount to a drift of few pixel rows over the course of days, may ruin measurement sequences which take a long time to complete. Cameras without moving parts are preferable to those that need to move a mirror prior to exposure or use mechanical shutters. The measurement lab should ideally be equipped with air conditioning to provide for constant ambient temperature and air humidity.

It should go without saying that in virtually all designs for measuring light transport, a high dynamic range pipeline is required so as to obtain exact reflections (Section 1.5). If highlights on a mirroring surface are clipped, for instance, reflections of dim light situations will come out much too dark.

6.7 Summary

When the task is to resolve the light transport the method of choice depends on the specific global illumination effect that should be resolved. Capturing diffuse or specular reflection of known surfaces is simpler than resolving the material properties contributing to interreflections, subsurface scattering or caustics. Though the light transport can be modeled as a linear operator its complexity in many dimensions still prevent a dense sampling approach—approximating models and reconstruction algorithms are therefore most often employed.

Part II

Reconstruction—Data Processing Techniques

7
Camera Registration from Images and Video

Jan-Michael Frahm and Enrique Dunn

7.1 Introduction

Nowadays, cameras are ubiquitously available and billions of photos and videos are uploaded every year to photo and video sharing sites. Combined with the recent progress in computer vision, this has lead to the development of large-scale 3D modeling from these images with approaches proposed by Snavely et al. [Snavely et al. 06], Agarwal et al. [Agarwal et al. 11], and Frahm et al. [Frahm et al. 10]. Figure 7.1 illustrates a dense 3D model obtained from a photo collection modeling of Frahm et al. [Frahm et al. 10]. These reconstruction methods all rely on the registration of the cameras (determination of their relative/absolute poses at capture and their internal camera parameters) in a common coordinate system. Once the registration is known, the cameras can be leveraged within dense scene geometry estimation (see Chapter 8 for more details on dense depth estimation) to obtain a dense 3D model. Beyond dense depth estimation, camera registration is also required for a broad range of applications like sensor fusion with camera images (see Chapter 9), reconstruction of dynamic structure (see Chapters 11, 12), and more. The automated process of jointly esti-

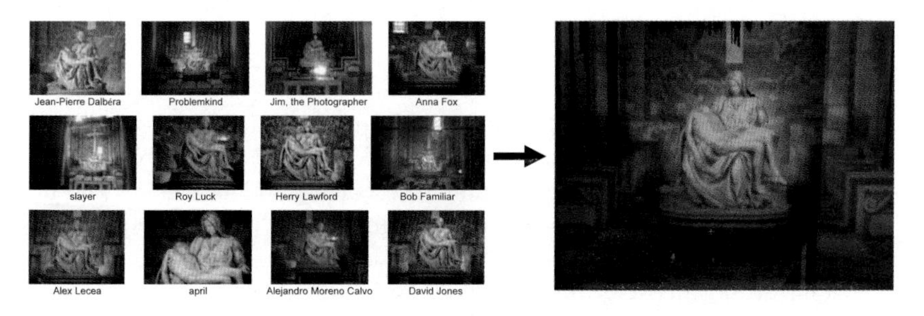

Figure 7.1: An example dense 3D model from an Internet photo collection of Rome, Italy, computed by the method of Frahm et al. [Frahm et al. 10].

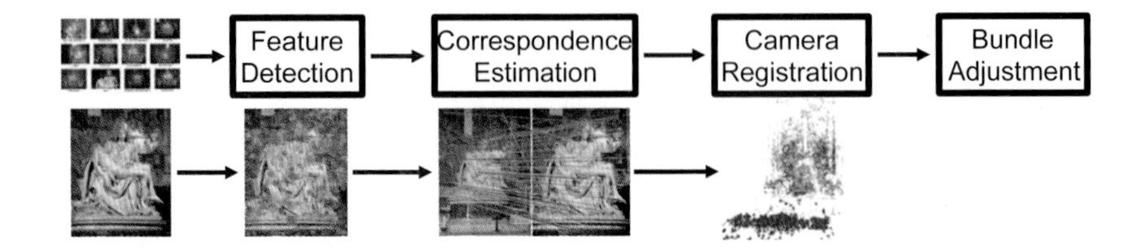

Figure 7.2: Overview of the general processing steps of structure-from-motion methods.

mating sparse scene structure and camera viewing parameters from a set of input imagery, is known as structure from motion. Camera registration is a prevalent task of each part of the structure-from-motion pipeline.

7.2 Structure from Motion Pipeline Overview

Structure from motion algorithms take as input a set of images or video frames and, if available, their associated camera calibration, and determine the relative registration of the cameras during capture. The relative camera registration consists of camera poses of all cameras in a common reconstruction coordinate system, which is related to the world coordinate system by a similarity transformation consisting of a rotation, translation, and scaling of the reconstruction coordinate system. While images and videos are captured in a wide variety of scenes and capture configurations, there are common steps to all structure from motion methods. Most structure from motion techniques perform the following steps (Figure 7.2):

- **Feature detection** determines salient points (features) in the images or video frames. These points are generally expected to be reliably detected across different viewpoints and zoom levels to ensure the detection of the same salient features across different frames. Each feature has an associated descriptor to quantize the characteristics of the feature.
- **Feature matching or tracking** determines the features corresponding to the same 3D point in multiple different images or frames. Matching or tracking determines the correspondence using the descriptors of the detected features in the images and delivers a set of putative correspondences. These putative correspondences typically contain both correct and incorrect correspondences.
- **Robust camera registration** simultaneously determines the correct correspondences from the putative set of correspondences and

the camera poses within the reconstruction coordinate system. Additionally, the 3D points for the correct correspondences are typically triangulated. The set of 3D points is referred to as a sparse point cloud.

- **Bundle adjustment** is an optional non-linear refinement to optimize the camera poses along with the sparse point cloud to obtain more accurate camera poses and sparse 3D features. While this step is optional in small-scale reconstructions, it is recommended for large-scale reconstructions in order to control drift.

Most structure from motion processes follow the above general steps to obtain a camera registration. The differences between various structure from motion systems [Wu 11, Frahm et al. 10, Snavely et al. 06, Agarwal et al. 11, Pollefeys et al. 04] typically result from accounting for different characteristics of captured data (unordered images vs. frames of a video with temporal order) or differences in the desired levels of accuracy or performance. This chapter focuses on the efficient registration of unordered image sets as the most general case of structure from motion.

Local Feature Detection and Description

Estimation of the camera motion between two given images, $\mathbf{I}$ and $\mathbf{I}'$, is at the core of structure from motion methods. While it is easy for humans to perform this task (e.g., if everything in the image moves to the left, the camera moved to the right), for the computer this task is hard as it does not have a global understanding of the image content. Hence, the computer cannot directly infer motion from the captured images $\mathbf{I}$ and $\mathbf{I}'$. Instead, camera motion estimation algorithms rely on the motion of salient 3D points observed in the images to reveal the motion of the capturing cameras. To support this task, structure from motion first detects these salient image features.

The detection of salient features is a long-standing research topic in computer vision and photogrammetry. The desired properties for feature points are:

- **Repeatability:** The same feature can be found in several images despite geometric and photometric transformations.
- **Saliency:** Each feature has to be distinctive and robust to clutter and partial occlusion.
- **Compactness and efficiency:** There are many fewer features than image pixels in each image.
- **Locality:** A feature occupies a relatively small area of the image.

Figure 7.3: Left: Original image of the Pieta in Rome (courtesy David McSpadden). Right: SIFT features detected for the image of the Pieta. The position, orientation and scale of each of the detected features is depicted by an overlayed box.

The above properties may be at odds with each other and will require the design of a feature detection mechanism to consider various performance trade-offs. For example, while achieving repeatability entails invariance/robustness with respect to a given set of image transformations, the presence of such transformations within a single image will hinder the distinctiveness of the detected features given the afforded invariance.

The output of a feature detection process is a sparse set of image pixel positions corresponding to the detected features. In order to perform feature association, a representation describing the (local) appearance properties of each detected feature is required. The set of properties desirable for feature detection is also applicable to feature description mechanisms. There is a wide variety of salient feature detectors/descriptors available in the literature [Lowe 04, Bay et al. 06, Harris and Stephens 88, Matas et al. 04]. One of the most commonly used features is the SIFT feature [Lowe 04], which is robust against in-plane image rotation, changes in the scale of the observed features, and small illumination changes of the scene (Figure 7.3).

While SIFT features are very robust [Mikolajczyk and Schmid 05] their detection is computationally expensive and methods to improve the computational performance have been proposed. One example of a more computationally efficient feature is the SURF feature [Bay et al. 06], which exercises a similar robustness to rotation, scale, and illumination as the SIFT feature [Heinly et al. 12]. Alternatively, the use of commodity graphics cards has been successfully proposed to perform the feature detection [Sinha et al. 11]. SIFT and SURF features compute a 128-dimensional or 64-dimensional feature descriptor, respectively, for each salient feature point in the image. Given that an image often has several thousand features, the amount of storage required for the feature descriptors is often comparable to the image size. Hence, for large-scale datasets, it is required to reduce the amount of storage to both conserve disk space and to reduce the bandwidth required during matching when the feature descriptors of the images are used to identify potentially corresponding features.

Binary features aim to address the memory efficiency for the descriptors by representing the feature descriptor as a binary vector [Calonder et al. 10, Leutenegger et al. 11, Rublee et al. 11, Alahi et al. 12, Heinly et al. 12]. These binary descriptors typically encode the sign of the local intensity gradient in the image around the 2D feature point for various directions and locations in the vicinity of the feature point. These are then encoded as binary strings to describe the feature. Typical binary descriptors are 128-512 bits and can perform well if chosen appropriately for the application. Heinly et al. [Heinly et al. 12] investigate the performance and the correct choice of binary features for a variety of visual tasks. They show that while no binary feature measures up to the overall performance of SIFT in the large variety of tested scenarios, with careful consideration for the specific needs of any application, binary features can perform similarly well in the appropriate limited scenario.

Once the appropriate features with respect to the targeted application scenario are chosen, these features build the foundation for the next steps. After feature detection and their descriptor computation, the correlation of features across images has to be tackled by the matching or tracking algorithm, which is discussed in more detail in the next section. Popular currently available implementations of the above techniques include [Bradski 00][1] and [Vedaldi and Fulkerson 08].[2]

Feature Matching and Tracking

After detecting the salient feature points in multiple images, the features need to be correlated to obtain the putative matches, which are then used in camera pose estimation. In principle, this correspondence information is obtained through two major classes of techniques. The first class are

[1] http://opencv.org/

[2] http://www.vlfeat.org/

the matching techniques, which detect salient features independently in each image and then correlate their descriptors to obtain the correspondences. The second class are the tracking techniques that detect salient image points in one image and then search for the best corresponding location in the second image. Matching techniques are the more often used for image collections while feature tracking methods are better suited to leverage the temporal correlation observed in video sequences.

Given that for unordered image collections there is typically no prior information about the overlap of any pair of images, no meaningful spatial prior for matching features can be established. Hence, for each feature in the first image, the corresponding feature can be anywhere in the second image, and the correspondence search has to be performed against all features in the image. Given that the feature descriptor is associated with a similarity metric $\mathcal{S}$, the correspondence search obtains the most similar feature point in the second image as the maximum similarity match. For the popular SIFT feature [Lowe 04], with its 128-dimensional feature descriptor, the angle between two feature descriptors $\mathbf{m}_i$ and $\mathbf{m}_j$

$$\mathcal{S}(i,j) = \frac{< \mathbf{m}_i, \mathbf{m}_j >}{\|\mathbf{m}_i\| \|\mathbf{m}_j\|}$$

is used as a similarity metric $\mathcal{S}$. In practice, it is more efficient to normalize the feature descriptors m_i beforehand and use the inner product $< m_i, m_j >$ directly as a similarity metric. For binary feature descriptors the similarity metric is typically the Hamming distance of the feature descriptors [Calonder et al. 10, Leutenegger et al. 11, Rublee et al. 11, Alahi et al. 12].

Using the similarity metric $\mathcal{S}$, the correspondence search then determines the putative match as the feature most similar to the first image's feature (Figure 7.4). Repetitive or duplicate structures in the scene, which have highly similar feature points cause inherent ambiguities in the matching and frequently lead to mismatches (Figure 7.4). To avoid this disturbance, many feature matching methods measure the similarity distance $\mathcal{S}_2$ to the second best match. Lowe [Lowe 04] proposed to remove a match from the pool of putative matches if the ratio $a = \frac{1-\mathcal{S}_1}{1-\mathcal{S}_2}$ of the best similarity $\mathcal{S}_1$ to the second best similarity $\mathcal{S}_2$ measures is too close to one, which means the feature descriptors are very similar and can easily be confused. It is typical to exclude matches whose ratio is higher than 80% [Lowe 04]. Similar ratio tests can be defined for the binary feature's similarity using the Hamming distance [Heinly et al. 12]. Additionally, it is common to test the consistency of the match by matching the feature from the first image to the features in the second image and vice versa. The match is only accepted when the matches from both matchings are consistent. Please note, this doubles the computational effort during matching and is in practice

Figure 7.4: Left: Matches for the Pieta in Rome. Right: Disturbing effect of repetitive structures in putative matching (images courtesy of Flickr users David McSpadden, Megan Allen, Luca Semprini, YougoPL).

often not performed to attain a higher processing speed. The remaining putative matches are then used in robust camera estimation (Section 7.2). In uncontrolled image collections it is expected that there are potentially high levels (>70%) of incorrect putative matches. This is due to several contributing factors, such as significant illumination changes, their larger angles of out-of-plane camera motion ($> 30°$), and the occurrence of wider baselines between cameras causing the viewing angle to change significantly.

For videos, there are additional constraints available to further improve the putative matches or reduce the computational effort of the correspondence recovery. In video, the motion between frames is limited and the appearance variation of the scene is typically very limited. Moreover, in practice, the scene illumination can only change slightly between two video frames. Accordingly, in video, feature tracking is exploited for the frame-to-frame correspondence search. In contrast to matching, feature tracking detects the salient feature points in one image, and then actively searches for the corresponding point in the second image, for example using a gradient based search, like it is done in the well-known KLT tracker [Shi and Tomasi 94, Lucas and Kanade 81]. The KLT tracker assumes a small motion (theoretically less than 1 pixel) between images and a constant appearance of the object between two consecutive frames of the video. The stable appearance assumption leads to the widely used brightness constancy assumption

$$\mathbf{I}^{t+1}\left(\mathbf{x} + \frac{\mathbf{dx}}{2}\right) - \mathbf{I}^{t}\left(\mathbf{x} - \frac{\mathbf{dx}}{2}\right) = 0, \tag{7.1}$$

where $\mathbf{I}^t$, $\mathbf{I}^{t+1}$ are the frames at time t and $t+1$ and $\mathbf{dx} = (dx, dy)$ is the motion of pixel $\mathbf{x} = (x, y)$ in image $\mathbf{I}^t$ with respect to the image $\mathbf{I}^{t+1}$, i.e., pixel $\mathbf{x}$ in $\mathbf{I}^t$ corresponds to pixel $\mathbf{x} + \mathbf{dx}$ in image $\mathbf{I}^{t+1}$. Leveraging the small motion assumption, Eq.(7.1) can be linearized around the pixel location $\mathbf{x}$. Equation (7.1) considers the brightness of a single pixel, which is subject to ambiguities due to the similarity of the pixels and due to disturbance by noise. Hence, the KLT tracker assumes a consistent motion of the pixels in a patch $\mathcal{P}$. This leads to the following linear equation system for solving for the unknown motion $\mathbf{dx}$ of the pixel $\mathbf{x}$

$$\sum_{\mathbf{x}\in\mathcal{P}} \begin{pmatrix} \mathbf{I}_x^2 & \mathbf{I}_x\mathbf{I}_y \\ \mathbf{I}_x\mathbf{I}_y & \mathbf{I}_y^2 \end{pmatrix} \begin{pmatrix} dx \\ dy \end{pmatrix} = 2\sum_{\mathbf{x}\in\mathcal{P}} \begin{pmatrix} \left(\mathbf{I}^t(\mathbf{x}) - \mathbf{I}^{t+1}(\mathbf{x})\right)\mathbf{I}_x \\ \left(\mathbf{I}^t(\mathbf{x}) - \mathbf{I}^{t+1}(\mathbf{x})\right)\mathbf{I}_y \end{pmatrix}, \quad (7.2)$$

with $\mathbf{I}_x = \frac{\partial \mathbf{I}^{t+1}(\mathbf{x})}{\partial x} + \frac{\partial \mathbf{I}^{t}(\mathbf{x})}{\partial x}$, $\mathbf{I}_y = \frac{\partial \mathbf{I}^{t+1}(\mathbf{x})}{\partial y} + \frac{\partial \mathbf{I}^{t}(\mathbf{x})}{\partial y}$. Eq.(7.2) can be solved to obtain the motion $\mathbf{dx}$ of pixel $\mathbf{x}$. Since this is only valid for small motion in the image, in practice, tracking is performed in a hierarchical way to enable the tracking of larger motions between frames. Kim et al. [Kim et al. 07] proposed an extension of KLT that also accounts for the change in illumination and thus overcomes the strict constraints posed by brightness constancy (Eq.(7.1)).

Due to the explicit correspondence search and the smaller motion, tracking typically produces significantly higher rates of correct correspondences and smaller feature position uncertainties. The uncertainty is, in most cases, reduced to less than a pixel compared to the positional uncertainty of the SIFT based matching, which is typically assumed to be accurate within about four pixels.

After determining the set of putative correspondences between the images or video frames, these correspondences are used to determine the camera pose. The OpenCV library [Bradski 00][3] includes an implementation of the KLT tracker as well as a variety of feature matching frameworks.

Camera Registration

Camera registration within the context of structure from motion entails the estimation of the viewing parameters of a given set of images (Chapter 2). This is achieved through the geometric analysis of their jointly observed scene structure. Intuitively, camera registration strives to determine both the camera pose and internal camera *parameters* that best explain the available image feature *measurements* with respect to a given geometric *model* relating 3D structures and their 2D image observations. The sought internal camera parameters are focal length, principal point, and the skew of the camera (Chapter 1). In general, the level of abstraction suitable for geometric analysis varies according to the image set's cardinality as well as the availability of parameter and scene priors. While pairwise camera analysis based on epipolar geometry can be used to bootstrap projective 3D scene modeling and camera registration, the availability of scene and/or camera

[3] www.opencv.org

intrinsic knowledge can simplify the camera registration problem to one of 3D Euclidean resection [Pollefeys et al. 04, Hartley and Zisserman 03].

These insights are leveraged by structure from motion modules implementing incremental camera pose estimation [Wu 11][4] [Snavely et al. 08b].[5] Specifically, an initial pair or triplet of cameras establishes a common camera coordinate system and the 3D positions of the correct salient feature point matches, enabling subsequent cameras to be registered with respect to the initial reconstruction. There are a few approaches in the literature treating structure from motion as a global problem [Sinha et al. 12]), but for large-scale camera registration this is computationally prohibitive in practice. Camera registration algorithms differ for the cases of uncalibrated and calibrated cameras (i.e., unknown vs. known internal parameters), as in the latter case stronger constraints are available. Camera registration for uncalibrated cameras is more demanding than registration of cameras with known internal camera calibration. Intuitively, the motion of the corresponding features reveals insights on the camera motion between the two images, i.e., if the feature moved to the right the camera moved to the left and similar constraints can be formulated for all degrees of freedom of the camera motion.

Uncalibrated camera registration exploits the fact that an image feature with homogeneous pixel coordinates $\tilde{\mathbf{x}} = (u, v, 1)^\top$, under the pinhole camera model, corresponds to a viewing ray r passing through the camera's optical center. The ray then intersects the image plane at 3D coordinates $\mathbf{K}^{-1}\tilde{\mathbf{x}}$, where $\mathbf{K} \in \mathbb{R}^{3\times3}$ is an upper triangular matrix describing the camera's internal calibration parameters, Eq.(1.10). The camera coordinate system is assumed to be aligned with the first camera, which is at the origin in canonical orientation. To obtain the position of the second camera of the initial pair, structure from motion models the relation between the salient feature points in the first image and its corresponding points in the second image.

The fundamental matrix $\mathbf{F} \in \mathbb{R}^{3\times3}$ linearly maps the viewing ray r of pixel $\tilde{\mathbf{x}}$ in the first camera into a homogeneous 2D line in a second camera, $\mathbf{l}' = \mathbf{F}\tilde{\mathbf{x}}$. The fundamental matrix can be composed as

$$\mathbf{F} = \mathbf{K}'[\mathbf{e}]_\times \mathbf{R}\mathbf{K}^{-1}, \tag{7.3}$$

where the epipole $\mathbf{e}$ is the projection of the second camera's center into the

[4] http://ccwu.me/vsfm/
[5] http://www.cs.cornell.edu/ snavely/bundler/

first camera, $[\cdot]_\times$ represents the matrix formulation of the cross product,[6] $\mathbf{R} \in SO(3)$ is a rotation matrix describing the orientation of the second camera, while $\mathbf{K}$, $\mathbf{K}'$ describe the internal camera parameters for the first and second cameras, respectively. The fundamental matrix has rank two and seven degrees of freedom [Hartley and Zisserman 03]. The seven degrees of freedom result from the fact that a 3×3 matrix has at most nine degrees of freedom. The rank 2 constraint removes one degree of freedom. Additionally, the fact that the fundamental matrix is a projective transformation means that it is scale-invariant, which reduces the degrees of freedom to seven. For a feature point $\tilde{\mathbf{x}}$, the corresponding matching point $\tilde{\mathbf{x}}'$ in the second image is constrained to lie on the line $\mathbf{l}'$, which is the projection of the viewing ray r. This relation is expressed in the epipolar constraint

$$(\tilde{\mathbf{x}}')^\top \underbrace{\mathbf{F}\tilde{\mathbf{x}}}_{\mathbf{l}'} = 0. \tag{7.4}$$

Equation (7.4) provides one constraint for each correspondence $\tilde{\mathbf{x}}$, $\tilde{\mathbf{x}}'$ between a pair of images, enabling the estimation of the fundamental matrix $\mathbf{F}$ from a set of seven or more feature correspondences in general configuration [Hartley and Zisserman 03]. The fundamental matrix can be exclusively computed from feature correspondences in the absence of any priors on the camera parameters. The process of estimating the epipolar geometry between two views is commonly referred to as pairwise geometric verification.

Equation (7.4) is linear with respect to the matrix values $\mathbf{F} = \{F_{jk}\}$ and can be estimated through least squares methods. Moreover, given two corresponding feature positions, $\tilde{\mathbf{x}}_i$ and $\tilde{\mathbf{x}}'_i$, each epipolar constraint can be expressed in linear form as $\mathbf{a}_i\mathbf{f} = 0$, where $\mathbf{f} = \mathbf{vec}(\mathbf{F})$ is a row-major vectorization of the $\mathbf{F}$ matrix into a column vector, while $\mathbf{a}_i^\top = \mathbf{vec}\left(\tilde{\mathbf{x}}'_i\tilde{\mathbf{x}}_i^\top\right)$ is the vectorization of the outer product of vectors $\tilde{\mathbf{x}}_i$ and $\tilde{\mathbf{x}}'_i$. Given multiple feature correspondences, the fundamental matrix can be estimated from a linear system of equations of the form

$$\begin{bmatrix} \mathbf{a}_1 \\ \vdots \\ \mathbf{a}_N \end{bmatrix} \mathbf{f} = \mathbf{A}\mathbf{f} = 0. \tag{7.5}$$

Solving Eq.(7.5) should leverage the rank deficiency and scale ambiguity of $\mathbf{F}$. Hartley addressed input normalization in over-constrained estimation,

[6] The cross product matrix for vector $\mathbf{e} = [e_1, e_2, e_3]$ is given by

$$[\mathbf{e}]_\times = \begin{bmatrix} 0 & -e_3 & e_2 \\ e_3 & 0 & -e_1 \\ -e_2 & e_1 & 0 \end{bmatrix}.$$

with eight or more features, to improve numerical stability [Hartley 97]. The fundamental matrix can then be used to obtain the projection matrix of the second camera [Hartley and Zisserman 03]. Note that due to the projective nature of the epipolar geometry formulation, the estimation of the fundamental matrix enables 3D reconstruction only up to a projective transformation of the scene.

Solving Eq.(7.5) assumes strictly error-free feature correspondence estimation, which is rarely attained in practice for scenes captured under uncontrolled settings. Accordingly, robust estimation frameworks like RANSAC [Fischler and Bolles 81] are employed for fundamental matrix estimation.

Robust model estimation is important for obtaining correct reconstructions in the presence of corrupted input measurements. One of the most commonly used robust estimation frameworks is RANSAC, which enables robust parametric model fitting through the joint estimation of model parameters (here the fundamental or essential matrix) and the classification of input data into model compliant (inliers) and non-compliant data (outliers). Algorithm 7.1 provides the pseudo-code for RANSAC.

On a high level the RANSAC algorithm iterates two steps, the *hypothesis generation* and the *hypothesis verification*. In the hypothesis generation RANSAC produces hypotheses for the model leveraging data samples to explore the space of models, i.e., it selects a random sample of the data and computes the model from this data. During the hypothesis verification phase RANSAC uses all data to verify if they support the model and it counts the number of supporting points for each model. After a sufficient number of iterations of hypothesis generation and verification, the best seen model obtained so far is returned.

Hypothesis generation draws sample sets $\mathcal{J}$ of size s from the data $\mathcal{D}$. Then the sample set $\mathcal{J}$ is used to compute a model $M_{\mathcal{J}}$. This model will be the correct model $M_{\mathcal{J}}$ if the sample set $\mathcal{J}$ only contains correct data, called *inliers*, which are compliant with the true underlying model and are only slightly corrupted by noise. If the sample set $\mathcal{J}$ contains erroneous data, called *outliers*, the model $M_{\mathcal{J}}$ will be wrong.

Hypothesis verification aims at determining, which of the generated hypotheses/models $M_{\mathcal{J}}$ is supported by the data. In this context, the support of an individual data point to a model $M_{\mathcal{J}}$ corresponds to a thresholding on the magnitude of the observation residuals with a value Θ (inlier threshold). The points below the threshold Θ form the inlier data points $\mathcal{I}_{\mathcal{J}}$. The set of outliers $\mathcal{O}$ is the complement of the inlier set $\mathcal{I}_{\mathcal{J}}$.

The number of inliers $|\mathcal{I}_{\mathcal{J}}|$ is then used as the criterion for model selection from the generated hypotheses $M_{\mathcal{J}}$ by keeping the best hypothesis seen so far as the current estimate of the model M. The sequential sampling is terminated once the current number of iterations guarantees that, with probability ρ, a good model has been sampled. The number of required samples h is defined by:

$$h = \frac{\log(1-\rho)}{\log(1-\epsilon^s)}, \tag{7.6}$$

with $\epsilon = \frac{|\mathcal{D}|}{|\mathcal{I}|}$ being the fraction of inlier data points in the total data $\mathcal{D}$. Hence, the required number of iterations h is a function of the fraction of inliers in the input data, the size of the sample set $\mathcal{J}$ and the desired level of confidence. Moreover, lower inlier ratios ϵ will reduce the probability of each sample to be a correct sample of data points not corrupted by noise. Similarly, a large data sampling set $\mathcal{J}$ will reduce the probability of finding an ensemble of data points exclusively comprised by inliers. Accordingly, the use of minimal sampling sets is required for efficiency-driven applications. For example, the fundamental matrix estimation based on the minimal sample of $n = 7$ points (sampling subset) is generally preferred over the more numerically stable 8-point method in the context of RANSAC-based robust estimation.

As explained above, by leveraging the fundamental matrix a projective camera registration can be established, which can then be upgraded into a Euclidean reconstruction. If the camera's internal calibration parameters are known, a Euclidean reconstruction can be established directly.

Data: dataset $\mathcal{D}$ for model fitting, confidence level ρ
Result: $M, \mathcal{I}$
$j = 0,\ h = \infty,\ \mathcal{I} = \emptyset$
repeat
  Hypotheses generation
  - chose random minimal sample set $\mathcal{J}$ of size s from data $\mathcal{D}$
  - compute model $M_{\mathcal{J}}$ from minimal sample $\mathcal{J}$
  Hypothesis evaluation
  - determine inlier set $\mathcal{I}_{\mathcal{J}}$
  if $|\mathcal{I}_{\mathcal{J}}| > |\mathcal{I}|$ **then**
    $M = M_{\mathcal{J}},\ \mathcal{I} = \mathcal{I}_{\mathcal{J}}$
    $\epsilon = \frac{|\mathcal{D}|}{|\mathcal{I}|},\ h = \frac{\log(1-\rho)}{\log(1-\epsilon^s)}$
  end
  $j+ = 1$
until $h < j$;

Algorithm 7.1: RANSAC algorithm

Calibrated camera registration is based on the more constrained essential matrix $\mathbf{E}$ instead of the fundamental matrix. The essential matrix $\mathbf{E}$ describes the relationship between a pair of views in a Euclidean setting, instead of the projective context provided by a fundamental matrix $\mathbf{F}$. The essential matrix is given by

$$\mathbf{E} = \mathbf{K}'^{\top}\mathbf{F}\mathbf{K} = [\mathbf{t}]_{\times}\mathbf{R}, \tag{7.7}$$

where $\mathbf{t}$ describes a translation vector between cameras. Moreover, $\mathbf{E}$ has only five degrees of freedom and can be estimated from five correspondences [Nistér 04], which leads to a more efficient RANSAC-based estimation. Given that the essential matrix avoids the degeneracy of the fundamental matrix for 3D points that are on a plane, it is always advised to employ essential matrix estimation for calibrated cameras. Even if there is no accurate calibration available, many digital cameras nowadays provide an estimate of their internal calibration through the EXIF data embedded into the image data file. In practice, approximating the principal point to be located at the image pixel center and stipulating a focal length within 50% of the ground truth nominal value is generally sufficient for attaining estimates of the essential matrix [Nistér 04]. Alternatively, the work of Bougnoux [Bougnoux 98], Sturm [Sturm 01], and Hartley [Hartley 93] has explored the estimation of focal lengths strictly from analysis of the fundamental matrix, encompassing a diverse set of scenarios.

Given an essential matrix estimate, the pairwise relative camera poses can be estimated by a matrix decomposition combined with oriented geometry [Hartley and Zisserman 03]. In this way, a registration of both cameras in the same coordinate system is attained. One camera defines the origin of the coordinate system in canonical orientation, while the displacement vector and the relative orientation of the other camera are denoted by $\mathbf{t}$ and $\mathbf{R}$ in Eq.(7.7). Please note that the resulting coordinate system is of arbitrary scale with respect to the world coordinate system, yielding an ambiguity in the magnitude of the baseline vector $\mathbf{t}$.

Two-view triangulation is the process of inferring the 3D structure of the image features, given the estimated relative camera motion described by the essential matrix. Given a pair of image measurements $\tilde{\mathbf{x}}$ and $\tilde{\mathbf{x}}'$ in each image, and the pairwise essential matrix $\mathbf{E}$, the geometry of both viewing rays r and r' in a common 3D reference frame can be determined, as well as their 3D intersection. In practice, image feature positions are corrupted by measurement errors, and thus rays do not intersect in 3D. This is why 3D triangulation is usually phrased as residual minimization. Hartley and Sturm [Hartley and Sturm 97] proposed a closed-form formulation to determine the optimal image corrections compliant with a specified epipolar geometry. Alternatively, Kanatani et al. [Kanatani et al. 08] and more recently Lindstrom [Lindstrom 10], have proposed iterative solutions based on non-linear optimization. In their work, feature measurements are used as initialization priors. The cost function being optimized is the sum of squared residuals, and the search space is (implicitly) defined over candidate 3D positions subject to the constraints defined by the essential matrix.

Incremental 3D reconstruction systematically augments an initial sparse 3D model obtained from a camera pair using the fundamental or the essential matrix. The sparse 3D reconstruction employs the feature correspondences and intersects their associated viewing rays to compute the position of sparse features in 3D. The 3D reconstruction process has so far 1) defined a Euclidean 3D reference coordinate system (up to scale), 2) identified a set of geometrically consistent 3D landmarks (i.e., triangulated features), and 3) determined the spatial relationships between the input images (i.e., epipolar geometry). The augmentation of our existing reconstruction can now leverage the estimated Euclidean structure and perform pairwise camera calibration by solving the perspective three point problem (P3P) [Haralick et al. 94]. More specifically, given a set of three 3D landmarks and their projections (i.e., the corresponding 2D feature points) to a calibrated camera with unknown pose, it is possible to use the known 3D positions and the angle among the corresponding viewing rays in the new camera to solve for the distance along each viewing ray and estimate the rigid motion transformation of the camera. The typical framework is to sequentially execute different RANSAC instances attempting to register a new input image against each of the existing registered cameras.

Bundle Adjustment

In the following it is assumed that the data used for camera registration, comprising input feature correspondence observations, estimated output camera registrations, and sparse 3D points are denoted by $\mathcal{O}$, $\mathcal{C}$, and $\mathcal{S}$, respectively. Initial estimates for an individual camera $\mathbf{c}_i \in \mathcal{C}$ and a 3D scene point $\mathbf{s}_j \in \mathcal{S}$ are typically obtained through pairwise camera registration techniques. The incremental nature of these registrations can lead to global estimation inconsistencies. The process by which sets of camera estimates $\mathcal{C}' \subset \mathcal{C}$ and structure estimates $\mathcal{S}' \subset \mathcal{S}$ are jointly refined through non-linear optimization techniques is known as bundle adjustment. In practice, bundle adjustment can alternatively serve as 1) a post-processing step with the goal of refining a given camera registration, and/or 2) a systematic geometric consistency enforcement module during incremental camera registration. More detailed discussions within the context of photogrammetric measurements can be found in [McGlone 13], while the survey presented by Triggs et al. [Triggs et al. 00] provides a discussion from the perspective of the computer vision community.

An observation residual is defined as $\mathbf{r}_{ij} = \mathbf{o}_{ij} - f(\mathbf{c}_i, \mathbf{s}_j)$, where $f(\cdot)$ describes the image formation model evaluated for a given camera and structure parameter instance, while the camera and 3D feature indices are denoted by i and j, respectively. To minimize the image reprojection errors a weighted least squares optimization problem is defined as follows:

$$\min_{\mathbf{c},\mathbf{s}} \sum_{\forall\{i \times j\}:\exists \mathbf{o}_{ij}} \mathbf{r}_{ij}^{\top} \mathbf{W}_{\mathbf{o}_{ij}} \mathbf{r}_{ij}, \tag{7.8}$$

where $\mathbf{W}_{\mathbf{o}_{ij}}^{-1}$ approximates the measurement covariance matrix. Equation (7.8) can be iteratively solved using an approximated second-order Gauss–Newton refinement step $\Delta_{\mathbf{c},\mathbf{s}}$ defined by

$$\left(\mathbf{J}^{\top}\mathbf{W}_{\mathbf{o}}\mathbf{J}\right)\Delta_{\mathbf{c},\mathbf{s}} = -\mathbf{J}^{\top}\mathbf{W}_{\mathbf{o}}\,\mathbf{r}, \tag{7.9}$$

where $\mathbf{J} = \left[\frac{\partial f(\mathbf{c}_i,\mathbf{s}_j)}{\partial \mathbf{c}} \; \frac{\partial f(\mathbf{c}_i,\mathbf{s}_j)}{\partial \mathbf{s}}\right]$ denotes the Jacobian of the image formation model with respect to the camera and structure parameters. The above formulation entails the inversion of the normal equations defined by $\mathbf{J}^{\top}\mathbf{W}_{\mathbf{o}}\mathbf{J}$, for which efficient solvers leveraging the equation's block (and possibly sparse) structure can be utilized [Agarwal et al. 10, Wu et al. 11]. Nevertheless, performing such global refinement in an incremental setting may become a limiting computational overhead for large 3D reconstructions. In this respect, the concept of *windowed* (i.e., reduced local neighborhood) camera subset selection (i.e., $\mathbf{c} \subset \mathbf{C}$, where $|\mathbf{c}| \ll |\mathbf{C}|$) enables a trade-off among estimation reliability and computational efficiency. Accordingly, temporal windowing is generally applied to video-based reconstructions, while spatial windowing is generally applied to unordered input image sets.

After introducing the general building blocks of structure from motion, the next section will discuss some of the modifications necessary for large-scale structure from motion based on these building blocks.

7.3 Scalable Structure from Motion

This section discusses some of the challenges and considerations when doing structure from motion of large-scale image sets. In this context, the main (often competing) objectives are scalability and robustness. While the former entails the ability to operate on Internet-scale input image sets (currently comprising thousands to millions of images), the latter implies the ability to operate on heterogeneously captured input data. In practice, both of these objectives are tightly coupled and hinge on diverse design and implementation considerations. Moreover, the relative maturity and effectiveness of state-of-the-art structure from motion technologies shifts the focus of large-scale implementation to developing effective data association and process management modules. Given that the use of an incremental structure from motion framework (as described in Section 7.2) is the *de facto* design choice for large-scale image sets, the remainder of this section discusses the different data association considerations within this framework.

Before discussing scalability, two major considerations are discussed that improve the robustness of structure from motion: the selection of the first pair of images from which to initialize, and the selection of the next best view to add to the reconstruction.

initial pair selection is critical for a robust and well-behaved structure from motion process. The initial pair needs to have a sufficient baseline to ensure that any triangulated 3D point has sufficient accuracy. On the other hand, if the baseline is too large, the triangulation produces very accurate 3D points but the matching ability degrades due to the appearance change. Moreover, the optimal initial pair would produce a high and well-distributed number of 3D points that match with a large number of other images in the image collection. In practice, the initial pair has to find a compromise between these competing goals of matching, triangulation, and scene structure. Beder and Steffen [Beder and Steffen 06] consider the uncertainty of the triangulated points for selecting the initial pair, which leads to stable structure from motion. Hence, their method focuses on increasing the baseline of the initial pair. In contrast, Snavely et al. [Snavely et al. 06] propose to initialize structure from motion by choosing the pair with the maximal number of corresponding points whose correspondences are not explained by a planar scene or a camera rotation. Hence, they emphasize the importance of the scene structure over the accuracy of the triangulated points, which is improved through their subsequent processing steps. A combination of these criteria is proposed by Raguram et al. [Raguram et al. 11]. They optimize the number of inliers of a pair and the uncertainty of the triangulated points simultaneously. After selecting the initial pair, structure from motion needs to decide in which order to register the remaining views, which is done by the next best view selection.

Next best view selection for incrementally growing the structure from motion is important to achieve a stable reconstruction. For example, selecting a weakly connected view could lead to an erroneous registration for the next best view. This corrupted reconstruction can then influence the succeeding registrations through perturbed 3D points used in the registration.

Similar to the selection of the initial pair, the next best view should have a sufficiently high number of 3D points that correspond to the 2D salient feature points visible in the camera view. Additionally, the uncertainties of those 3D points should be as small as possible to allow accurate registration of the view. For example, in Raguram et al. [Raguram et al. 11] and Snavely et al. [Snavely et al. 06] the next best view is chosen as the view with the highest number of visible 3D points. The 2D-3D point correspondences are obtained by using the pairwise viewing registrations and their 2D-2D matches to predict the visibility of the 3D points (Section 7.2). This criterion favors high scene overlap over enforcing higher accuracy of the 3D points.

Scalability is critical for large-scale structure from motion in order to reduce the computational complexity. Please note that most of the above selection strategies consider the availability of pairwise registrations of all images. These pairwise registrations are, for example, described by the essential matrix of the pair and its feature inliers. Computing these relations and inliers for all possible pairs of images in a collection is computationally prohibitive, even for collections of a few thousand images. One such example is the seminal work of Snavely et al. [Snavely et al. 06], where the registration of approximately three thousand images required about two weeks of computation time. It can be observed that in larger photo collections, where a large number of pairs does not overlap at all [Frahm et al. 10] or are not required for a stable reconstruction [Snavely et al. 08c]. Hence, Agarwal et al. [Agarwal et al. 11] propose to use a vocabulary tree search [Nistér and Stewenius 06] for identifying overlapping images and combine the search with query expansion [Chum et al. 07] to increase the number of overlapping images returned by the search. Frahm et al. [Frahm et al. 10] propose the iconic scene graph to limit the search for overlapping pairs. They employ iconic images, i.e., representative images for subsets of the database, to represent the viewpoints in the scene. Both Agarwal et al. [Agarwal et al. 11] and Frahm et al. [Frahm et al. 10] improve the scalability with the latter technique, reaching overall linear complexity for image registration, which enables the scaling to the processing of millions of images on a single day using a single PC.

7.4 Summary

This chapter discussed the underlying principles of structure from motion, which is an important enabling technique for image-based reconstruction-from photo and video collections. These principles are common to the large body of state-of-the-art algorithms. The discussed methods span from small-scale structure from motion to methods for large-scale structure from motion for crowd sourced data. The latter have to overcome more significant challenges with respect to robustness and scalability to meet the demand of reconstructing from thousands [Snavely et al. 06] to millions of images [Frahm et al. 10]. Yet, the discussed methods only represent a small fraction of the state-of-the-art structure from motion methods and do not represent the full range of methods [Pollefeys et al. 04, Hartley and Zisserman 03, Beardsley et al. 97, Lhuillier and Quan 05, Dellaert et al. 00, Wilson and Snavely 13].

8

Reconstruction of Dense Correspondences

Martin Eisemann, Jan-Michael Frahm, Yannick Remion, and Muhannad Ismaël

8.1 Introduction

This chapter concentrates on dense image correspondence estimation with a special focus on stereo. Images are the basic input for a vast majority of algorithms dealing with the reconstruction of the real world. To analyze a scene from a collection of images it becomes inevitable to put these images into correspondence. These correspondences then form the basis for many subsequent analyses, including camera calibration, stereo and 3D reconstruction, motion information, scene flow, and others. While some of these tasks like camera calibration require only sparse correspondences between the images (Chapter 7), others require per-pixel correspondence, also known as dense correspondence estimation.

Humans are extremely good at solving the correspondence problem which most of them do all the time during depth perception. Basically, the eyes serve as two cameras, slightly displaced, with respect to each other, that capture the surroundings from two different viewpoints. When focusing on an object at a certain distance one has already computed an estimate of the distance in the brain and therefore of the object's position in space. It turns out the same problem is quite difficult for a computer and has been researched for several decades now.

The difficulty in correspondence estimation is caused by several factors: images are often corrupted by sensor noise, e.g., when recorded in a poorly lit environment (Section 1.1); the captured scene signal is discretized and represented by some finite image resolution; not every pixel actually has a corresponding partner in the other views as it might be occluded; and ambiguities due to the absence of texture are difficult to solve.

If one can solve the dense correspondence problem a variety of different applications becomes possible especially in the field of computer vision. Robot navigation and autonomous cars require depth perception to avoid obstacles [Giachetti et al. 98, Kastrinaki et al. 03]. Quality assurance in industrial applications is often based on stereo algorithms to detect cracks and ridges in manufactured products. Reconstruction of urban environments from images has recently gained a lot of interest in the research community [Gallup et al. 07, Frahm et al. 10]. The dense correspondences allow for video editing [Adobe Systems Inc. 13, The Foundry 13], super-resolution [Irani and Peleg 91], video stabilization [Matsushita et al. 06], to interpolate between images [Chen 95, Lipski et al. 10a], e.g., to create bullet time effects made famous in the blockbuster movie *The Matrix* and for specific tracking applications in graphics, e.g., the local pose optimization for texture correspondence matching in Chapter 11 is related. Disparity remapping based on the correspondences and reconstructed depth becomes important to avoid visual fatigue in stereoscopic cinema [Devernay and Beardsley 10].

The following will give a hands-on guide on how to compute dense correspondences between images. After a short overview of current state-of-the-art approaches (Section 8.2), a robust solution to the correspondence problem is described and extended (Section 8.3). It is described how to compute correspondences from multiple images (Section 8.4), and means to speed up the computations using graphics hardware are presented (Section 8.5).

8.2 Overview

This section gives a brief overview of different approaches dealing with the dense correspondence problem. The goal is to find the best corresponding (sub-)pixel position in neighboring views for every pixel of a reference image, if such corresponding positions exist.

The algorithms dealing with the correspondence problem can be broadly classified into two categories: stereo and optical flow. Intrinsically the problem is the same for both of them, finding good correspondences between the views, but they differ in the premises. Stereo can be seen as a special case of optical flow, where correspondences are searched along the same scanline (or epipolar line), reducing the solution space from 2D to 1D. The following will give a short overview of the most seminal papers in both categories and their contributions.

Stereo In analogy to the human eyes, the input to classic binocular stereo algorithms are two images $\mathbf{I}^l$ and $\mathbf{I}^r$, a left and a right one. The task is to find for every pixel $\mathbf{p}$ with pixel coordinates (x, y) in the left image a corresponding pixel $\mathbf{q}$ in the right view with pixel coordinates $(x - d_{\mathbf{p}}, y)$.

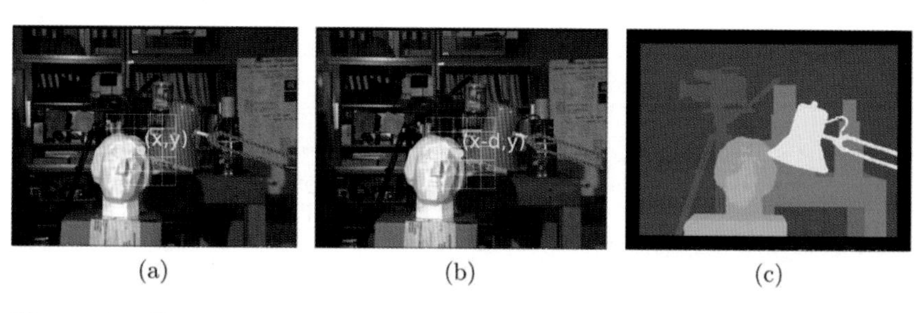

(a) (b) (c)

Figure 8.1: Dense correspondence estimation in stereo for the Middlebury Tsukuba dataset [Scharstein and Szeliski 02]. (a) The task is to find for each pixel at any position (x, y) in the left view (b) a corresponding position $(x - d, y)$ in the right view and encode the result in (c) a disparity map from which 3D coordinates can be reconstructed. In the stereo setting, the corresponding pixels lie on the same scanline, whereas in the more general problem of optical flow estimation the correspondence can be any position within the right view. Instead of comparing single pixel values, comparing neighborhoods of pixels (shown as the overlaid grid) results in higher robustness.

$d_{\mathbf{p}}$ is called the disparity of pixel $\mathbf{p}$. The disparity information is typically saved in an intensity image, the so-called disparity map $\mathbf{D}$, where low/dark values encode low disparity and high/bright values encode high disparity (Figure 8.1(c)).

In stereo one generally distinguishes between local and global methods. In the first category local areas of one image are matched to local areas in the corresponding view, often called support regions. The difficulty lies in the choice of the support region as matching single pixels is highly ambiguous in most scenes. Simple rectangular windows around the pixel under consideration can be efficiently implemented [Hirschmüller et al. 02, Mühlmann et al. 02] but it can be difficult to choose the right size. By shifting the center position of the window and testing different sizes [Fusiello et al. 97] or by deactivating parts of the support region [Hirschmüller et al. 02, Veksler 02] one can hope that at least one constellation does not overlap with a depth discontinuity. This otherwise poses a matching problem as in many cases a depth discontinuity marks the separation line between two objects with different disparities and, therefore, a different amount of motion in image space from the left to the right view. The research community has thus investigated methods to find a good support region, with different criteria on how much influence each pixel inside this region should have on the final result [Hosni et al. 13].

One key component, and a breakthrough for local methods in recent years, has been the introduction of adaptive support weights [Yoon and

Kweon 05]. The idea is to adjust the influence of neighboring pixels on the final matching cost based on a similarity metric, most often color and spatial similarity. [Yoon and Kweon 05]'s bilateral weighting scheme is based on a Gaussian distribution depending on the spatial proximity and proximity of intensity values. To overcome the problem of spatially close but distinct objects influencing each other, the spatial proximity can be exchanged with a geodesic distance [Hosni et al. 09].

Unfortunately, the computation of adaptive support weights is costly if implemented in a naive way. To speed up the aggregation step it can be converted to an image filtering procedure. It turned out that the bilateral weighting scheme of [Yoon and Kweon 05] is equivalent to applying a cross-bilateral filter or derivations of it to the x, y-slices of a cost volume [Hosni et al. 11b, Richardt et al. 10, Zhang et al. 10a, Ju and Kang 09]. To further speed up the computation, the pixel-wise matching for fixed disparities can be elegantly formulated as a plane-sweeping algorithm on the GPU [Yang and Pollefeys 03, Gallup et al. 07, Zach et al. 08] allowing for real-time stereo implementations.

An implicit assumption made by the aforementioned techniques is that each local support region is basically a patch with fronto-parallel orientation to the image plane of the reference view. Treating the slices in the cost volume not as virtual planes representing a certain disparity but as real 3D planes in the scene one can easily use rotated versions of these slices to compute the matching cost for slanted surfaces [Gallup et al. 07]. The computation times, however, increase linearly with the number of orientations used. Therefore, [Zhang et al. 08] propose to iteratively refine the disparities and orientations in a feedback loop. Another alternative is to initialize each pixel with a random orientation and disparity and propagate good matches to neighboring pixels based on a PatchMatch update scheme [Bleyer et al. 11a].

The second category of stereo algorithms forms the so-called global methods. Global stereo methods pose the matching problem as an energy minimization problem which is usually of the following form:

$$E(\mathbf{D}) = E_{\text{data}}(\mathbf{D}) + \alpha \cdot E_{\text{smooth}}(\mathbf{D}) \quad , \tag{8.1}$$

where $\mathbf{D}$ is the current estimate of the disparity map. The goal is to find $\mathbf{D}$ that produces the lowest energy. E_{data} in this context is a photo-consistency measure that can be equal to the matching function of the local methods but is traditionally simpler. Instead of implicitly stating a smoothness function in the form of a support region, as in the local approaches, here the smoothness is explicitly expressed within the error formulation as E_{smooth}. This regularization of the solution can be especially useful for textureless regions as it basically smoothes out the solution.

Several optimization approaches have been proposed to minimize Eq.(8.1) through dynamic programming [Veksler 05, Bleyer and Gelautz 08], graph-cuts [Boykov et al. 01, Hong and Chen 04, Bleyer and Gelautz 07] or belief propagation [Sun et al. 03, Yang et al. 06b, Taguchi et al. 08].

Interestingly, the usage of tree-reweighted message passing (TRW) and a comparison to ground truth results revealed that modern optimization algorithms yield energies that are actually lower than that of the ground truth solution [Szeliski et al. 08]. This indicates that the model in Eq.(8.1) is actually a limiting factor. Further advances, therefore, need to extend the model. Explicit occlusion handling or enforcing symmetrical matches between the input images was used, e.g., in [Kolmogorov and Zabih 01, Lin and Tomasi 04, Sun et al. 05, Woodford et al. 09]. Truncating the smoothness term to a user-defined maximum value favors large jumps in the disparity map instead of many small changes [Hirschmüller 05, Sun et al. 05, Yang et al. 06a]. Segmentation-based methods presegment the image into patches of coherent color and match whole segments at once [Deng et al. 05, Hong and Chen 04, Zitnick et al. 04]. The idea is that in many cases depth discontinuities coincide with segment borders. An extension of segmentation-based stereo is object-based stereo which matches semantic objects instead of single colored patches. In this way it becomes possible to handle even semi-occluded surfaces [Bleyer et al. 11b]. Extending the idea of object-based stereo one can estimate simple 3D approximations for the different objects [Bleyer et al. 12]. On the basis of these higher semantic concepts one can add sophisticated additional constraints to the optimization, for instance to prevent intersections between the objects or to add a gravity constraint.

Optical flow The problem of optical flow estimation is strongly related to the stereo problem and several of the aforementioned algorithms are applicable to both. Basically, optical flow estimation is a generalization of the stereo problem from a 1D solution space, the disparity map, to a 2D solution space, the flow or motion field. While stereo algorithms aim at reconstructing correspondences between images captured at the same instance in time, optical flow allows to track the motion of pixels also across the time dimension, e.g., in a video.

During the last 30 years, hundreds of research papers have been published in the field of optical flow and various surveys and benchmarks cover and compare the state-of-the-art [Barron et al. 94, Baker et al. 11]. The seminal work of [Horn and Schunck 81] and [Lucas and Kanade 81] laid the foundations for the algorithms to follow. Interestingly, similar to stereo, one can distinguish global and local approaches to the optical flow problem, explicitly enforcing smoothness in the solution [Horn and Schunck 81] and assuming local constancy within a window around each pixel [Lucas and Kanade 81]. Neither assumption of smoothness holds at motion boundaries for which robust [Black and Anandan 96, Zach et al. 07] and anisotropic regularizers [Nagel and Enkelmann 86, Werlberger et al. 09, Sun et al. 10, Zimmer et al. 11] have therefore been proposed. To reduce the influence of outlier pixels caused by brightness changes and sensor noise the simple data terms based on color-constancy assumption are mostly replaced by robust penalizer functions [Black and Anandan 96, Brox et al. 04, Zach et al. 07] or pixel-descriptors [Mileva et al. 07, Liu et al. 08].

To cope with fast motion, scale-space approaches [Anandan 89] and iterative warping schemes [Alvarez et al. 00, Brox et al. 04] make use of image pyramids to find corresponding pixels. As downsampling only works well for sufficiently large objects several search schemes have been proposed in the literature that either perform a full search [Steinbrücker et al. 09, Linz et al. 10a, Lipski et al. 10b, Hosni et al. 11b] or use tracked features as reliable priors for the optimization [Brox and Malik 11]. In a more hardware-based approach [Lim et al. 05] make use of a high-speed camera to reduce the per pixel displacement to less than a pixel.

Probably due to its success in stereo, explicit occlusion handling has been introduced to optical flow estimation as well. The occlusion detection thereby is either based on the optimization residual and divergence of the flow [Xiao et al. 06, Sand and Teller 06], the symmetry of forward and backward flow [Alvarez et al. 07, Linz et al. 10a, Lipski et al. 10b], or is integrated in the image formation model making use of alternate exposure images [Sellent et al. 11] by alternate capturing of long- and short-exposed images in a video.

8.3 Dense Correspondence Estimation

In the following, an approach is described to compute dense correspondences between two images. The algorithm is mainly based on the fast cost-volume filtering by [Hosni et al. 11b] which is one of the top ranked local methods for stereo and which is also applicable to the more generalized optical flow problem.[1] For simplicity it is assumed that the images have already been rectified, i.e. corresponding points lie on the same scanline (Figure 8.1). Otherwise, it is assumed that standard rectification algorithms are applied first [Hartley and Zisserman 03].[2] These are generally based on the camera registration procedures described in Chapter 7. These constraints will be loosened in the later part of this chapter (Section 8.5).

[1] Code is available at https://www.ims.tuwien.ac.at/publications/tuw-210567

[2] Code is available at http://www.robots.ox.ac.uk:5000/~vgg/hzbook/code/.

Figure 8.2: Different dissimilarity functions. (a) Pixel-wise matching solely based on color/intensity is highly ambiguous. (b) A 3×3 correlation window is still noisy. (c) A 21 × 21 correlation window results in edge fattening. (d) The cost filter method of [Hosni et al. 11b].

The basic task of estimating a disparity map $\mathbf{D}$ can be formulated for each pixel $\mathbf{p}$ as

$$d_{\mathbf{p}} = \underset{0 \leq d \leq d_{\max}}{\operatorname{argmin}}\ c(\mathbf{p}, \mathbf{p} - d) \ . \tag{8.2}$$

The term $d_{\max}$ is a user-defined constant which must be larger than the expected maximum disparity. Note that due to rectification d is always a positive value or 0. In case the disparity map for the right image is to be computed $\mathbf{p} - d$ in Eq. (8.2) is replaced by $\mathbf{p} + d$. For simplicity, only disparity computations for the left image are considered. The simplified notation $\mathbf{p} - d$ denotes the pixel 2D pixel position $(x_p - d_p, y_p)$ where (x_p, y_p) are the pixel coordinates of pixel $\mathbf{p}$. The symbol c denotes a cost / dissimilarity function.

Dissimilarity functions To find an appropriate disparity $d_{\mathbf{p}}$ for each pixel $\mathbf{p}$ one needs to find a suitable dissimilarity function c in Eq.(8.2). The probably most simple one would be a naive per-pixel matching, that is, $c(\mathbf{p}, \mathbf{q}) = |\mathbf{I}^l_{\mathbf{p}} - \mathbf{I}^r_{\mathbf{q}}|_2$ where $\mathbf{I}^l_{\mathbf{p}}$ denotes the pixel intensity of $\mathbf{I}^l$ at pixel position $\mathbf{p}$ and $|\cdot|_2$ is the Euclidean distance between the two vectors. But matching only simple intensity values is highly ambiguous and leads to very noisy results (Figure 8.2(a)).

Instead of matching single pixels one can match small image patches centered at $\mathbf{p}$. In this case the cost function becomes

$$d_{\mathbf{p}} = \underset{0 \leq d \leq d_{\max}}{\operatorname{argmin}} \sum_{\mathbf{q} \in \mathbf{W}_{\mathbf{p}}} c(\mathbf{q}, \mathbf{q} - d) \ , \tag{8.3}$$

where $\mathbf{W}_{\mathbf{p}}$ is a square window centered at $\mathbf{p}$, and c as defined above. Figures 8.2(b) and 8.2(c) show the resulting disparity maps using correlation windows of the size 3×3 and 21×21 pixels, respectively. The choice of a right size has a notable influence on algorithmic performance, and no single window size generally works for all cases. While smaller window sizes capture finer details, matching scores can be highly ambiguous. Larger window sizes, on the other hand, lead to edge fattening around discontinuities and oversmooth results. What is needed is an adaptive support weight that adjusts the shape of the window or, in other words, reduces the influence of pixels that do not belong to the same object as pixel $\mathbf{p}$.

Adaptive support weights Adaptive support weights adjust the influence of each individual pixel considered in the matching process. This can be formulated as a simple extension to Eq.(8.3)

$$d_{\mathbf{p}} = \underset{0 \leq d \leq d_{\max}}{\operatorname{argmin}} \sum_{\mathbf{q} \in \mathbf{W_p}} w(\mathbf{p}, \mathbf{q}) \cdot c(\mathbf{q}, \mathbf{q} - d) \quad , \tag{8.4}$$

where $w(\mathbf{p}, \mathbf{q})$ is a weighting function which should return a value of 1 if $\mathbf{q}$ has the same disparity as $\mathbf{p}$, and 0 otherwise. As this disparity is not known, the weight is usually based on some heuristic that represents the probability that pixel $\mathbf{q}$ exhibits the same disparity as $\mathbf{p}$. The most common assumption made is that pixels close to $\mathbf{p}$ are more likely to belong to the same object and have a more similar disparity than pixels farther away. Additionally, pixels with similar color are more likely to belong to the same object than those with dissimilar color values. The bilateral weighting scheme proposed in [Yoon and Kweon 05] expresses this correlation as

$$w_b(\mathbf{p}, \mathbf{q}) = exp\left(-\left(\frac{c_c(\mathbf{p}, \mathbf{q})}{\sigma_c} + \frac{c_s(\mathbf{p}, \mathbf{q})}{\sigma_s}\right)\right) \quad . \tag{8.5}$$

The function $c_c(\mathbf{p}, \mathbf{q})$ denotes the similarity in color defined as the Euclidean distance of pixels at position $\mathbf{p}$ and $\mathbf{q}$ in RGB space, whereas $c_s(\mathbf{p}, \mathbf{q})$ is the spatial component defined as the Euclidean distance of $\mathbf{p}$ and $\mathbf{q}$'s pixel coordinates. The terms σ_c and σ_s are user-defined constants that control the spread of each term similar to the window size before.

The computation of the bilateral weights for each pixel in the input image is time consuming. A fast and qualitatively even better alternative to the bilateral weighting scheme in Eq.(8.5) is the guided image filter [He et al. 10].[3] While the output is similar to the bilateral weighting, the computation is different

$$w_g(\mathbf{p}, \mathbf{q}) = \frac{1}{|\mathbf{W}|} \sum_{\mathbf{k}:(\mathbf{p},\mathbf{q}) \in \mathbf{W_k}} (1 + (\mathbf{I}^l_{\mathbf{p}} - \mu_{\mathbf{k}})^\top (\Sigma_{\mathbf{k}} + \epsilon \mathbf{U})^{-1} (\mathbf{I}^l_{\mathbf{q}} - \mu_{\mathbf{k}})) \quad , \tag{8.6}$$

with $\mu_{\mathbf{k}}$ and $\Sigma_{\mathbf{k}}$ being the mean vector and covariance of $\mathbf{I}^l$ in a squared window $\mathbf{W_k}$ of user-defined size, centered at and being constant for each

[3] Code is available at http://research.microsoft.com/en-us/um/people/kahe/eccv10/

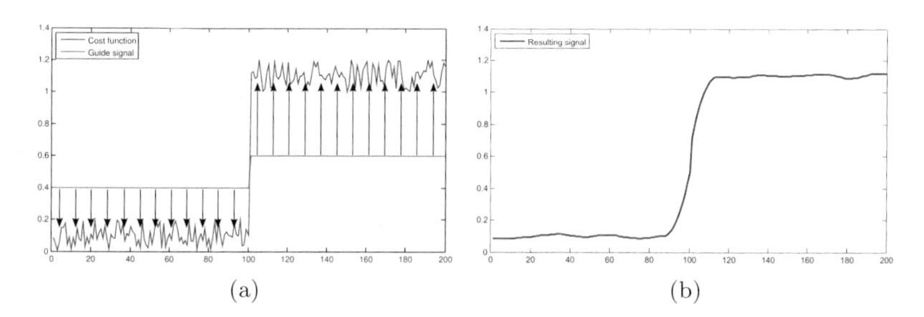

(a) (b)

Figure 8.3: 1D example of the guided image filter [He et al. 10] for a 1D signal. (a) The filter takes a guide signal (green) and fits it locally to the given, potentially noisy, cost function (blue) resulting in (b) a smoothed but edge preserving signal (red).

pixel $\mathbf{k}$. $|\mathbf{W}|$ denotes the number of pixels in the window. $\mathbf{U}$ is the identity matrix and ϵ a smoothness parameter. While, at first glance, Eq.(8.6) appears highly complex in comparison to Eq.(8.5), it turns out that the computation requires only running a series of box filters which can be computed in constant time, independent of the window size. For details see [He et al. 10].

Intuitively, the guided image filter takes a guide image, in this case the input image $\mathbf{I}^l$, and tries to fit it locally to the cost function $\mathbf{C}^d$ that encodes pixelwise costs for a certain disparity d.

$$\mathbf{C}^d_{\mathbf{p}} = c(\mathbf{p}, \mathbf{p} - d) = |\mathbf{I}^l_{\mathbf{p}} - \mathbf{I}^r_{\mathbf{p}-d}| \quad ,$$

This results in a smoothed version of $\mathbf{C}^d$ which is equal to aggregating the weighted costs in a given window around each pixel. For this, the best fitting linear transformation for local windows $\mathbf{W}_{\mathbf{k}}$ around each pixel is computed, i.e., a scaling and an offset of the guide image, to get from $\mathbf{I}^l$ to $\mathbf{C}^d$. In a second step the linear transformation coefficients of all windows overlapping at a pixel are averaged. An example for a single-channel input is given in Figure 8.3. Other commonly used matching techniques and pixel descriptors can also be found in Chapter 7.

Cost volume filtering Stacking the functions $\mathbf{C}^d$ for all disparities d onto each other into a 3D array $\mathbf{C}$ creates the so-called cost-volume. The filtered cost-volume can be extracted by filtering each x, y-slice that belongs to a fixed disparity d with the guided image filter as described above. The final disparity for each pixel $\mathbf{p}$ is then defined in Winner-Takes-All manner as

$$d_{\mathbf{p}} = \underset{0 \leq d \leq d_{\max}}{\operatorname{argmin}} \; \mathbf{C}^d(\mathbf{p}) \quad .$$

Occlusion Occluded pixels are detected using a left-right cross checking procedure. Once the disparities for image $\mathbf{I}^l$ to image $\mathbf{I}^r$ are computed, one can exchange both images and additionally compute the disparities from $\mathbf{I}^r$ to $\mathbf{I}^l$. A pixel $\mathbf{p}$ is marked as invalid, i.e., occluded, if $d^l_{\mathbf{p}} \neq d^r_{\mathbf{p}-d_{\mathbf{p}}}$ where $d^l_{\mathbf{p}}$ is the disparity at pixel $\mathbf{p}$ with reference image $\mathbf{I}^l$. Again note that the disparity is always positive and the sign in Eq.(8.2) is changed according to whether the disparity for the left or right image is computed.

One cannot assign disparities to pixels being occluded in one of the input images. If the application demands such an assignment, it has to be based on some kind of sensible heuristic. In [Hosni et al. 11b] a weighted median filter is used for filling invalidated pixels.

Extensions An advantage of the presented framework is that it naturally extends to higher dimensional and more fine-grained solution spaces at the cost of higher computation times. In the previous example each slice in the cost volume corresponds to a certain integer-valued disparity. One can easily increase the precision to fractional values by increasing the number of slices and assigning each slice to a certain fractional disparity. More generally speaking, each slice of the cost volume can be considered to be a distinct label l from a set $\mathcal{L} = \{1, \ldots, L\}$. The user only needs to specify how these labels are mapped to semantically meaningful parameters for the algorithm. That means one is not bound to interpret l only as integer-valued disparities but could extend the label space to fractional disparities as well, e.g., [Gehrig et al. 12]. Alternatively, a set of slanted windows could be included to better handle slanted surfaces that are not fronto-parallel, e.g., [Gallup et al. 07].

By defining a mapping from a 2D solution vector (u, v) to the label space $\mathcal{L}$ one can directly extend the presented stereo approach to optical flow problems by exchanging $d_{\mathbf{p}}$ and d in Eq.(8.2) by $(u_{\mathbf{p}}, v_{\mathbf{p}})$ and (u, v), respectively. In this context it should be noted that modern optical flow methods mostly use a more sophisticated cost function including not only color- but also gradient-similarity; details for the presented approach can be found in [Hosni et al. 11b].

Limitations A principal limitation of all local methods, such as the one presented, is their inability to cope with highly ambiguous data such as unicolored walls or objects. Depending on the application this may not be crucial, e.g., for image interpolation, as no visible artifacts will occur if objects of the same color are interpolated incorrectly. In other applications, such as autonomous driving vehicles or robot navigation, this may pose a high risk, because there, accurate disparity, which means accurate depth, is crucial. Imagine similar stone pillars standing next to each other. Matching the right ones is highly ambiguous. In such cases more complicated global correspondence estimation algorithms are required, a good overview is given in [Bleyer and Breiteneder 13].

Another limitation of the presented problem formulation is its discrete nature, which means it can only produce a solution that consists of combinations of preset labels. Even though labels may represent fractional values and the solution is therefore sub-pixel precise, it is always limited by the label space. The quality of any correspondence algorithm also depends highly on the scene content. While local methods are ranked high in the famous Middlebury benchmark [Scharstein and Szeliski 02], they are not always as successful in other benchmarks, e.g., [Geiger 12]. The reason could be a higher sensitivity to noise or ambiguities occurring more often in natural scenes. And finally occlusion handling can usually not be integrated into the matching process directly with local methods. Once the cost-volume has been created one could exchange the Winner-Takes-All strategy by a more sophisticated global label selection algorithm that could handle such cases by a better or more robust disparity assignment even for pixels occluded in one view. Therefore, the presented algorithm is a good starting point for further investigation of dense correspondence algorithms.

Section 8.4 extends the stereo correspondence estimation to multiple input cameras and deals with appropriate camera layouts and scene representations. Section 8.5 describes the plane-sweeping stereo algorithm that is easily portable to the graphics card to allow even real-time correspondence estimation.

8.4 Multi-View Stereo

The following section gives an overview of multi-view stereovision. The term multi-view stereovision (MVS) refers to stereovision-based reconstruction from $n > 2$ views, $\mathbf{I}^0$ to $\mathbf{I}^{n-1}$, and is sometimes called *multiocular stereovision* in contrast to *binocular stereovision* from one pair of views (Section 8.3). MVS has been an active field of research for several decades and more than seventy algorithms are listed on the Middlebury Multi-View Benchmark website [Seitz et al. 06]. This benchmark provides a commonly accepted test suite to evaluate the quality of multi-view stereo algorithms.

An important assumption of any MVS method lies in its required, compatible, or intended camera layout since various possibilities exist and may have an impact on the 3D reconstruction strategy (Section 2.2).

Most MVS methods (notably among those on the Middlebury list) are designed for n cameras freely laid out in space. Some apply binocular stereovision (as previously discussed in Section 8.3) on different couples of views

$(\mathbf{I}^i, \mathbf{I}^j)$ and then merge their separate binocular results. The main difficulty in such approaches concerns regularizing the union of separate results, especially in scene areas where reconstructions overlap. Common problems to be solved in such areas are to reduce too high point density and to resolve ambiguities/inconsistencies. This task pertains to point cloud merging and is discussed in Chapter 10 in more detail. Another type of MVS approach for a free camera layout consists in fitting some form of geometric model of the scene in order to maximize its local photo-consistency in available views [Furukawa and Ponce 10].

Some other MVS methods, sometimes called *multi-baseline stereovision* methods, are designed for the "parallel" or "decentered parallel" camera layouts discussed in Section 2.2 and especially fitted for 3DTV content capture. Those layouts are characterized by aligned, evenly distributed and parallel cameras. As will be demonstrated below, such restrained settings induce a set of geometrical constraints on corresponding pixels from different views arising from the so-called *simplified multi-epipolar geometry*. Those constraints enable searching correspondences over every view at once as a multi-view matching process [Okutomi and Kanade 93, Szeliski and Golland 99, Niquin et al. 10, Ismael et al. 14]. This is also called *multi-scopic stereo matching* and consists in matching n-tuples of pixels instead of couples which yields a consistent and more robust reconstruction.

In the following, the main concepts behind multi-baseline stereovision are reviewed. The "parallel" layout of Section 2.2 implies the optical centers $\mathbf{o}_0 \ldots \mathbf{o}_{n-1}$ to be aligned and evenly distributed on the baseline, parallel optical axes orthogonal to this baseline, cameras of same focal and darkroom depth, sensors of same size and resolution $nc \times nl$ centered on their optical axes with rows parallel to the baseline. The "decentered parallel" layout (Figure 8.4), generalizes this setting by translating the sensor centers off their optical axis in such a way that the *line of sight* of all the views, defined for each camera by its sensor center and optical center, now converge on a chosen 3D *convergence point* $\mathbf{c}$, possibly at finite distance [Prévost et al. 13]. The convergence point is of utmost importance in 3DTV content shooting as it will be displayed exactly on the center of the 3D display and thus controls how captured scene space is mapped in the perceived 3D space (Section 2.2). One should note that, in this layout, the convergent lines of sight no longer coincide with the parallel optical axes. The off-axis translation of the "sensor" may be achieved both at hardware design stage as sensor chip physical/mechanical decentering and/or, to a given extent, at software post-processing stage as region of interest (ROI) cropping. In the following, the term sensor ROI is used to denote all of the above-mentioned possibilities.

Another benefit of such a layout is usually achieved thanks to rectification of the n views from aligned and evenly distributed cameras with

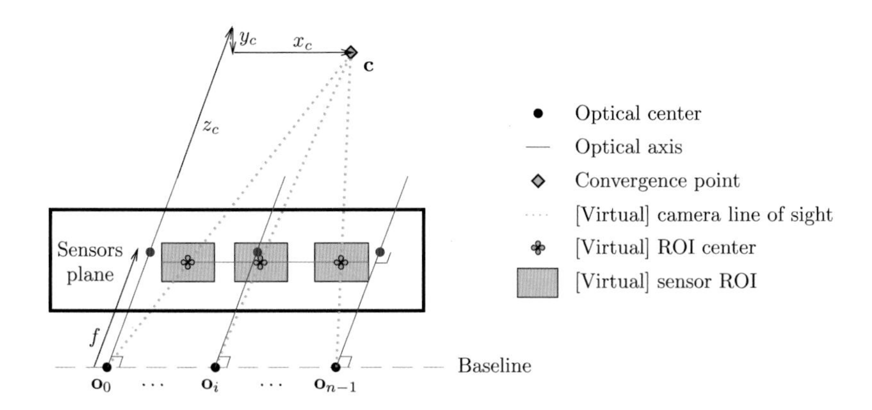

Figure 8.4: Decentered parallel camera layout.

(approximately) convergent optical axes. Similar to its binocular counterpart, the multiocular rectification consists in intersecting pixel rays by a plane at distance f (virtual sensors' plane) parallel to the common baseline connecting the optical centers (Figure 8.4). The rectified virtual sensor ROI in which the rectified views will be computed are then virtually laid in this sensors' plane with same size and orientation, so that the rows are parallel to the baseline. Furthermore, their centers are chosen to make every line of sight converge at the chosen 3D convergence point $\mathbf{c} = (x_\mathbf{c}, y_\mathbf{c}, z_\mathbf{c})$ (coordinates expressed in reference frame of camera 0) (Figure 8.4). One should note that there is not as much freedom in the layout of the actual cameras as in the binocular case as the rectification process relies on actual optical centers being aligned and rather evenly distributed.

For n images $\mathbf{I}^i$ recorded or rectified in "(decentered) parallel" layout and numbered $i \in \{0, n-1\}$ from left to right, the *epipolar constraint*, previously discussed for the binocular case (Figure 8.5), states that any pixel at $\mathbf{p}_i$ in any image $\mathbf{I}^i$ represents the actual 3D scene point projected onto $\mathbf{p}_i$. Pixel $\mathbf{p}_i$ and the optical center $\mathbf{o}_i$ of the camera are aligned on $\mathbf{p}_i$'s "pixel ray." Considering that pixel rays of corresponding pixels at $\mathbf{p}_i$ and $\mathbf{p}_j$ in two views $\mathbf{I}^i$, $\mathbf{I}^j$ have to intersect at their common 3D point $\mathbf{p}$, a straightforward derivation yields that optical centers $\mathbf{o}_i$, $\mathbf{o}_j$ and corresponding pixels at $\mathbf{p}_i$, $\mathbf{p}_j$ have to be coplanar (they belong to 2 intersecting and yet different lines). An *epipolar plane* is then defined for a couple of views (i, j) by both optical centers and any studied pixel in one of these views. The corresponding pixel in the other view has thus to be searched for within the *epipolar segment* defined as the intersection of this plane

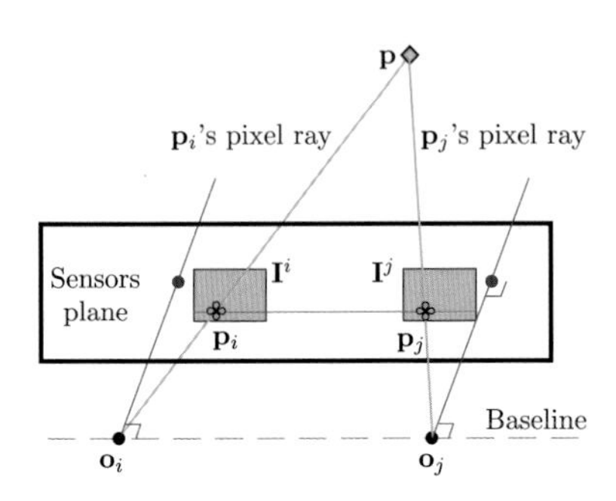

Figure 8.5: Simplified epipolar geometry.

with the other image (black horizontal line in Figure 8.5). When those two (rectified) cameras are set in "(decentered) parallel" layout, the epipolar segment in $\mathbf{I}^j$ defined by pixel $\mathbf{p}_i$ in $\mathbf{I}^i$ is part of the scanline of $\mathbf{I}^j$ of the same rank as the one holding $\mathbf{p}_i$ in $\mathbf{I}^i$.

In the multi-baseline context, because the optical centers are aligned, epipolar planes defined for a given pixel $\mathbf{p}_i$ in $\mathbf{I}^i$ and any other view $\mathbf{I}^j$ coincide. Successive pairwise binocular epipolar constraints thus ensure that corresponding pixels have the same y-value in every view $\mathbf{I}^i$. Hence, any 3D point $\mathbf{p} = (x_\mathbf{p}, y_\mathbf{p}, z_\mathbf{p})$ is projected into the [rectified] views $\mathbf{I}^i$, $\mathbf{I}^j$ onto corresponding pixels whose coordinates are respectively $\mathbf{p}_i$ and $\mathbf{p}_j = \mathbf{p}_i - (d^{i,j}_{\mathbf{p}_i}, 0)$, where $d^{i,j}_{\mathbf{p}_i} = u_i - u_j$ is called *horizontal disparity*.

A relationship between the horizontal disparity $d^{i,j}_{\mathbf{p}_i}$ and $\mathbf{p}$'s depth $z_\mathbf{p}$ can be established based on scale ratios between similar triangles. Let us consider the scale ratios in two pairs of such triangles, with apices on $\mathbf{c}$ and $\mathbf{p}$, respectively, and the camera centers $\mathbf{o}_i$ and $\mathbf{o}_j$ (Figure 8.6). Let $e_{i,j}$ be the distance between the center of the sensor ROI of camera i and j, defined by the triangle with apex on $\mathbf{c}$, and $e_{i,j} - d^{i,j}_{\mathbf{p}_i}$ be the distance between the corresponding pixels $\mathbf{p}_i$ and $\mathbf{p}_j$, in the sensors' plane, defined by the triangle with apex at $\mathbf{p}$. The relation between these two triangles yields the disparity-to-depth relation:

$$\left.\begin{aligned} e_{i,j} &= b_{i,j} \cdot (z_\mathbf{c} - f) \cdot z_\mathbf{c}^{-1} \\ e_{i,j} - d^{i,j}_{\mathbf{p}_i} &= b_{i,j} \cdot (z_\mathbf{p} - f) \cdot z_\mathbf{p}^{-1} \end{aligned}\right\} \Rightarrow d^{i,j}_{\mathbf{p}_i} = f \cdot b_{i,j} \cdot (z_\mathbf{p}^{-1} - z_\mathbf{c}^{-1}) \ .$$

When the optical centers are evenly distributed (i.e., $\forall i, j \in \{0, \dots, n-1\}, b_{i,j} = (j-i) \cdot b$), disparity values are scaled by $(j-i)$:

$$\forall i, j \in \{0, \dots, n-1\} \quad d^{i,j}_{\mathbf{p}_i} = (j-i) \cdot f \cdot b \cdot (z_\mathbf{p}^{-1} - z_\mathbf{c}^{-1}) = (j-i) \cdot d_\mathbf{p} \ .$$

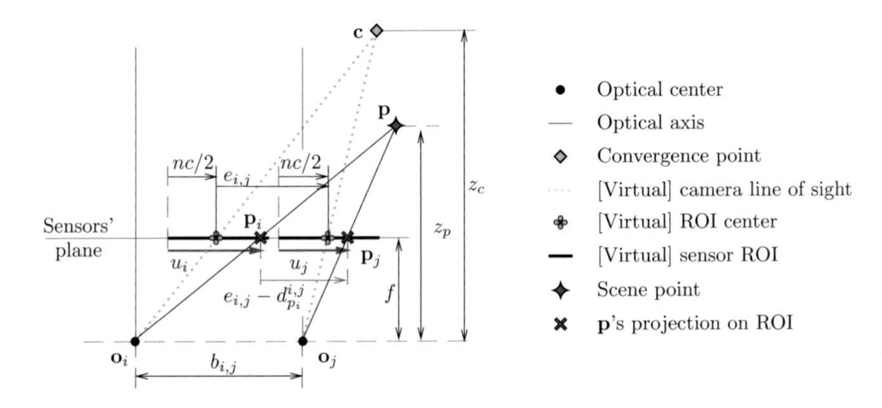

Figure 8.6: Projective geometry in off-axis simplified epipolar geometry (top view).

and disparity values for successive views are identical:

$$\forall i \in \{0, \ldots, n-2\} \quad d^{i,i+1}_{\mathbf{p}_i} = d_{\mathbf{p}}.$$

This common disparity $d_{\mathbf{p}}$ among each pair of successive views is conveniently used for each disparity assumption d for a pixel at $\mathbf{p}_i$ in $\mathbf{I}^i$. Instead of testing only two corresponding pixels for photo-consistency, one builds an associated geometrically consistent n-tuple in the multiscopic stereo matching process:

$$\forall j \in \{1, \ldots, n-1\} \quad \mathbf{p}_j = \mathbf{p}_i + (i-j) \cdot (d, 0) \quad . \tag{8.7}$$

These n-tuples contain one pixel per image, ordered according to their image number. Furthermore, thanks to Eq.(8.7), they all lie in the same epipolar plane and a common horizontal disparity assumption d is assigned to them. As such, pixels of a single n-tuple are corresponding projections on every view of a single 3D point.

To summarize, the presented multi-baseline stereovision paradigm reformulates the dense correspondence reconstruction problem as the answer to the question: "which of the geometrically consistent n-tuples correspond to actual 3D points in the scene according to their photo-consistency in the n views?".

Aside from the differences in the camera layout they are intended to handle, MVS methods may also be categorized according to what data or representation of the world they operate on [Seitz et al. 06]:

Scene-based methods employ a 3D scene model whose projections on views are checked for photo-consistency. As they are designed for a general freely arranged camera layout, they often use a 3D volume and rely on photometric similarity measures of the projections of the voxel cells, and remove others from the volume. Voxel coloring [Seitz and Dyer 99] preserves voxels whose cost is below a threshold. Space carving [Kutulakos and Seitz 00] progressively removes the photo-inconsistent voxels from an initial volume. More recently, a different category of methods has been proposed that uses a scene model composed of a collection of planar patches or surfels whose depth and orientation are separately optimized to maximize their photo-consistency. For representing patches, such methods rely on planar polygons [Habbecke and Kobbelt 06], circular disks [Habbecke and Kobbelt 07], or pre-segmented superpixels [Micusik and Kosecka 10]. The seminal work of [Furukawa and Ponce 10] fits patches on pixels around detected sparse features, then expands them in order to fill gaps between their projections, and afterwards reconstructs and refines a mesh.

Some multi-baseline methods make use of the disparity space introduced by [Yang et al. 93] for reconstruction instead of working on the standard 3D scene. Making use of photo-consistency and visibility reasoning, [Ismael et al. 14] optimize a so-called *materiality map* in this space for improved multi-view reconstruction.

Image-based methods compute a set of depth or disparity maps which are merged later [Narayanan et al. 98, Goesele et al. 06] or to which they apply constraints [Gargallo and Sturm 05, Szeliski 99] to ensure a consistent 3D scene reconstruction. Some methods that expect a more restrictive camera layout, typically multi-baseline, directly match n-tuples as multiscopic pixel sets [Niquin et al. 10, Kang and Szeliski 04], as described above. Among methods intended for a free camera layout, some computationally more intensive techniques are dedicated to MVS from community photo collections (CPC) and have gained an increasing interest. They have to handle a large number of uncalibrated views of a scene [Goesele et al. 07]. New difficulties then arise as such views are typically shot at different times, with differing acquisition geometries (viewpoints, angles, focal lengths, resolutions), and usually differing environmental conditions (weather, exposure, occlusions). This makes it necessary to restrict the matching to subsets of views sharing similar exposure, and empower the methods to deal with significant baselines (distances) between the cameras.

Feature-based methods compute dense correspondences by first matching feature points which can be more robustly estimated than a complete disparity map. In a second step a surface model is fitted to the reconstructed features [Taylor 03].

Image-based methods that rely on multiscopic matching of n-tuples share an important advantage with scene-based methods: implicit consistency of the reconstruction. Furthermore, both take full advantage of pixel redundancy to avoid as many false matches as possible while enabling smart occlusion handling schemes. The photo-consistency cost implied in those methods is often defined, for each 3D point of interest, as the aggregation of dissimilarity costs of its corresponding pixels over a set $\mathcal{R}$ of several pairs of views

$$c(\mathbf{p}) = \sum_{(i,j)\in\mathcal{R}} c(\mathbf{p}_i, \mathbf{p}_j) \quad . \tag{8.8}$$

Here $c(\mathbf{p}_i, \mathbf{p}_j)$ is the same cost function as used before in the binocular case and $\mathbf{p}_i$ is the pixel position of the backprojected 3D point $\mathbf{p}$ into the i-th view $\mathbf{I}^i$. Commonly employed pair sets $\mathcal{R}$ consist of:

- successive views $\mathcal{R} = \{\ (i, i+1) \mid \forall i \in \{0, \ldots, n-2\}\ \}$,
- every available pair $\mathcal{R} = \{\ (i, j) \mid \forall i, j \in \{0, \ldots, n-1\}, i < j\ \}$,
- pairs specifically selected according to geometrical considerations and/or similar recording conditions.

The first option is often preferred in a rectified layout as it makes the stereo method less sensitive to colorimetric shifts among the image set. Contrarily, the third is used when a very large number of views is available with widely spread viewpoints.

Using all views to compute the dissimilarity cost in Eq.(8.8) rarely leads to high-quality reconstructions, as a scene point $\mathbf{p}$ may be occluded in some of the cameras. However, as multiple views are available visibility may be reconstructed as well during the correspondence estimation [Kolmogorov and Zabih 02, Kutulakos and Seitz 00, Seitz and Dyer 99]. This visibility information can be used to improve the correspondence reconstruction by:

- restricting to a useful set of image pairs $\mathcal{R}_{\mathbf{p}} = \{(i,j) \in \mathcal{R} \mid$ with $\mathbf{p}$ visible in both i and $j\}$,
- weighting the dissimilarity costs in Eq.(8.8) according to $\mathbf{p}$'s visibility in the images, or
- replacing the dissimilarity cost by a predefined, heavy, penality cost for pairs for which $\mathbf{p}$ would occlude some already reconstructed 3D point.

Multi-view stereovision methods vary strongly with respect to methodology and tend to be computationally more complex than their binocular

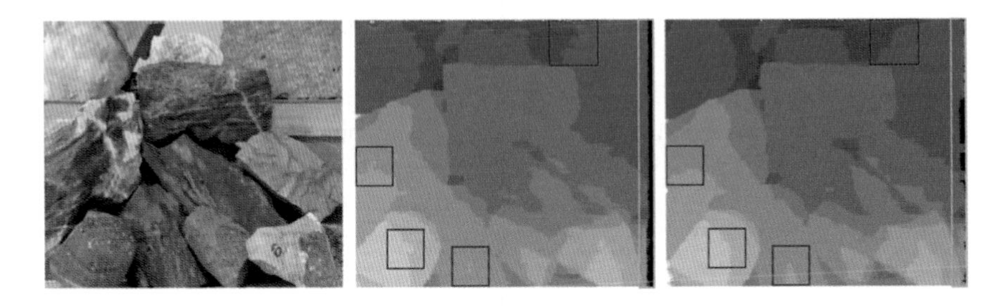

Figure 8.7: Results from a multi-baseline, scene-based method [Ismael et al. 14] on the Middlebury dataset "Rocks2": left, one source view; center, disparity map computed from two views only; right, disparity map computed from a set of 4 views. Green rectangles highlight an area more completely reconstructed from 4 views; red rectangles focus on some areas more regularly reconstructed from four views; the blue rectangle points to a region with higher accuracy in the 4-view case.

counterparts, as they have to deal with more data. Nevertheless, they tend to exploit the increased redundancy to achieve more robust reconstructions: Figure 8.7 shows that, with similar context, data, method, and parameters, results computed from four views are more regular and complete and contain fewer outliers than those using two views.

8.5 Stereo on the GPU

Stereo estimation is typically a significant computational expense and improving the computation time of stereo has long been in the focus of research. One characteristic that has been leveraged is that the depth/disparity estimation in local stereo for pixel (x, y) has no dependencies to any other pixel (Sections 8.2 and 8.3). Hence, parallel computation has long been explored for improving the computation time. Commodity graphics hardware (GPUs) nowadays provides a massively parallel processing platform with thousands of parallel compute cores and a significantly higher memory bandwidth than CPUs have at their disposal. Yang and Pollefeys [Yang and Pollefeys 03] proposed to leverage these highly parallel architectures to improve the computational performance utilizing the plane-sweeping stereo algorithm [Collins 96]. Besides leveraging the parallelism of GPUs, their method further leverages the high efficiency of texture mapping in GPUs.

Plane-sweeping stereo [Collins 96] is a multi-view stereo method with $n > 2$ views, $\mathbf{I}^0$ to $\mathbf{I}^{n-1}$ which does not rely on multiocular rectification

(Section 8.4). It can use any set of multiple overlapping views to perform stereo estimation. Plane-sweeping stereo estimation only requires the camera calibration of the views $\mathbf{I}^0$ to $\mathbf{I}^{n-1}$, as for example computed by structure from motion (Chapter 7). The core idea of plane-sweeping stereo is to perform the dense correspondence estimation by testing a series of plane hypotheses Π^i with $i = 1, \ldots, K$ for the scene, i.e., it assumes the scene is on a plane and then tests this hypothesis. Once all K plane hypotheses have been tested, the best plane is chosen for each pixel. This, however, does not mean that the scene has to be planar, as the plane only represents a local planar approximation of the scene for the function used to compute the matching cost. Please note that each plane is basically a slice in the cost-volume introduced in Section 8.3.

In plane-sweeping stereo the depth map relative to one of the images $\mathbf{I}^0, \ldots, \mathbf{I}^{n-1}$ is computed. This image is referred to as the reference view $\mathbf{I}^{ref}$ and its camera projection matrix is transformed to be $\mathbf{P}^{ref} = [\mathbf{U}_{3\times3} \quad 0_{3\times1}]$ where $\mathbf{U}$ is again the identity matrix. All other images' camera projection matrices are transformed as well to be in the same coordinate system as the reference image $\mathbf{I}^{ref}$. Given this unified coordinate system, the plane hypotheses Π^i are chosen with respect to $\mathbf{I}^{ref}$ to sample the depth interval $[d_{near}, d_{far}]$ with K steps[4] from a closest distance d_{near} to a farthest distance d_{far}. The plane hypotheses Π^i in Yang and Pollefeys [Yang and Pollefeys 03] were chosen to be fronto-parallel to the reference image $\mathbf{I}^{ref}$, i.e., their normals n_i are equal to $[0\ 0\ 1]^T$ and their distance d_i corresponds to the depth (distance from the reference camera to the plane). Conceptually, to test any specific plane hypothesis Π^i all views can be warped onto the plane and their photoconsistency[5] can be evaluated using any of the cost functions from Section 8.3. This would require the definition of a raster on the plane hypothesis Π^i, which is a challenge. Equivalently, a hypothesis Π^i can be tested with respect to the reference view $\mathbf{I}^{ref}$ by warping all other images $\{\mathbf{I}^0, \ldots, \mathbf{I}^{n-1}\} \backslash \mathbf{I}^{ref}$ to the reference view $\mathbf{I}^{ref}$. Photoconsistency at each pixel (x, y) in the reference image $\mathbf{I}^{ref}$ can then be tested using the warped images. The warp of pixel (x, y) from the reference view $\mathbf{I}^{ref}$ to image $\mathbf{I}^j$ over plane Π^i is a planar mapping and can be described by a planar homography $\mathbf{H}_{\Pi^i, \mathbf{P}^j}$. Here, $\mathbf{P}^j = \mathbf{K}_j \left[\mathbf{R}_j^T \ -\mathbf{R}_j^T \mathbf{C}_j\right]$ is the camera projection matrix of the camera corresponding to image $\mathbf{I}^j$ with $\mathbf{K}_j$ being the camera calibration matrix and $\mathbf{R}_j$, $\mathbf{C}_j$ representing the rotation of the camera and the camera center, respectively [Hartley and Zisserman 03].

[4]Please note that the steps are typically not equidistant steps. They are chosen to have equal disparity sampling in the reference view. For more details on the hypotheses generation see Gallup et al. [Gallup et al. 07].

[5]Photoconsistency is the color similarity, i.e., the highest photo consistency is achieved when all views have the same color.

Figure 8.8: Left: Plane-sweeping stereo's reference image from an outdoor video sequence. Right: Depth map computed by plane-sweeping stereo using a total of eleven views.

The homography $\mathbf{H}_{\Pi^i,\mathbf{P}^j}$ is given by:

$$\mathbf{H}_{\Pi^i,\mathbf{P}^j} = \mathbf{K}_j \left(\mathbf{R}_j^T + \frac{\mathbf{R}_j^T \mathbf{C}_j n_i^T}{d_i} \right) \mathbf{K}_r^{-1}. \tag{8.9}$$

Then, the location (x_j, y_j) in image $\mathbf{I}^j$ of the warped pixel (x, y) in the reference image $\mathbf{I}^{ref}$ can be computed by

$$(x' \; y' \; w')^T = \mathbf{H}_{\Pi^i,\mathbf{P}^j}(x \; y \; 1)^T \text{ and } x_j = \frac{x'}{w'}, \quad y_j = \frac{y'}{w'} \quad . \tag{8.10}$$

If the scene point that projects into pixel (x, y) is on the plane Π^i then the colors of pixel (x, y) in the reference image $\mathbf{I}^{ref}$ and the color of pixel (x_j, y_j) in image $\mathbf{I}^j$ should be very similar. Similar to the binocular case, their similarity can be measured by a variety of measures, as explained in Section 8.3.

Using the GPU, the warping between the reference image $\mathbf{I}^{ref}$ and image $\mathbf{I}^j$ with the homography $\mathbf{H}_{\Pi^i,\mathbf{P}^j}$ can be performed using projective texture mapping [Segal et al. 92], i.e., a projection of an input image onto a geometric primitive. This projection is highly efficient on GPUs. The similarity evaluation and the selection of the best plane hypothesis for each pixel in the reference view are independent for each pixel and hence can be performed in parallel as well. This high degree of parallelism provides speedup factors of one hundred and more for stereo estimation on GPUs [Gallup et al. 07]. In [Gallup et al. 07] it is further proposed to improve the local approximation of the scene geometry with the plane hypotheses Π^i by using multiple plane orientations. This indeed improves the accuracy of the stereo estimation by better modeling planes that are seen under oblique viewing directions, for example the ground plane. Figure 8.8 shows an image and its example depth map computed by multiway plane-sweeping [Gallup et al. 07].

8.6 Summary

This chapter only touched the tip of the iceberg that represents the field of dense correspondence estimation. Nevertheless, the knowledge provided here poses a useful basis for understanding any of the other current state-of-the-art correspondence techniques and provides a flexible basic framework upon which to build. The simplicity of local methods makes them attractive from a beginner's perspective as well as a computational view. Choosing the right dissimilarity function proves crucial for the quality of the algorithm, but with well-chosen adaptive weights, state-of-the-art results are achievable. Posing the aggregation step as a filtering process can dramatically improve the speed of the correspondence algorithm. To handle occlusions, a symmetry check between the input images can be used and several extensions including higher dimensional solution spaces and slanted surfaces can be easily incorporated at the cost of higher computation times.

The finding that current state-of-the-art dense correspondence algorithms achieve energies in the cost function that are below that of ground truth scenes [Szeliski et al. 08] may appear dissatisfying for researchers starting in this area. It should be mentioned, though, that this opens the door for more creative approaches that do not follow the standard paths but try to come up with novel ideas and complete new algorithms that differ in more than the choice of a new data or regularization term. Maybe the way to go is also specialized algorithms specifically targeting certain scenes or applications. While from a vision perspective the goal is to find the best automatic algorithm for dense correspondence matching, it may make sense to have an algorithm that one can improve through additional user input [Klose et al. 11, Ruhl et al. 13], e.g., for multimedia applications like image interpolation. In such cases, these interactive correction tools are of utmost importance. Further on, if massive amounts of images are to be matched, speed may be of highest interest [Frahm et al. 10].

9
Sensor Fusion

Andreas Kolb, Jiejie Zhu, and Ruigang Yang

9.1 Introduction

Many means to recover scene depth have been introduced so far. They range from specialized sensors that can directly return depth (e.g., LiDAR sensors and ToF sensors described in Chapter 4) to the most widely used stereo matching algorithms in Chapter 8 that require just two or more images as input. As with many real-world situations, for example in the context of cultural heritage (Chapter 22), there is no single method that can claim to be the panacea for all and every application that relies on depth data. As shown in Table 9.1, these different depth sensing approaches all have their advantages and disadvantages with respect to several criteria. Some of the disadvantages are inherent to the principle of operations, such as the quadratically growing depth error in stereo, while others are due to design

Table 9.1: Characteristics of different depth sensing methods.

	LiDAR	ToF	Stereo	Light Field	Photometric Stereo
Range‡	long	near	mid	mid	near
Depth accuracy	high	high	variable*	variable*	high†
Cost	high	low	low	high	low
Robustness	high	high	low	medium	medium
Resolution	single	low/medium	high	high	high
Dynamic scene handling	limited	yes	yes	yes	yes
Computational overhead	little	little	high	high	medium

‡: long > 20m; mid < 20m; near < 5m;
* The depth error grows quadratically with respect to range.
†Photometric stereo estimates the surface normal direction, not absolute metric depth.

choices or practical and technological considerations, such as the limited measurement range of ToF sensors. Furthermore these different depth sensing methods all have different systematic errors. ToF sensors, for example, cannot reproduce sharp concave corners due to the multi-path propagation of light. Results from stereo matching usually show the "fattening" effect around discontinuities (often due to unavoidable regularization). Naturally, researchers have therefore developed techniques for combining different sensor observations or modalities to increase the sensing performance. This is the central topic of this chapter.

Strictly speaking, *sensor fusion* is the combination of sensory data from disparate sources to improve the resulting data quality, usually in terms of accuracy and robustness. In the scope of this chapter, sensor fusion methods can be roughly divided into two categories: *multi-sample* fusion and *multi-modal* fusion approaches. Multi-sample fusion takes advantage of the redundancy in the input data to significantly reduce the noise in individual sensor readings, generating much cleaner output. It will be discussed in detail in Section 9.2 using several specific algorithmic examples. Multi-modal fusion takes advantage of the often complementary nature of different sensing modalities; for example, it combines the ability of photometric stereo to capture detail with the metric reconstruction accuracy of stereo, in order to reduce systematic errors in the fused data. Several typical examples of multi-modal fusion are presented in Section 9.3. This chapter concludes with a summary in Section 9.4.

9.2 Multi-Sample Fusion

Multi-sample fusion is a research topic with a long history in computer vision. The necessity for multi-sample fusion arises if several independently acquired range scans have to be fused into a single consistent model. Historically, the *Digital Michelangelo Project* [Levoy et al. 00] can be seen as a project involving the full pipeline from laser-scanner based range acquisition via alignment to postprocessing. At that time, the acquistion of a full historical site took 30 nights, and the full processing of the dataset required 1080 man hours resulting in 32 GB model data.

Multi-Sample Fusion Pipeline

Having new real-time range sensing cameras at hand (Section 4), the quest for easy online scene acquisition techniques has gained momentum. Here, a system is desired that is capable of processing a stream of range images in a sequential manner. A first system that allowed for online acquisition of 3D models using a structured light scanner has been presented by [Rusinkiewicz et al. 02]. In this work, a scanner with up to 60 Hz range image acquisition rate is used and the registration is based on an optimized version of the *Iterative Closest Point (ICP)* algorithm [Besl and McKay 92] (Chapter 10).

[Schuon et al. 09] present an approach for superresolution for range images acquired from time-of-flight (ToF) cameras. Here, only very small motions are performed, so that the range image alignment can be done using optical flow. By further extending this idea, [Cui et al. 10] present a fusion approach using a ToF camera. Due to the low lateral (x/y) resolution of the ToF camera used, their approach is formulated as a simultaneous superresolution and a probabilistic registration technique. The major drawback is the complex optimization approach which is very time intensive and cannot be performed in real-time.

The generic data processing pipeline presented by [Rusinkiewicz et al. 02] is still used in many of nowaday's approaches for free-hand model acquisition based on range cameras and contains the following steps (Figure 9.1):

- **Range Image Acquisition and Data Pre-Processing**: Several sensor and algorithm alternatives are available to acquire range data (Chapter 4). Depending on the device, several pre-pocessing operations may be needed: For instance flying pixels in ToF camera data may need to be removed (Section 4.3). Also data may need to be further filtered and smoothed, e.g., using bilateral filtering.

- **Camera Pose Estimation/Registration**: Any sequential range image acquisition device is assumed to produce redundant data, i.e., overlapping regions between subsequent range images. Using this redundancy, the individual range images can be registered against each other. This operation is usually achieved with an ICP algorithm.

- **Model Update**: After a new frame has been registered into a model, any new range image has to be fused with the already existing scene

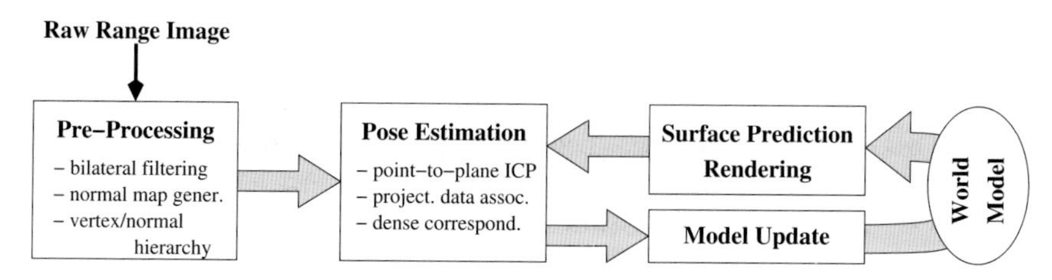

Figure 9.1: The basic workflow of the KinectFusion algorithm.

model. In this step, the redundancy that was needed for the registration step should be removed in order to reduce sensor noise and obtain a compact scene model.

- **Rendering**: Depending on the target application, the model is further processed. In most cases, some kind of visualization, e.g., for visual inspection, is provided, i.e., the model needs to be rendered.

KinectFusion

In the context of modern high-speed range imaging cameras, the KinectFusion system presented by [Izadi et al. 11, Newcombe et al. 11] is the first realization of the pipeline sketched above, that entirely runs in real-time.

Whenever designing a system that is able to process range data at this high throughput, i.e., 512×424 pixel for the Kinect 2™ at 30 FPS, the key question is, what is the right representation of the model. This question is tied to the solution of the associated challenges of how to use this model representation for online registration, merging, and rendering. KinectFusion adopts the *volumetric model* and fusion method of [Curless and Levoy 96]. The KinectFusion algorithm utilizes a *regular voxel grid* in order to store the depth measurements of the range camera device and the world model as *truncated signed distance fields*, where the zero-set corresponds to the scene's surfaces S. The truncated distance of a 3D point $\mathbf{x}$ is defined using a distance threshold μ

$$\text{tsdf}(\mathbf{x}) = \begin{cases} \max(-\mu, -\min_{\mathbf{y} \in S} \|\mathbf{x} - \mathbf{y}\|) & \text{if } \mathbf{x} \text{ is inside } S \\ \min(\mu, \min_{\mathbf{y} \in S} \|\mathbf{x} - \mathbf{y}\|) & \text{if } \mathbf{x} \text{ is outside } S \end{cases}.$$

The motivation for the truncation is mainly the fact that signed distance values with respect to a surface are estimated using the polar distance given in the range image, which is only reliable close to the surface. Transforming the input range data into a truncated signed distance field and merging it with the current representation in the voxel grid is done by sweeping the volume (Algorithm 9.1).

In a first step, the proper alignment of the input range map to the data accumulated in the voxel grid so far needs to be computed. Therefore, for each individual depth map $Z^i(\mathbf{u}) \in \mathbb{R}$ with pixel coordinates $\mathbf{u} = (x, y)$, a map of 3D vertex coordinates, henceforth called *vertex map* $\mathbf{v}^i(\mathbf{u}) \in \mathbb{R}^3$, is generated using the intrinsic camera matrix $\mathbf{K}$. An additional *normal map* $\mathbf{n}^i(\mathbf{u}) \in \mathbb{R}^3$ is computed using a bilateral filtered version of the vertex map. Both maps are further processed to get a multi-scale representation, which is required for the fast ICP method (Section 10.3). The ICP correspondence finding and alignment is subsequently done in a coarse-to-fine manner, starting with the coarsest hierarchy level. A fixed amount of iterations is performed per level. On each hierarchy level, ICP uses a dense correspondence map for the alignment, which is explained next.

For the alignment of the new input data frame i, it is assumed that the camera pose of the previous frame $\mathbf{T}^{i-1} = \begin{bmatrix} \mathbf{R}^{i-1} & \mathbf{t}^{i-1} \\ 0\ 0\ 0 & 1 \end{bmatrix}$ consisting of the rotation matrix $\mathbf{R}^{i-1}$ and the translation $\mathbf{t}^{i-1}$, is known. One also assumes that the input vertex map $\mathbf{v}^i$, normal map $\mathbf{n}^i$ and distance map Z^i with polar distances for frame i are given (Figure 9.1). The KinectFusion approach follows a frame-to-model alignment, where each vertex in the input vertex map is a potential correspondence to the current truncated signed distance field accumulated in the voxel grid which represents the current *model*. To identify the corresponding points in the model, the previous camera pose $\mathbf{T}^{i-1}$ is used in order to synthesize a *model range map* $\mathbf{v}_m^{i-1}$ and a *model normal map* $\mathbf{n}_m^{i-1}$. To serve this purpose, the so far accumulated truncated signed distance field of the model is rendered using ray casting. From the previously estimated camera pose $\mathbf{T}^{i-1}$, rays are cast through the volume searching for the first zero crossing [Parker et al. 98]. Aligning the input map $\mathbf{v}^i$ to the model map $\{\mathbf{v}_m^{i-1}, \mathbf{n}_m^{i-1}\}$ requires the estimation of the relative rigid transformation $\mathbf{T}^{i\to(i-1)}$, which enables the computation of the new camera pose $\mathbf{T}^i = \mathbf{T}^{i-1} \cdot (\mathbf{T}^{i\to(i-1)})^{-1}$. As mentioned earlier, $\mathbf{T}^{i\to(i-1)}$ is determined using an iterative closest point (ICP) approach. Initializing $\mathbf{T}_k^{i\to(i-1)} = [\mathbf{I}_{3\times3}|\mathbf{0}]$ for $k = 0$, the iterative solution minimizes the following error metric

$$\mathbf{T}_{k+1}^{i\to(i-1)} = \min\arg \sum_{\mathbf{u}\in\mathcal{M}} \left((\mathbf{T}_{k+1}^{i\to(i-1)} \tilde{\mathbf{v}}^i(\mathbf{u}) - \tilde{\mathbf{v}}_m^{i-1}(\mathbf{u}_k)) \cdot \mathbf{n}_m^{i-1}(\mathbf{u}_k) \right)^2 . \tag{9.1}$$

Here, $\mathcal{M}$ denotes the set of input range map pixels $\mathbf{u}$ with valid depth, i.e., depth values masked out by the depth sensor as invalid and correspondences with strongly deviating normals $\mathbf{n}^i(\mathbf{u}) \cdot \mathbf{n}^{i-1}(\mathbf{u}_k) < 1 - \epsilon$ are discarded. Furthermore, $\mathbf{u}_k$ is the pixel coordinate in the model map corresponding to $\mathbf{u}$ in the k-th iteration when applying the previous estimate $\mathbf{T}_k^{i\to(i-1)}$ of the relative transformation. Using the known intrinsic camera matrix $\mathbf{K}$, $\mathbf{u}_k$ is computed in homogeneous coordinates as

$$\tilde{\mathbf{u}}_k = \mathbf{K} \cdot (\mathbf{T}_k^{i\to(i-1)})^{-1} \tilde{\mathbf{v}}^i(\mathbf{u}).$$

The computation of $\mathbf{T}^{i\to(i-1)}$ is based on the *small angle* assumption between the iterative steps $\mathbf{T}_k^{i\to(i-1)}$ and $\mathbf{T}_{k+1}^{i\to(i-1)}$. Based on this assumption, the incremental update matrix between the iterative steps can be linearly approximated, i.e.,

$$\mathbf{T}_{k+1}^{i\to(i-1)} = \mathbf{\Delta}\cdot\mathbf{T}_{k}^{i\to(i-1)} \text{ with } \mathbf{\Delta} \approx \begin{pmatrix} 1 & -\gamma & \beta & t_x \\ \gamma & 1 & -\alpha & t_y \\ -\beta & \alpha & 1 & t_z \\ 0 & 0 & 0 & 1 \end{pmatrix} = \begin{pmatrix} \mathbf{R}^{\Delta} & \mathbf{t}^{\Delta} \\ 0\,0\,0 & 1 \end{pmatrix},$$

where α, β, and γ are the rotation angles around the $x-, y-$, and $z-$axis. After applying the following separation, $\mathbf{R}^{\Delta} = \mathbf{I}_{3\times 3} + \mathbf{S}^{\Delta}$, and after applying a linear approximation to Eq.(9.1), yields

$$\begin{aligned} &(\mathbf{T}_{k+1}^{i\to(i-1)}\tilde{\mathbf{v}}^{i}(\mathbf{u}) - \tilde{\mathbf{v}}_{m}^{i-1}(\mathbf{u_k}))\cdot\mathbf{n}_{m}^{i-1}(\mathbf{u_k}) = \\ &(\mathbf{\Delta}\,\underbrace{\mathbf{T}_{k}^{i\to(i-1)}\tilde{\mathbf{v}}^{i}(\mathbf{u})}_{=\tilde{\mathbf{v}}_{k}^{i}(\mathbf{u})} - \tilde{\mathbf{v}}_{m}^{i-1}(\mathbf{u_k}))\cdot\mathbf{n}_{m}^{i-1}(\mathbf{u_k}) \approx \\ &(\mathbf{S}^{\Delta}\mathbf{v}_{k}^{i}(\mathbf{u}) + \mathbf{v}_{k}^{i}(\mathbf{u}) + \mathbf{t}^{\Delta} - \tilde{\mathbf{v}}_{m}^{i-1}(\mathbf{u_k}))\cdot\mathbf{n}_{m}^{i-1}(\mathbf{u_k}). \end{aligned}$$

Since $\mathbf{S}^{\Delta}$ is the skew matrix $[\mathbf{a}^{\Delta}]_{\times}$ for vector $\mathbf{a}^{\Delta} = (\alpha, \beta, \gamma)$ one finally gets

$$(-[\mathbf{v}_{k}^{i}(\mathbf{u})]_{\times}\mathbf{a}^{\Delta} + \mathbf{t}^{\Delta} + \mathbf{v}_{k}^{i}(\mathbf{u}) - \tilde{\mathbf{v}}_{m}^{i-1}(\mathbf{u_k}))\cdot\mathbf{n}_{m}^{i-1}(\mathbf{u_k}).$$

This expression is linear in the unknowns $\mathbf{a}^{\Delta}, \mathbf{t}^{\Delta}$, thus minimizing Eq.(9.1) results in solving a linear system in six unknowns. It should be noted that the resulting angles α, β, γ are used to set up a correct rotation matrix instead of using the linearized version.

In KinectFusion [Newcombe et al. 11], a fixed number of ICP iterations is performed on each level, i.e., $4, 5, 10$ from the coarsest to the finest level. Also the ϵ-threshold to discard correspondences due to normal deviation should be adjusted while moving through the hierarchy, i.e., less variation should be allowed on finer levels.

Assuming several registered signed distance fields in a common voxel grid representing different overlapping regions of a given object, the resulting fused implicit model is simply achieved via averaging the distance fields. Thus, fusion is a very efficient and simple step. KinectFusion uses truncated signed distance fields, which carry valid information only in a small band around the measured surface. The thickness of this band is a user-defined parameter μ and needs to be chosen according to the noise level of the sensor and the resolution of the voxel grid.

Averaging needs to be done in such a way that the range inaccuracy or sensor noise is handled properly. In case a grid voxel stores information with high reliability, i.e., many input frames have "seen" this voxel, new points should not have a strong influence on the resulting signed distance value. Vice versa, an input point should be weighted stronger if the voxel

```
Data: Current depth map Z^i parameterized over pixel coordinates;
Voxel volume; voxel g stores weight g.w and signed distance g.s;
Current transformation matrix [T_i|t_i];
foreach (x, y) volume slice in parallel do
    while sweeping voxel g from slice z_front to z_back do
        v^g ← convert g from voxel to global 3D coordinates;
        v ← T_i^{-1} v^g // voxel in camera space
        u_v ← perspectively project vertex v onto image plane;
        if v in camera view frustum then
            sdf ← ||t_i − v^g|| − Z^i(u_v) // compute signed distance
            // truncate signed distance
            if sdf > 0 then
                tsdf ← min(μ; sdf);
            else
                tsdf ← max(−μ; sdf);
            end
            // avg signed dist. and weight & assign to grid
            g.s ← (g.w · g.s + w_i(u_v) · tsdf)/(g.w + w_i(u_v));
            g.w ← min(weight_max, g.w + w_i(u_v));
        end
    end
end
```

Algorithm 9.1: Updating the truncated signed distance field represented in the voxel volume by averaging a new registered input depth map.

has been observed only a few times so far, i.e., the stored surface information is less reliable. KinectFusion stores in each voxel g the truncated signed distance field value $g.s$ and a weight $g.w$. The weight might be a simple counter, i.e., each new point averaged within the truncation range increases the weight by 1. Alternatively, the input depth map carries additional per-distance weights $w_i(\mathbf{u})$ for each range pixel $\mathbf{u}$. These weights can, for example, represent the accuracy of the range value.

Algorithm 9.1 explains the update process of the signed distance field of the model's volume grid. Each voxel sequence $(x, y, z_{\text{front}}, \ldots, z_{\text{back}})$ is processed in parallel on the GPU. After transforming the current voxel to camera space $\mathbf{v}$ and identifying the corresponding range pixel $\mathbf{u}_v$, the signed distance with respect to the range map sdf is computed (note, that $\mathbf{t}_i$ represents the coordinates of the camera's focal point in world space). The last step is the update of the voxel's distance and weight values.

Improvements

There are several shortcomings of the KinectFusion approach. The first issue is the constrained work space and resolution, which is predefined by the fixed volume grid. As the grid has to be processed on the GPU, 512^3 voxels is a typical resolution that can be handled given the memory constraints of current GPUs. In case of a desired accuracy of 5 mm, the working volume is restricted to some 2.5 m in each dimension. Another costly step is the transformation of the input point set into a truncated signed distance field. A third issue is the difficulty to handle dynamic scenes.

Several approaches have been proposed to overcome constraints imposed by the restricted working volume. One solution strategy is simple volume moving methods, as proposed by [Roth and Vona 12]. They stream out voxels from the GPU in order to make space for new voxels; however, this approach is lossy. [Chen et al. 13a] present a hierarchical GPU data structure which is capable of lossless streaming of subvolumes between GPU and host, decoupling the active volume from a predefined physical space. [Nießner et al. 13] present a method to restrict the volumetric model representation to regions that contain input points. This is achieved by subdividing space into a coarse, regular and virtually infinite voxel block grid. In case input points get assigned to one of these coarse voxel blocks the first time, the "real" voxels on resolution level are allocated and referenced in a hash-table. The associated hash function transforms the voxel block ID into the hash table ID. Thus, voxel blocks without any surface points do not consume any memory. [Nießner et al. 13] demonstrate the acquisition of very large scenes, e.g. a courtyard of $16 \times 12 \times 2$ meters in extent (Figure 9.2).[1]

An alternative KinectFusion approach that uses a point-based rather than a voxel grid representation is presented by [Keller et al. 13]. Their major challenges are the rendering of the point-based model from a given camera pose, and the compactification of the points in the model. Every model point stores the point location, the normal, and a radius. Here, the radius describes the surface portion covered by a model point. Using this information, [Keller et al. 13] explicitly render the model by means of a simple surface-splatting technique which employs the per-point information to generate disk-shaped surface splats. Point fusion is applied in order to prevent memory overflow. First, an image-based correspondence finding establishes corresponding model points for each input point. This is achieved by rendering a high-resolution index map, which stores a single model point index in each pixel. Usually 4×4 pixels (indices) are associated with each input point. In case a model point is sufficiently close to the input point under consideration, the normal is sufficiently similar and the model

[1] http://graphics.stanford.edu/~niessner/niessner2013hashing.html

Figure 9.2: Large courtyard scene acquired with the method presented by [Nießner et al. 13].

point's radius is larger, the model and the input points are merged. Similar to the original KinectFusion algorithm, the number of merges is stored as a weight which represents the reliability of the point; model points with low weights are removed from the model. The radius of an input point is given as the projected pixel size scaled with respect to the distance. Thus, larger points implicitly model higher distances from the camera and, at the same time, they have a higher uncertainty due to sensor noise. Therefore, if an input point has a larger radius than the corresponding model point, it is less reliable and should not be merged. In contrast to the original KinectFusion, this approach explicitly merges normal and radius information.

In order to handle dynamic scenes to a certain extent, the original KinectFusion system [Izadi et al. 11, Newcombe et al. 11] can simply be extended by incorporating a temporal blend weight, resulting in a fading out of model information over time. Thus, if a surface point is not observed for a longer period, it is removed from the model. [Keller et al. 13] propose a more explicit approach to handle dynamic scene objects. The idea is to mark the input points in the dense ICP correspondence map as potentially moving objects, in case they have not been successfully assigned to a model point. By further morphological operations and region growing, connecting parts of potentially moving objects are identified. In case marked input points get a model point as merge partner in the merging stage, the weight of the corresponding model point is set to a low value. If the object is really moving, it is very likely that this model point will not be observed again and thus it is automatically removed from the model later on. Figure 9.3 shows some sample results obtained with this strategy for handling of dynamic scenes. [Keller et al. 13] further show, that the segmentation

Figure 9.3: Sample scenario capturing a scene with dynamic objects. Merged model normals of the static scene (top left), current input normals after players start to interact with the ball (top center), and the regions detected as potentially dynamic (top right). The lower sequence shows the corresponding renderings of the model (bottom left), superimposed with the dynamic scene parts (bottom center and right).

of potentially moving objects has a very positive influence on the ICP stability, since potentially moving parts are suppressed from correspondence finding during ICP.

9.3 Multi-Modal Fusion

In the previous section, the fusion has been carried out by using multiple samples from the same sensor to achieve better results than any single frame can deliver. However, some sensing modalities have an inherent system bias, which it can be difficult to remove by considering data only from the sensor itself. Fortunately, some sensor modalities have complementary characteristics. In these situations, fusion using two or more distinctive sensor modalities is thus a natural choice. In this section two sets of widely used fusion schemes will be discussed. They are fusion of stereo with shape-from-shading and fusion of time-of-flight sensors with stereo.

Fusion of Stereovision and Shape-from-Shading

Stereovision is based on the principle of *triangulation* (Chapter 8). Given two or more images of the same object, taken from known and different viewpoints, one can triangulate the depth of a 3D point from its corresponding pixels in the input images. The earliest attempt to solve the stereo problem, by Marr and Poggio, dates back to 1976 [Marr and Poggio 76]. The foundations of shape-from-shading are described in the pioneering work of Woodham [Woodham 80], in which the surface normal of a Lambertian object is estimated using images taken under illumination from three distant light sources. The surface normal field can then be integrated to obtain a 3D surface [Horn and Brooks 86, Agrawal et al. 06].

The complementary nature of stereovision and shape-from-shading has been recognized for many years. Stereovision systems are simple to set up and can generate metric measurements, but their accuracy is inversely proportional to the object distance from the cameras, and stereo reconstructions often lack fine scale detail. On the other hand, photometric stereo is known for capturing surface details, but the integrated surface is typically not metrically accurate and often times suffers from global non-uniform distortions due to the inaccuracy in normal estimation and unknown boundary conditions. Therefore, several methods have been developed to fuse (stereo) depth maps with normal maps [Cryer et al. 95, Nehab et al. 05, Anderson et al. 11]. Early methods [Cryer et al. 95, Mostafa et al. 99] focused on integrating the normal field and the depth map in a postprocessing step. Chen et al. [Chen et al. 03] formulate the reconstruction as a nonlinear optimization; Vogiatzis et.al [Vogiatzis et al. 06] apply the non-linear optimization on vertices of a closed mesh. Nehab et al. [Nehab et al. 05] presented the first paper to combine positional and orientational information in a linear way. Their positional measurement was obtained via active stereo with a projected random pattern and then depth values are assigned to all image pixels. By taking the normals from the photometric stereo technique into account, the recovered shape in the depth map is augmented with high-frequency detail, and the low-frequency reconstruction bias is successfully countered. In the following, a state-of-the-art fusion scheme that explicitly models discontinuities [Zhang et al. 12] is presented as an example.

Suppose the 3D surface $S(u,v) = [x,y,z]^T$ is parameterized in a 2D field $\Omega = [u,v]$ and the initial measured surface is S^0. In this section the superscript ^{0} is used to denote the initial measurement. Assume the field $[u,v]$ coincides with the image grid $\mathbf{u} := [i,j]$, and that per-pixel normal and depth are denoted as $\mathbf{n}(\mathbf{u}) = N_\mathbf{u}$ and $z(\mathbf{u}) = Z_\mathbf{u}$, respectively. Considering the perspective projection of 3D positions into pixels, the surface can be represented in terms of the depth map $Z_\mathbf{u}$:

$$S(\mathbf{u}) \quad = \quad \mu(\mathbf{u}) Z_\mathbf{u}, \tag{9.2}$$

$$\mu_\mathbf{u} \quad := \quad \begin{bmatrix} \frac{1}{f_x} & & -\frac{p_x}{f_x} \\ & \frac{1}{f_y} & -\frac{p_y}{f_y} \\ & & 1 \end{bmatrix} \begin{bmatrix} \mathbf{u} & 1 \end{bmatrix}^T, \tag{9.3}$$

where $[f_x, f_y]$ and $[p_x, p_y]$ are the camera's focal length and principal point

(Section 1.6). Then the distance of the surface to be reconstructed from the measurement is represented using the depths, as follows:

$$E_p = \sum_{\mathbf{u}} \|\mu_{\mathbf{u}}\|^2 \left(Z_{\mathbf{u}} - Z^0_{\mathbf{u}}\right)^2 . \tag{9.4}$$

To use the normal information, the cost function is defined as

$$E_n = \sum_{\mathbf{u}} \left(N^0_{\mathbf{u}} \cdot \frac{\partial S}{\partial u}\right)^2 + \left(N^0_{\mathbf{u}} \cdot \frac{\partial S}{\partial v}\right)^2 . \tag{9.5}$$

The corresponding derivatives of the surfaces are

$$\frac{\partial S}{\partial u}(\mathbf{u}) = \mu_{\mathbf{u}} \frac{\partial Z}{\partial u}(\mathbf{u}) + \left[\frac{Z_{\mathbf{u}}}{f_x}, 0, 0\right]^T , \text{ and} \tag{9.6}$$

$$\frac{\partial S}{\partial v}(\mathbf{u}) = \mu_{\mathbf{u}} \frac{\partial Z}{\partial v}(\mathbf{u}) + \left[0, \frac{Z_{\mathbf{u}}}{f_y}, 0\right]^T , \tag{9.7}$$

where $\frac{\partial Z}{\partial u}$ and $\frac{\partial Z}{\partial v}$ are discrete derivatives computed through finite differences. Due to the truncated error and the Gibbs phenomenon[2] arising near the gradient discontinuities, a 3-point derivative formula similar to [Harker and O'Leary 08] was used.

Combining all terms above, the desired surface is obtained by minimizing the following function in terms of depth values:

$$\begin{aligned} E &= \lambda_p \sum_{\mathbf{u}} \|\mu_{\mathbf{u}}\|^2 \left(Z_{\mathbf{u}} - Z^0_{\mathbf{u}}\right)^2 \\ &+ \lambda_n \sum_{\mathbf{u}} \left(\left(N^0_{\mathbf{u}} \cdot \mu_{\mathbf{u}}\right) \frac{\partial Z}{\partial u}\Big|_{\mathbf{u}} + \frac{N^0_{\mathbf{u},x}}{f_x} Z_{\mathbf{u}}\right)^2 \\ &+ \lambda_n \sum_{\mathbf{u}} \left(\left(N^0_{\mathbf{u}} \cdot \mu_{\mathbf{u}}\right) \frac{\partial Z}{\partial v}\Big|_{\mathbf{u}} + \frac{N^0_{\mathbf{u},y}}{f_y} Z_{\mathbf{u}}\right)^2 , \end{aligned} \tag{9.8}$$

where $\lambda_p + \lambda_n = 1$ and $\lambda_p, \lambda_n \geq 0$ are blending weights for each penalty. Since the differential operator over the depth map boils down to a matrix multiplication, the total energy is a quadratic form of depth values, which can be solved by an over-constrained linear least square system:

$$\begin{bmatrix} \lambda_p I \mu \\ \lambda_n \left[(N^0 \cdot \mu) \frac{\partial}{\partial u} + \frac{N^0_x}{f_x}\right] \\ \lambda_n \left[(N^0 \cdot \mu) \frac{\partial}{\partial v} + \frac{N^0_y}{f_x}\right] \\ \lambda_s \nabla^2 \end{bmatrix} [Z] = \begin{bmatrix} \lambda_p Z^0 \mu \\ 0 \\ 0 \\ 0 \end{bmatrix} , \tag{9.9}$$

[2]When a signal with a sharp discontinuity is approximated by its Fourier series, the values around a discontinuity are always oscillating. This is referred to as the *Gibbs phenomenon*.

 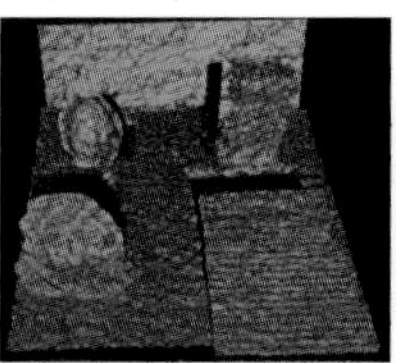 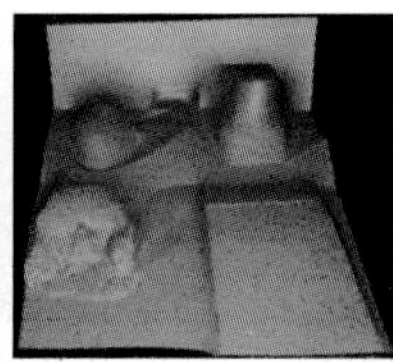 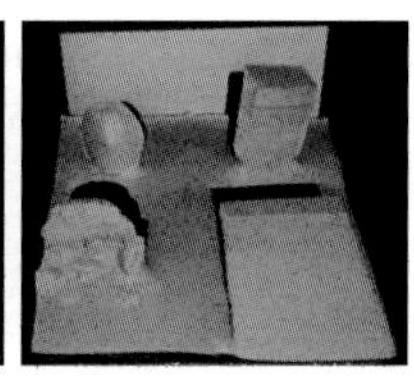

Figure 9.4: Fusion results of shape-from-shading and stereovision [Zhang et al. 12]. From left to right: color input image; depth map obtained via stereo reconstruction rendered in 3D; reconstructed 3D surface after [Nehab et al. 05]; 3D surface reconstructed after [Zhang et al. 12].

where $[Z]$ symbolizes stacking all the depth variables into a column vector, and the multiplications taken in the left-hand side are arranged in the order of each corresponding pixel. The smoothness term is meant to suppress artifacts due to quantization and due to the Gibbs phenomenon, where ∇^2 denotes the Laplacian operator on the 4-neighbor image grid and the weight λ_s is set to a small value. The weights λ_p and λ_n adjust how the depth and normal affect the final reconstructed surface. In practice, a large λ_n and small λ_p are chosen to ensure high-quality details as long as the depth bias of the result is comparable to the original one.

Fusion of Time of Flight (ToF) and Stereo

ToF cameras provide independent range estimates for each pixel in real-time, and are now becoming available from companies such as Microsoft, SwissRanger, and PMDTechnologies at commodity prices. A basic principle of ToF measurement is the timing of the roundtrip of a pulse of light in order to estimate the depth (Chapter 4). Given an array of such pulsed light emitters and a properly designed and synchronized pixel grid, a ToF sensor can return an array of depth measurements. In addition, by calculating the amplitude between the emitted and the returned signal, a ToF sensor can return an intensity image.

Unfortunately, while having the advantage of obtaining depth from texture-less areas, which is difficult with passive 3D reconstruction methods, the ToF sensing principle also has several disadvantages. Each of the pixel range estimates typically has a measurement bias which is dependent on the albedo of the recorded scene. Pixel measurements are also corrupted by noise which usually follows a Poisson distribution around the true depth. In addition, ToF sensors often have low resolution, both in x and y dimensions. In order to overcome these issues, recent works have shown improved depth estimates by modeling ToF and stereo depth as probability functions, and using a global optimization framework to fuse both data sources to yield an improved reconstruction [Zhu et al. 11].

Fusion Framework The estimation of the depth map $Z_{\mathbf{u}}$ for an image can be viewed as a conditional probability. Its posterior probability can be written as

$$P(Z|o) = \frac{P(Z,o)}{P(o)} \propto P(Z,o), \tag{9.10}$$

where o is evidence or observation, $P(o)$ is the normalization factor and usually assumed to be a constant.

To solve for the joint distribution $P(Z,o)$, the true depth map is modeled as a Markov Random Field (MRF). Each node in the MRF represents an independent pixel depth value, and each edge represents this node's relation to a neighboring node. According to the Hammersley–Clifford theorem, the joint distribution of $P(Z,o)$ can be written as

$$P(Z,o) \propto \prod_i \phi(Z_i, o_i) \prod_{\substack{i,j \\ j \in N(i)}} \varphi(Z_i, Z_j), \tag{9.11}$$

where $N(i)$ represents the neighborhood of the node i. Function $\phi(Z_i, o_i)$ is called the evidence function. It encodes the local evidence for the node i. Function $\varphi(Z_i, Z_j)$ is called the compatibility function. It is a smoothness function measuring the depth difference between the node i and its neighboring node j. Applying the negative logarithm to both sides of the above equation yields

$$\begin{aligned} -\log(p(Z,o)) &= -\log\Big(\prod_i \phi(Z_i, o_i) \prod_{\substack{i,j \\ j \in N(i)}} \varphi(Z_i, Z_j)\Big) \\ &= \sum_i -\log(\phi(Z_i, o_i)) + \sum_{\substack{i,j \\ j \in N(i)}} -\log(\varphi(Z_i, Z_j)). \end{aligned} \tag{9.12}$$

To compute the optimal depth values for a frame is thus equivalent to minimizing the right-hand side

$$\mathbf{E(Z)} = \mathbf{min}\Big(\sum_i -\log(\phi(Z_i, Z_i)) + \sum_{\substack{i,j \\ j \in N(i)}} -\log(\varphi(Z_i, Z_j))\Big), \tag{9.13}$$

The first sum in the above equation represents a data term which calculates the deviation of the estimate from the observed data (evidence). The second sum is often referred to as a smoothness term, and it encodes prior knowledge of what depth values a node is likely to have given its neighboring pixel's depth values. The notation can be simplified in the above equation by using function $D(i)$ to represent the data term, and function $V(i,j)$ to represent the smoothness term. Node j is a neighbor of node i:

$$\mathbf{E(Z)} = \sum_i D(i) + \sum_{\substack{i,j \\ j \in N(i)}} V(i,j). \tag{9.14}$$

In order to fuse depth from the ToF sensor and that from the stereo, the observation O is modeled from two components x, y. The variable x denotes the depth evidence from stereo, and y is the depth evidence from ToF. By assuming that x and y are independent, and $P(x)$ and $P(y)$ are constants, the posterior probability from Eq.(9.10) becomes

$$P(Z|x,y) = \frac{P(Z,x,y)}{P(x,y)} = \frac{P(Z,x,y)}{P(x)P(y)} \propto P(Z,x,y), \tag{9.15}$$

An enhanced energy function is derived by introducing two weighting factors to allow for more flexibility in fusing ToF and stereo:

$$D(i) = w_s \cdot f_s(Z_i, x_i) + w_t \cdot f_t(Z_i, x_i). \tag{9.16}$$

Here, f_s and f_t are functions of the data cost for stereo and the ToF sensor, respectively. The factor w_s and w_t are weights that trade-off the relative robustness of depth measurements from passive stereo and ToF reconstruction, respectively. The following paragraphs described these two terms in more detail.

Stereo Matching Cost and Weighting The cost function of stereo matching f_s is designed as pixel dissimilarity between the left and right views (the two views are rectified in which matched pixels can be searched on a scan line) with an aggregation process. To weight on both smooth and discontinuous regions, an appropriate window should be selected during the cost aggregation. In a sense, the window should be large enough to cover a sufficient area in textureless regions, while small enough to avoid crossing regions with depth discontinuities. In our implementation, a color weighted aggregation is incorporated to obtain this reliable correlation volume.

The weights are computed using both color and spatial proximity to the central pixel of the support window. The color difference in this support window (in the same view) is expressed in RGB color space as:

$$\Delta C(x,y) = \sum_{c \in R,G,B} |I_c(x) - I_c(y)|, \tag{9.17}$$

where I_c is the intensity of the color channel c. The weight of pixel x in the support window of y (or vice versa) is then determined using both its color and spatial difference as:

$$w_{xy} = e^{-\left(\frac{\Delta C(x,y)}{\gamma_C} + \frac{\Delta G_{xy}}{\gamma_G}\right)}, \tag{9.18}$$

where ΔG_{xy} is the geometric distance from pixel x to y in the 2D image grid. The factors γ_C and γ_G control the shape of the weighting function, and their values are determined empirically.

The data term (cost in the left and right views) is then an aggregation with the soft windows defined by the weights as:

$$f_s(x_l, x_r) = \frac{\sum_{y_l, y_r \in W(x_l) \times W(x_r)} w_{x_l y_l} w_{x_r y_r} d(y_l, y_r)}{\sum_{y_l, y_r \in W(x_l) \times W(x_r)} w_{x_l y_l} w_{x_r y_r}}, \quad (9.19)$$

where $W(x)$ is the support window around x; $d(y_l, y_r)$ represents the pixel dissimilarity using Birchfield and Tomasi's approach [Birchfield and Tomasi 98]; x_l and y_l are pixels in the left view; x_r and y_r are pixels in the right view. Results show that f_s can provide both moderate smoothness and preserve boundary sharpness on depth.

The best depth candidate for a pixel should have the lowest cost value of all the possible depth candidates. matching reliability of pixel p is inutitively defined as how distinctive the best and the second best cost is:

$$Ratio_p = \begin{cases} 1 - \frac{c_p^{1st}}{c_p^{2nd}} & C_p^{2nd} > T_c \\ 0 & otherwise \end{cases}, \quad (9.20)$$

where c_p^{1st} and c_p^{2st} are the best (lowest) and the second best matching cost of all depth candidates at pixel p. T_c is a small threshold value to avoid division by zeros. $Ratio_p \in [0, 1]$ can be used to prevent the deteriorating effects of ambiguous matching in case of poor signal-to-noise ratio (SNR).

ToF Cost The cost function for the ToF data f_t measure the depth difference reported from the ToF sensor d_t and a vector of depth candidates from stereo triangulation d_s. The transformation between them is estimated by a pre-calibration step similar to the geometric calibration step in [Zhu et al. 11] where the three cameras (for calibration, the ToF sensor is regarded as a regular camera because it can return a grayscale image along a depth map) are registered into one coordinate system. The method is explained in the following.

For each pixel p in the left (stereo) view, given a vector of disparity candidates d_c, a list of 3D candidate point locations for pixel p after stereo triangulation is generated. For each depth candidate, the pixel's depth cost is computed as the distance to the point returned from the ToF sensor (Figure 9.5).

[Zhu et al. 11] observed that the ToF's measurement reliability depends on the distance to the scene and the reflection of the captured object. In order to mathematically model the relationship between reliability and a pixel's true geometric distance, they placed a white calibration board into the scene and captured multiple frames at different distances. They noticed that the further a depth pixel is radially away from the center of the image, the larger the variance in the depth measurements of multiple

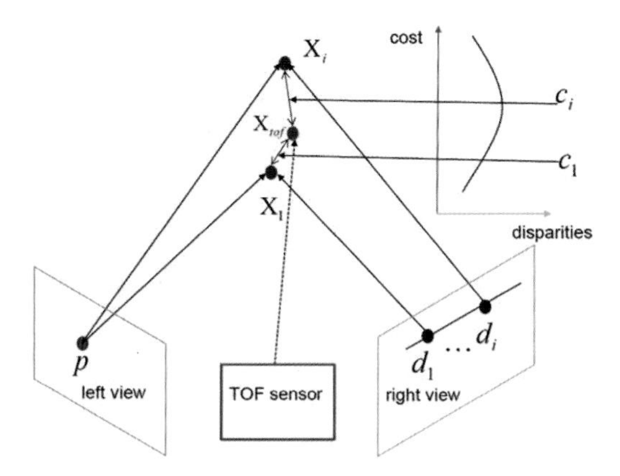

Figure 9.5: Example of how to calculate the ToF sensor cost. Disparity candidates for a pixel in the view of the left stereo camera and its disparity candidates are triangulated into a 3D space; ToF and RGB cameras are calibrated into a common coordinate system. The Euclidian distance between 3D candidates and 3D points captured by the ToF sensor determine the cost.

frames is. To calibrate the same dependency on an object's reflection, they took multiple depth images of boards with changing reflectances. They observed low deviations from the ground truth depth for highly reflective objects, and high deviations for low intensity objects. Based on these two observations, they were able to fit a reliability function for depth values measured by the sensor that depends on distance and intensity.

Figure 9.6 shows an example of ToF and stereo fusion results.

Other Fusion Methods The fusion framework introduced in Section 9.3 can be regarded as a *global method* which considers the depth from the ToF sensor through an additional data term, and then jointly optimizes for the depth values in a whole frame. In the literature, first methods for fusing ToF with stereo employed *local methods* for reconstruction. They optimize the per-pixel depth within a local window using evidences from either type of sensor modality [Kuhnert and Stommel 06, Beder et al. 07, Gudmundsson et al. 08, Hahne and Alexa 09, Bartczak and Koch 09, Chan et al. 08, Yang et al. 10] instead of over a whole frame. Recently, most of ToF and stereo fusion works have focused on *global methods*, because they yield and improved depth accuracy. For example, Kim et al. [Kim et al. 09] proposed a method for fusion of multi-view ToF and multi-view stereo to reconstruct dense 3D models by combining depth from ToF and multi-view stereo, and by using additional object silhouette constraints. Ruhl

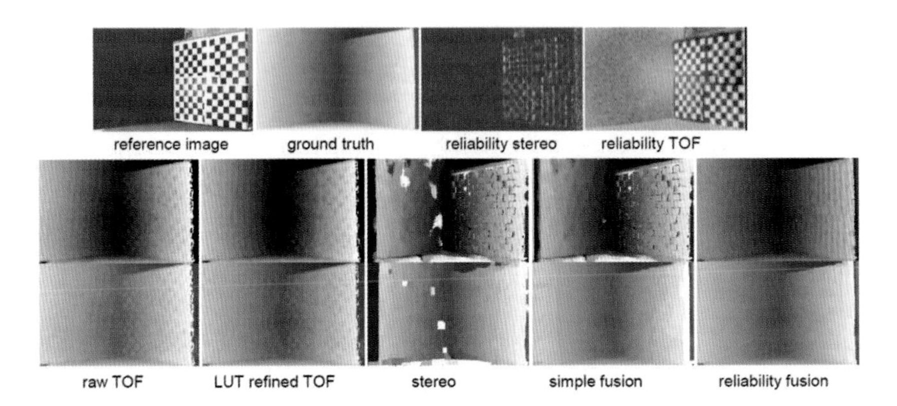

Figure 9.6: Example from [Zhu et al. 11]. Depth map from a simple scene with two planar boards. The first row shows the reference image, ground truth depth from a structured-light method and two maps of per-pixel reliability for stereo and the ToF sensor. The second row shows the depth map from local method. The third row shows the depth map from the global method. From left to right are: Raw depth from the ToF sensor, refined depth from the ToF sensor using a look up table, depth from stereo, depth from simple fusion, and depth from reliability fusion. Simple fusion results are obtained by setting f_s and f_t equal to 0.5. While the reliability fusion results are obtained by automatically computing f_s and f_t, as introduced in the text.

et al. [Ruhl et al. 12b] and Nair et al. [Nair et al. 12] formulated a variational model to fuse ToF and stereo for a sequence of frames by assuming a continuous image domain and continuous variables for estimation.

9.4 Summary

Sensor fusion has been proven to be a very effective approach to improve the quality of range sensing. When the depth map from a single measurement from a single sensor type is not of satisfactory quality, sensor fusion techniques can be considered for data enhancement.

10 Mesh Reconstruction from a Point Cloud

Tamy Boubekeur

10.1 Introduction

Typical 3D capture setups, such as *structure from motion*, Chapter 7, *multi-view stereo*, Chapter 8, and *RGBD sensors*, Chapter 9, provide a dense sampling of the measured 3D object or scene. In most cases, the measured data can take the form of a dense point cloud sampling the geometry of the scene and carrying surface attributes such as diffuse color and more advanced appearance parameters. However, the majority of visual computing applications require a continuous surface representation of their models which either interpolates or approximates the point data. The standard format of this representation is an *indexed triangle mesh* which consists of a list of vertices and a list of triangles indexed over the vertices.

Surface reconstruction algorithms [Hoppe et al. 92, Amenta et al. 01, Kazhdan et al. 06] convert 3D point clouds to meshes, "filling the holes" between the measured point samples and eventually removing the noise they embed. The pipeline supporting most of such algorithms contains several pieces:

1. **pre-processing stages**, including registration, outlier removal, and normal estimation,
2. **a surface model** defined from the processed sampling, which can take the form of a scalar field approximating the distance to the surface in 3D, a projection procedure folding the entire 3D space on the surface subspace, a binary indicator function defining an inside/outside segmentation of the surrounding space, or an energy measuring how close a given surface is from the point set,
3. **a mesh extraction method** discretizing the model in a polygon surface mesh,
4. **post-processing stages**, which are optional and act on the newly defined mesh structure by refining, simplifying, or filtering it.

Most of the numerous reconstruction methods can be described through this decomposition and formulated within an implicit, combinatorial or variational framework. They are characterized according to the kind of input point cloud they can handle (e.g., noisy, incomplete, large, varying in density) and the performance level they can reach (e.g., interactive, parallel scalable, local).

10.2 Overview

A complete review of existing reconstruction techniques goes beyond the scope of this chapter and can be found in the survey by Berger et al. [Berger et al. 14]. Essentially, these algorithms range from local surface approximation models to global and slower solutions. Local methods model the surface at a certain point in space by looking only at a subset of the input set to fit a simple object (e.g., radial basis function, analytical shape). These methods scale well and can still handle small-scale high-frequency noise in the input. Global methods compute the surface model at once using the entire input set and can cope with large holes, outliers, and high-quality connectivity in the output.

This chapter focuses on a simple solution which gives convincing results on dense point sets while being quite simple to implement: *moving least-squares* (or MLS) approaches [Levin 98, Levin 03]. Such reconstructions are local in the sense that they are mostly output-sensitive and do not require solving a large system of equations. They are also amenable to parallel implementations and streaming/out-of-core evaluations.

A typical MLS reconstruction pipeline is composed of three steps. First the point set is preprocessed by registering several partial subsets in the same frame, indexing the resulting sample set in a hierarchical data structure (e.g., kd-tree) to speed-up neighborhood queries, removing outliers, and estimating normals vectors (if not provided). Second, a particular MLS model is choosen, which typically takes the form of a simple geometric primitive, a fitting procedure and a projection operator. Last, a meshing machinery is executed using the MLS operator to locally detect the surface in 3D space. Optionally, a cleaning post-processing step can be performed to improve the connectivity quality, simplify/densify the mesh structure, or even compress it. When implementing parts of the full mesh reconstruction pipeline, a large number of the required operators can be supported by open source libraries such as the *Point Cloud Library*,[1] for instance.

[1] http://pointclouds.org

10.3 Registration

Mesh reconstruction algorithms assume that their input point cloud provides a reasonable sampling of a piecewise smooth real-world surface. This sampling usually corresponds to the union of several subsets which have been independently generated and individually provide only a partial covering of the object to reconstruct, as seen from a particular point of view in the real scene. Therefore, prior to reconstruction, these subsets are registered together. Essentially, this registration step aligns together different point sets measured on the same object by providing, for each of them, a specific space transformation function. When applied to each individual sample, this transformation brings it to the same global coordinate system used for the complete point set.

The *iterative closest point* (or ICP) algorithm [Besl and McKay 92] and its variants [Rusinkiewicz and Levoy 01] has become the standard method to perform this registration (Section 9.2). Assuming an initial rough alignment of two point subsets A and B, as well as some overlapping between them, the ICP algorithm computes a single rigid transformation for each individual subset by iterating with the following procedure:

1. select a subset of samples in A,
2. find, for each of them, the closest sample in B,
3. compute the rigid transformation that minimizes the distance of the so-defined pairs,
4. apply this transformation to B,
5. restart in 1 until convergence or a prescribed number of iterations.

The minimization is usually performed in the least-square sense and the numerous variants of this algorithm are either improving on the metric used to select closest samples, or they are improving the heuristic for selecting samples to match.

ICP provides an estimation of the extrinsic parameter of the sensors relative to a particular one, is simple to implement and fast to compute (Figure 10.1). Open source implementations, such as libICP[2] are available online. However, the intrinsic parameters and depth estimation biases may cause distortion in each subset already during the capturing step. In this case, the single per-subset rigid transformation of ICP is not suifficient and a non-rigid solution [Brown and Rusinkiewicz 07] may be required.

[2]http://www.cvlibs.net/software/libicp/

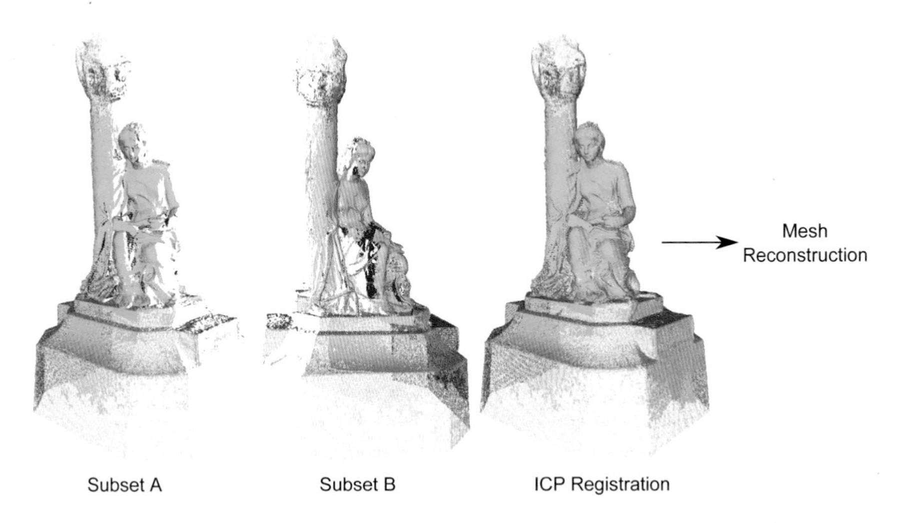

Figure 10.1: ICP registration of two point sets (in green and orange) coming from two distinct acquisitions (scans) of the same object.

10.4 Outlier Removal

While the smale-scale high-frequency noise is usually removed efficiently by mesh reconstruction algorithms, isolated samples located far away from the surface and corresponding to various errors occurring during the capturing process usually degrade strongly the quality of the reconstruction and shall be removed prior to the modeling and meshing stages.

Such samples are usually called *outliers*, and can be efficiently classified using local statistics that capture, for each sample, how far its closest neighbors are. A simple example for such statistics is the distance between the sample and the centroid of its k nearest neighbors: starting from a conservative threshold, which is application- and scale-dependent, the outlier classification is typically performed by iterating the computation of the statistics while decreasing the threshold progressively. An alternative solution casts the problem as a local density estimation by computing the *local outlier factor* [Kriegel et al. 09, Wang et al. 13a], classifying samples as outliers when the local density of their neighborhood is low.

The particular case of structured outliers is certainly harder to handle automatically. Although recent advances have shown efficiency for a number of cases [Giraudot et al. 13], this problem remains open in general, in particular in the context of high performance reconstruction pipelines.

10.5 Normal Estimation

Providing a normal vector $\mathbf{n}$ for each point sample can be done with at least two techniques. When the point cloud has been generated by registering several depth (2.5D) images, the gradient of the depth values in image space provides a good estimate of the local tangent plane of the geometry and therefore a simple (normalized) cross product of the horizontal and vertical gradients at the depth pixel $\{x, y\}$ gives a good normal vector for the equivalent 3D sample pretty much everywhere:

$$\mathbf{n} = \frac{G^X_{x,y} \otimes G^Y_{x,y}}{\|G^X_{x,y} \otimes G^Y_{x,y}\|}.$$

This estimate is indeed wrong only when there is a local high variance in depth (e.g., object contour). In such cases, the normal is corrupted and can be considered as noise, which will usually be filtered out in the upcoming stage of the pipeline. If the sensor topology is not known in the point cloud (unorganized point clouds), Hoppe et al. [Hoppe et al. 92] have proposed a simple solution to compute the normal: given a point sample with position $\mathbf{p}$ and $\mathcal{N}_\mathbf{p}$ its k nearest neighbors, the covariance matrix M of $\mathcal{N}_\mathbf{p}$ is computed as:

$$M = \Sigma_{\mathbf{q} \in \mathcal{N}_\mathbf{p}} (\mathbf{q} - \mathbf{o}) \bullet (\mathbf{q} - \mathbf{o})$$

with $\mathbf{o}$ the centroid of $\mathcal{N}_\mathbf{p}$ and $\bullet$ the outer product vector operator. The estimated normal $\mathbf{n}$ is taken along the eigenvector associated to the smallest eigenvalue of M. This eigenvector provides a good estimate of the orientation of the normal vector, but is defined locally and therefore cannot help deciding on the direction of the normal, i.e., two possible normals depending on which side of the local tangent plane is inside the object. To cope with this problem, the direction of the normals requires building a minimum spanning tree to be consistently oriented, unless the original position of the sensor which generated the sample is known, which avoids this direction ambiguity. In both cases, the resulting normal field carried by the point set can eventually be improved using a bilateral filter.

10.6 Point Set Surface

A simple and elegant way to define a surface model from the pre-processed sample set $\mathcal{S}$ is to use *point set surfaces* [Alexa et al. 01, Guennebaud and Gross 07, Alexa and Adamson 09], or (PSS), which are defined as the stationary set of $\mathbb{R}^3$ under a MLS projection.

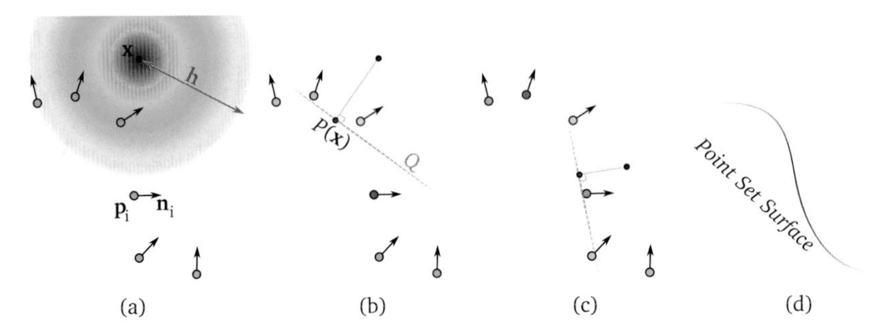

Figure 10.2: PSS from MLS projection illustrated in 2D. (a) The weighting kernel (in gradient color) is set at the evaluation position $\mathbf{x}$ (dark red) with support h. (b) A primitive (here a simple plane) Q is optimized to fit, in the weighted least square sense, the input point set and their normals, with the per-sample weight defined by the $\mathbf{x}$-centered kernel. The resulting projection $P(\mathbf{x})$ is in light red. (c) Illustration of the "moving" effect when evaluating the projection from a different point in space. (d) resulting PSS.

MLS projection Let us consider an input point sample set $\mathcal{S} = \{\mathbf{s_i}\}$, for which each sample $\mathbf{s_i} = (\mathbf{p_i}, \mathbf{n_i})$ carries a 3D spatial position $\mathbf{p_i} \in \mathbb{R}^3$ and a normal estimate $\mathbf{n_i} \in \mathbb{S}^2$. Given any point $\mathbf{x} \in \mathbb{R}^3$, the MLS projection of $\mathbf{x}$ with respect to $\mathcal{S}$ is defined as:

$$MLS_{\mathcal{S}}^{Q,\omega_h}(\mathbf{x}) := P_{\mathcal{S}}^{Q,\omega_h\,\infty}(\mathbf{x}) = P_{\mathcal{S}}^{Q,\omega_h} \circ \ldots \circ P_{\mathcal{S}}^{Q,\omega_h}(\mathbf{x})$$

with $P_{\mathcal{S}}^{Q,\omega_h} : \mathbb{R}^3 \to \mathbb{R}^3$ being a projection onto the geometric primitive Q (e.g., plane), fitted to $\mathcal{S}$ using a weighting kernel ω_h centered at $\mathbf{x}$ with support h.

This MLS projection procedure is rather easy to implement and can be summarized as follows (see Figure 10.2):

1. gather $\mathcal{N}_{\mathbf{x}}$, a local set of neighboring samples of $\mathbf{x}$ in $\mathcal{S}$,
2. fit Q to this set, weighting the contribution of each neighboring sample by its distance to $\mathbf{x}$ using ω_h,
3. project $\mathbf{x}$ onto Q once fitted,
4. restart in 1 until convergence or a maximum number of iterations is reached.

The neighborhood query is usually the most expensive step and is greatly sped up if a kD-tree [Bentley 75] has been pre-built over $\mathcal{S}$. The weighting kernel ω_h typically drives the low-pass filtering effect of this procedure. Typical choices for the weighting kernel are Gaussian or compact polynomial ones, such as, for instance, the Wendland quartic kernel [Wendland 95]:

$$\omega_h(t) = (1 - \frac{t}{h})^4(\frac{4t}{h} + 1) \text{ if } 0 < t < h, \; 0 \text{ otherwise }.$$

The support size h of ω tailors how much high frequency will be removed in the reconstruction, which corresponds to both noise and surface details: large values for ω better remove noise but may lead to over-smoothed surfaces.

Beyond the choice of this kernel, the main difference between the various flavors of MLS projections resides in the choice of the fitted primitive Q: a plane, a sphere or other simple shapes are usually good choices because they allow using simple fitting procedures. The most simple solution is probably to compute a "mean plane" [Alexa et al. 04] defined by a center $\mathbf{c}$ and a normal $\mathbf{n}$, for which the fitting procedure boils down to the weighted average of the neighoring sample positions and normals:

$$\mathbf{c}_{\mathcal{S}}^{\omega_h}(\mathbf{x}) = \frac{\Sigma_{\mathbf{s_j} \in \mathcal{N}_\mathbf{x}} \omega_h(||\mathbf{x} - \mathbf{p_j}||)\mathbf{p_j}}{\Sigma_{\mathbf{s_j} \in \mathcal{N}_\mathbf{x}} \omega_h(||\mathbf{x} - \mathbf{p_j}||)} \quad \mathbf{n}_{\mathcal{S}}^{\omega_h}(\mathbf{x}) = \frac{\Sigma_{\mathbf{s_j} \in \mathcal{N}_\mathbf{x}} \omega_h(||\mathbf{x} - \mathbf{p_j}||)\mathbf{n_j}}{||\Sigma_{\mathbf{s_j} \in \mathcal{N}_\mathbf{x}} \omega_h(||\mathbf{x} - \mathbf{p_j}||)\mathbf{n_j}||}$$

and the projection is simply defined as:

$$P_{\mathcal{S}}^{\omega_h}(\mathbf{x}) = \mathbf{x} - ((\mathbf{x} - \mathbf{c}_{\mathcal{S}}^{\omega_h}(\mathbf{x})) \cdot \mathbf{n}_{\mathcal{S}}^{\omega_h}(\mathbf{x}))\mathbf{n}_{\mathcal{S}}^{\omega_h}(\mathbf{x}),$$

A number of extensions have been proposed to improve this basic framework. One can, for instance, fit algebraic spheres [Guennebaud and Gross 07] to better cope with poor input samplings, take into account the normals of $\mathcal{S}$ using a hermite interpolation [Alexa and Adamson 09], preserve sharp features [Fleishman et al. 05] or even exploit the input set self-similarity in a non-local scheme [Guillemot et al. 12]. Several alterative projection procedures have also been proposed, to ensure orthogonality or better control the low pass filtering.

Implicit form The MLS projection procedure also provides an implicit form $f_{\mathcal{S}} : \mathbb{R}^3 \rightarrow \mathbb{R}$ of the surface when simply looking at the signed distance between the original location $\mathbf{x}$ and the converged location under MLS projection. This distance can be approximated by estimating a local tangent plane at the converged location and computing the signed distance from $\mathbf{x}$ to this plane:

$$f_{\mathbf{S}} = -(\mathbf{x} - MLS(\mathbf{x}) \cdot \mathbf{n_x}).$$

The 0-contour (iso-contour for the value 0) of this implicit MLS form is often instrumental for meshing algorithms which compute a polygonal contour of the point set surface (Figure 10.3).

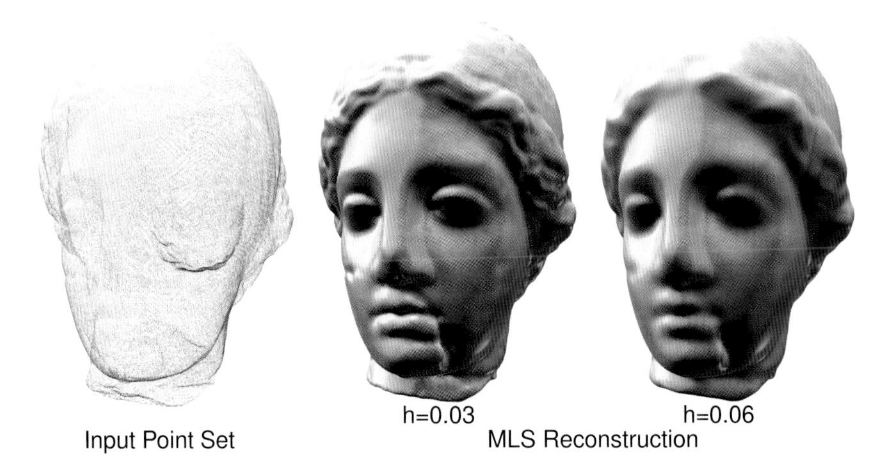

Figure 10.3: Mesh reconstruction from an unorganized point cloud using the MLS operator (here with a least square plane as the geometric primitive) and a mesh contouring algorithm.

10.7 Meshing

With the point cloud in hand and the MLS projection choosen, the output mesh structure can be extracted using several techniques. At least three classes of algorithms have proven to be effective here: *contouring*, *Delaunay meshing*, and *3D snakes*.

Contouring Contouring algorithms, such as *marching cubes* [Lorensen and Cline 87] or *dual contouring* [Ju et al. 02] methods, can be used on the implicit form of the surface. A number of open source implementations of such techniques are available online [Kobbelt et al. 01] and they can be quite easily translated to parallel (e.g., GPU) versions. Such approaches clearly have the advantage of being fast, but the resulting mesh quality is often low, and a remeshing post-process may be mandatory. Still, contouring is clearly the method of choice for most application scenarios requiring to mesh an implicit surface. Most of these techniques can be summarized with the following sequence:

1. Compute a bounding box of the surface to mesh.

2. Generate a 3D lattice inside the box, such as for instance a regular grid or an octree.

3. Evaluate the implicit surface either at the corners of the lattice cells (e.g., marching cube) or inside the cell (e.g., dual contouring).

4. Depending on the sign of the implicit surface at those evaluation locations, the cell containing the surface can be detected. The combinatorics of positive (inside) and negative (outside) implicit values induce the right local mesh structure to generate.

The original marching cube algorithm is deciding on the cell polygonization looking only at the 8 cell's corners sign (256 possibilities, boiling down to 15 when accounting for symmetry). The small set of polygons generated in each cell then approximates the geometry of the implicit surface in the cell, but solving for all cases requires looking also in the neighboring cells to ensure a watertight surface mesh in the end.

Delaunay meshing A better, yet slower, alternative is to use a *restricted Delaunay triangulation* over the implicit form of the PSS stemming from the point cloud. In this case, the connectivity is nearly optimal and the triangles are well-shaped. Such a meshing method can be implemented using the CGAL library[3] for robust computations.

3D snakes Last, a 3D *deformable model* can also be used together with the projective form of the PSS to generate the mesh while controlling its topology. The evolving mesh can be progressively remeshed to preserve well-shaped triangle and adapt to the geometric signal but such solutions are usually slow and hard to implement robustly.

10.8 Mesh Processing

The reconstructed mesh often has a resolution which is proportional to the input point set or related to the level of details captured in the original sampling. This mesh density may not fit the application scenario and the mesh then needs to be either simplified or subdivided. The simplification step can be performed using progressive *edge-collapses*, ordered using the quadric error metric [Garland and Heckbert 97], for which a number of open source implementations are available, such as QSlim[4] or MeshLab.[5] The subdivision can be performed using the Loop subdivision scheme, which has also a number of implementations available, such as OpenSubdiv.[6] A complete survey on how to process polygonal meshes can be found in the book of Botsch et al. [Botsch et al. 10].

[3]http://www.cgal.org

[4]http://www.cs.cmu.edu/afs/cs/Web/People/garland/quadrics/qslim.html

[5]http://meshlab.sourceforge.net/

[6]http://graphics.pixar.com/opensubdiv/

10.9 Summary

Using simple local operators with contouring extraction can cope with the meshing of dense point clouds exhibiting a restricted amount of noise. This solution is extremely fast and easy to run on GPU. However, point clouds exhibiting severe defects, such as large holes or unreliable normals, require more complex solutions. The Poisson Surface Reconstruction [Kazhdan et al. 06] is for instance quite efficient at filling large holes, while a combinatorial solution based on the *graph minimal cut* algorithm [Hornung and Kobbelt 06] can generate a high-quality surface without using the normal information.

The mesh reconstruction pipeline presented in this chapter allows to convert efficiently the raw 3D point sampling coming, for instance, from multiview stereo, laser range scanning, structure-from-motion data, or explicit surface models. The resulting mesh model can be enhanced with a parameterization to store surface attributes and support the basic intersection tests (with lines or volumetric primitives) which are at the root of physically based simulation and rendering. Last, it serves itself as input for numerous higher level geometry processing and analysis methods, including 4D face capture, full body reconstruction, animated cloth motion modeling, automatic rigging (Chapter 13), and 3D web compression and transmission (Chapter 22).

11
Reconstruction of Human Motion

Yebin Liu, Juergen Gall, Céline Loscos, and Qionghai Dai

11.1 Introduction

Motion capture is a technology to record and digitalize the motion information of living creatures. In recent years, with the emergence of new types of sensors and the improvement of computational performance, motion capture has started to play an important role in many fields, such as realistic character animation for games and movies, motion analysis for medical diagnostics and sport science, or virtual reality. While motion capture techniques can apply to general moving objects, they are mostly applied to capture full-body motion of humans or the motion of body parts like the face or hands. This chapter focuses only on human motion.

Motion capture techniques can be split into two categories, depending on if they use some form of fiducials or not. Marker-based systems require the attachment of different kinds of sensors or markers on the subject that is captured. By considering the types of sensors used, one can distinguish four categories: mechanical systems, electromagnetic systems, inertial systems, and optical systems (Figure 11.1). Among these, optical marker-based motion capture systems are the most popular ones, in particular in the movie industry, since they are very accurate and achieve real-time performance. However, marker-based systems suffer from widely known shortcomings, such as the need for a special controlled capture environment, errors due to broken marker trajectories, and their inability to capture motions of people wearing normal everyday apparel. Moreover, it is difficult to capture natural motions when a special marker suit or other technical equipment is disturbing. For example, when capturing the motion of hands interacting with objects, markers or sensors on the hands or the object might result in unnatural grasping poses. In general, the often considerable setup times with such invasive systems may also be a problem. Finally, there is a need for capture systems of detailed human motion, which also includes

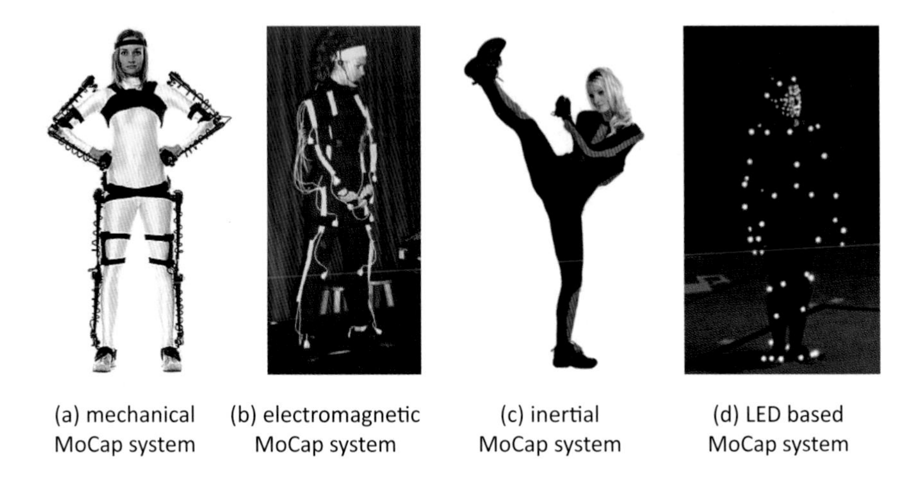

Figure 11.1: Examples of commercial marker-based and invasive motion capture systems: (a) Gypsy^{TM}—mechanical motion capture, (b) magnetic motion capture, (c) Xsens MVN—inertial motion capture, (d) LED-based optical motion capture.

non-rigid surface motion such as is caused by tissue or garment, while marker-based systems capture only articulated motion. Markerless motion capture approaches that use only video cameras or depth sensors (Chapter 4) provide a solution to these problems. Although they still lack the accuracy of marker-based systems, there has been substantial progress in recent years, promising less restrictive motion reconstruction.

This chapter concentrates on markerless motion capture using multi-view video cameras (Chapter 2), and multiple depth cameras (Chapter 4). After decades of research efforts, many kinds of markerless motion capture techniques have emerged for different application scenarios. From a technical aspect, markerless motion capture can be mainly classified into two categories: discriminative approaches [Ganapathi et al. 10, Shotton et al. 11] and generative approaches [Bregler et al. 04, Deutscher et al. 00, Gall et al. 10, Gall et al. 09].

Discriminative approaches rely on data-driven machine learning strategies to convert the motion capture problem into a regression or pose classification problem. Although learning approaches require sufficient training data and lack the accuracy of generative approaches, they are able to estimate the pose from a single frame in real-time and do not require an initial pose estimate. They are therefore currently the best choice for applications like human-computer interfaces or games where reliability is more important than accuracy.

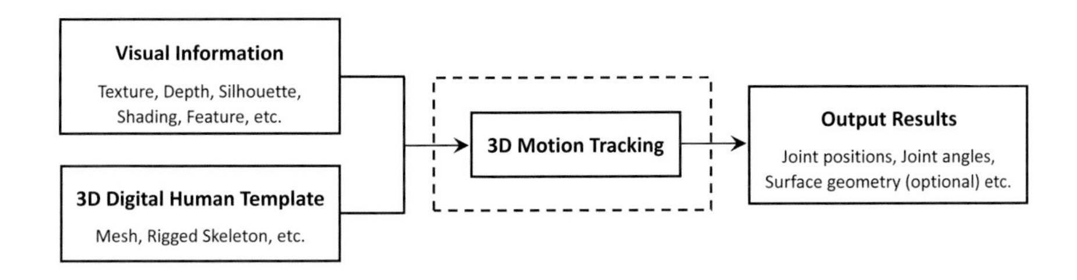

Figure 11.2: A general framework for template-based markerless motion capture. The module in the dashed box is the key processing step for motion reconstruction.

In contrast, generative approaches such as [Gall et al. 09], often rely on temporal information and solve a tracking problem. Many of these approaches parameterize the high dimensional digital human body by a low dimensional skeleton embedded in a template shape. Based on this template, an analysis-by-synthesis approach is started which optimizes the skeletal pose, and optionally the surface geometry, such that the synthesized human body images are consistent with the observed images. Figure 11.2 shows the generalized processing framework of template-based generative motion capture methods. Generative approaches are the preferred choice when very accurate results or motion details beyond joint positions are required. However, they require an initial pose, struggle to recover from tracking errors, and often rely on a user specific model. Consequently, some methods combine these two strategies [Wei et al. 12] where the initial pose is estimated by a discriminative approach and the estimated pose is then refined by a generative approach. This chapter mainly focuses on the discussion of generative approaches for markerless motion capture.

Despite a lot of progress in markerless motion capture in recent years, there are still many challenges remaining. The two major challenges for generative approaches to be addressed are how to capture motion of very high precision and how to achieve robust motion capture under general and uncontrolled capture settings.

One of the key factors to achieve high accuracy is the number and the precision of visual features that can be associated with the template model. While marker-based approaches only need to detect sparse but very reliable features, detailed fine-grained motion can only be captured by dense features. However, reliable dense features from images or depth data are very difficult to extract and the development of methods for this purpose is still a very active area of research. More details on 4D reconstruction using spatio-temporal features can be found in Chapter 12. Another important factor to achieve high accuracy is the quality of the template model [Liu et al. 13]. When it is not only the reconstruction of the skeletal motion that matters, but also the fine-scale rigid and non-rigid surface deformation, both surface and skeleton models need to be accurate and sufficiently fine-grained, and both need to be properly coupled with each other. Information on how to embed the skeleton into the template mesh model can be found in Chapter 13. The proper design and reliable solution of the objective function are undoubtedly the other two key factors to achieve high tracking accuracy and robustness. Usually, the objective function is defined as a data term measuring the consistency between the hypothesized pose and the observed image data. The data term is commonly extended by other terms that regularize the objective function to increase the robustness to noise or occlusions. An example of such a regularizer is a skeletal pose prior that prefers poses that are more likely to occur. The development of efficient algorithms for optimizing these analysis-by-synthesis energy functionals is an important research question in computer vision. Optimizers are challenged by the fact that these functions are usually non-convex and non-smooth, and for various reasons analytic derivatives may not be available. Commonly proposed optimization strategies for these functions can be split into local and global approaches. While local optimization is fast, it can easily erroneously converge to local optima if not properly initialized. Global optimization methods avoid this problem, but normally have a much higher computational cost.

Motion capture under general and uncontrolled settings presents another challenge. Early generative markerless motion capture approaches require tens of calibrated cameras, hardware camera synchronization, green screen background, and controlled indoor lighting [Moeslund et al. 06]. With the progress of camera technology, especially the emergence of depth cameras, outdoor markerless motion capture with more convenient setups becomes feasible and is thus intensively investigated in the research community. State-of-the-art motion capture techniques that have been shown to succeed under less controlled capture settings include motion capture algorithms using a single [Baak et al. 13, Wei et al. 12] or multiple handheld depth cameras [Ye et al. 12], methods using input from binocular video [Wu et al. 13], as well as approaches that can handle unsynchronized multi-view video [Elhayek et al. 12]. Other methods have demonstrated successful motion capture in outdoor environments [Hasler et al. 09b], or under time varying illumination in the scene [Wu et al. 12a, Li et al. 13]. These new techniques thus take important steps toward removing the need of complicated acquisition setups, and thus greatly advance practical usability of markerless motion capture. In the following, this chapter introduces fundamental algorithmic concepts used in core components of the above generative motion capture systems.

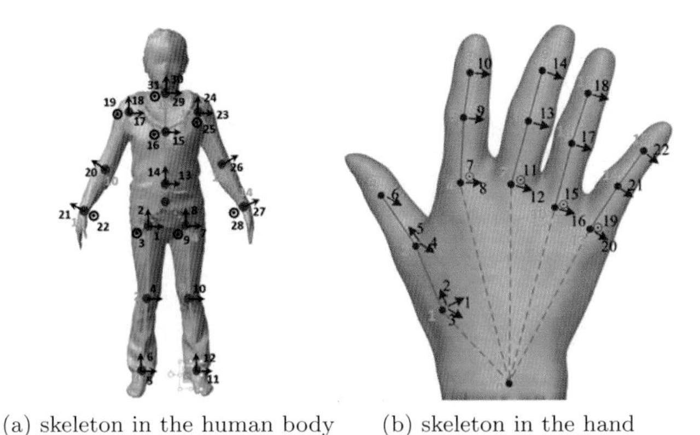

(a) skeleton in the human body (b) skeleton in the hand

Figure 11.3: Mesh templates and skeletons for the whole human body and the hand. The red dots are the joints and green numbers are the joint indexes; the black arrows and black numbers are the DoFs and their indices.

The chapter continues with a brief introduction of basic background concepts employed in generative markerless motion capture, including human skeleton parameterization and mesh skinning in Section 11.2. Two example algorithms for generative motion capture of one human actor, namely an approach based on local optimization and an approach based on global optimization, are presented in Section 11.3. Subsequently, Section 11.4 explains what challenges need to be solved for motion capture of multiple closely interacting subjects. In Section 11.5, it is explained how concepts developed for full body motion capture can be employed to capture the motion of hands interacting with objects.

11.2 Kinematic Skeleton and Skinning

A surface mesh template with its embedded skeleton is a widely used data structure to model the tracked subject in markerless motion capture. The mesh template can be a generic model, such as the SCAPE human shape model that is fitted to the subject, or it can be a laser-scanned mesh model with detailed surface geometry (Chapter 14). The process of embedding a skeletal structure into the mesh template, such that the surface deforms plausibly with the skeleton motion, is commonly referred to as "rigging" (Chapter 13).

Generally, the skeleton model is represented as a tree structure (Figure 11.3). In this figure, the nodes in the tree, which represent joints, are marked as red dots. Each joint represents a possible rigid body transformation, and it can have several degrees of freedom (DoFs). In the figure, these DoFs are marked as black arrows to indicate the rotation axes of each joint. For example, the joint of the wrist has 2 rotational DoFs while the joint of the neck has 3 rotational DoFs. Henceforth, the number of joints in the skeleton is denoted by N and the number of DoFs by M.

The rigid motion of a joint can be represented by a "twist" [Murray et al. 94], which can be mathematically expressed in two ways. The first representation is in the form of a 6D vector

$$\xi = (v_1, v_2, v_3, \omega_x, \omega_y, \omega_z)^T, \tag{11.1}$$

and the other representation takes the form of a 4×4 matrix

$$\hat{\xi} = \begin{bmatrix} 0 & -\omega_z & \omega_y & v_1 \\ \omega_z & 0 & -\omega_x & v_2 \\ -\omega_y & \omega_x & 0 & v_3 \\ 0 & 0 & 0 & 0 \end{bmatrix}. \tag{11.2}$$

In Eqs.(11.1) and (11.2), $\omega_\xi = (\omega_x, \omega_y, \omega_z)^T$ is the unit vector that points in the direction of the rotation axis. The vector $\mathbf{v}_\xi = (v_1, v_2, v_3)^T$ is the cross product of the rotation center and the rotation axis, i.e., $\mathbf{v}_\xi = \mathbf{p}_\xi \times \omega_\xi$. and $\mathbf{p}_\xi$ is the position of the joint in 3D space. If a joint has only one DoF, the 3D rigid motion transform can be described by the rotation angle θ, i.e., $\theta\hat{\xi}$, and similarly more rotation angles are needed to parameterize the transformation at a joint with more DoFs.

The twist obeys the cascade property [Murray et al. 94] in the skeleton, namely, for skeleton joint i, all of its parent joints (according to the tree hierarchy) affect its motion. Therefore, the combined transformation of joint i is represented as a matrix $\mathbf{T}_i$, which takes all the transformations of its parent joints into consideration,

$$\mathbf{T}_i = e^{\theta_0\hat{\xi}_0} \cdot \prod_{j \in Parent(i), j \neq 0} e^{\theta_j \hat{\xi}_j}, \tag{11.3}$$

where j is the index for the DoF, $Parent(i)$ the set of all the parent DoFs of joint i, θ_j the rotation angle of the jth DoF and $\hat{\xi}_j$ the rotation matrix corresponding to the jth DoF. For example, in Figure 11.3(a) the $10th$ joint is parameterized by the $20th$ DoF in the model, and $Parent(10) = \{13, 14, 15, 16, 17, 18, 19\}$. Here, note that $\theta_0\hat{\xi}_0$ is the global translation and rotation of the whole human body. Eq.(11.3) can be linearized by Taylor expansion to yield

$$\mathbf{T}_i = \mathbf{I} + \theta_0\hat{\xi}_0 + \sum_{j \in Parent(i), j \neq 0} \theta_j \hat{\xi}_j, \tag{11.4}$$

where $\mathbf{I}$ is the identity matrix.

Given a new pose parameter set (i.e., value assignment to each DoF) of the skeleton, the deformed surface geometry that corresponds to the new skeleton pose can be computed by a technique called "skinning." One widely used skinning technique is the linear blend skinning (LBS) operator [Lewis et al. 00]. The main idea of LBS is that the position of each vertex on the surface model depends linearly on the transformation matrices of the skeleton joints

$$\mathbf{v} = \left(\sum_{i=1}^{N} \omega_i \mathbf{T}_i\right) \tilde{\mathbf{v}} \tag{11.5}$$

Here, $\tilde{\mathbf{v}}$ and $\mathbf{v}$ are the positions of a vertex on the surface of the 3D model before and after the deformation; ω_i is the precomputed skinning weight of the surface vertex with respect to the skeleton joint i with $\sum_{i=1}^{N} \omega_i = 1$. Intuitively, a skinning weight describes if and how strongly the transformation of a joint i influences the deformation of the vertex on the surface. Assigning skinning weights to vertices on the surface is a process supported by most standard computer animation packages through some form of painting interface. But there are also automatic methods to compute the weights [Baran and Popović 07]. Using Eq.(11.4), Eq.(11.5) can be rewritten as

$$\mathbf{v} = \left(\mathbf{I} + \theta_0\hat{\xi}_0 + \sum_{m=1}^{M}\left(\sum_{j\in Children(m)} \omega_j\right)\theta_m\hat{\xi}_m\right)\tilde{\mathbf{v}}, \tag{11.6}$$

where $Children(m)$ is the set of all the children (joints) of the joint corresponding to the mth degree of freedom. By defining $\bar{\omega}_m = \sum_{j\in Children(m)} \omega_j$, Eq.(11.6) can be simplified to

$$\left(\theta_0\hat{\xi}_0 + \sum_{m=1}^{M} \bar{\omega}_m\theta_m\hat{\xi}_m\right)\tilde{\mathbf{v}} = \mathbf{v} - \tilde{\mathbf{v}}. \tag{11.7}$$

Assuming $\mathbf{v} = (x, y, z, 1)^T$, $\tilde{\mathbf{v}} = (\tilde{x}, \tilde{y}, \tilde{z}, 1)^T$, Eq.(11.7) can be further written as

$$\mathbf{A\Theta} = \mathbf{b} \tag{11.8}$$

where

$$\mathbf{\Theta} = (\theta_0 v_1^0, \theta_1 v_2^0, \theta_1 v_3^0, \theta_1\omega_x^0, \theta_1\omega_y^0, \theta_1\omega_z^0, \theta_1, \cdots, \theta_M)^T,$$

$$\mathbf{b} = (x - \tilde{x}, y - \tilde{y}, z - \tilde{z})^T,$$

and

$$A = \left[\begin{array}{cccccc} 1 & 0 & 0 & 0 & z & -y \\ 0 & 1 & 0 & -z & 0 & x \\ 0 & 0 & 1 & y & -x & 0 \end{array} \left[\begin{array}{c} \\ \mathbf{e}_1 \\ \\ \end{array}\right] \begin{array}{c} \cdots \\ \cdots \\ \cdots \end{array} \left[\begin{array}{c} \\ \mathbf{e}_M \\ \\ \end{array}\right]\right] \quad \text{with}$$

$$\mathbf{e}_1 = \begin{bmatrix} (v_{1,1} - \omega_{z,1}y + \omega_{y,1}z)\bar{\omega}_1 \\ (v_{2,1} + \omega_{z,1}x - \omega_{x,1}z)\bar{\omega}_1 \\ (v_{3,1} - \omega_{y,1}x + \omega_{x,1}y)\bar{\omega}_1 \end{bmatrix}, \mathbf{e}_M = \begin{bmatrix} (v_{1,M} - \omega_{z,M}y + \omega_{y,M}z)\bar{\omega}_M \\ (v_{2,M} + \omega_{z,M}x - \omega_{x,M}z)\bar{\omega}_M \\ (v_{3,M} - \omega_{y,M}x + \omega_{x,M}y)\bar{\omega}_M \end{bmatrix}.$$

In this way, all the vertices on the surface model can be linearized by M rotational degrees of freedom of the skeleton nodes. Here,

$$\xi_m = (v_{1,m}, v_{2,m}, v_{3,m}, \omega_{x,m}, \omega_{y,m}, \omega_{z,m}),$$

and

$$\mathbf{\Theta} = (\theta_0\xi_0, \theta_1, \cdots, \theta_M)^T, \tag{11.9}$$

Finally, Eq.(11.5) can be reformulated as

$$\mathbf{v} = \mathbf{T}(\Theta)\tilde{\mathbf{v}}. \tag{11.10}$$

where $\mathbf{T}$ maps an input pose $\mathbf{\Theta}$ to a linear deformation function, which further maps the vertex coordinates $\tilde{\mathbf{v}}$ to new coordinates $\mathbf{v}$.

11.3 Pose Optimization

The modeling of the energy function is essential to the success of the generative skeleton pose optimization. The energy function E to be optimized can be generally described by the sum of a data term E_D and a smoothness term E_S:

$$E(\mathbf{\Theta}) = E_D(\mathbf{\Theta}) + \gamma\, E_S(\mathbf{\Theta}). \tag{11.11}$$

Here, E_D is a data term that measures the difference between the synthesized human model and the observed image data. E_S is a pose prior that penalizes poses that are not physically possible (such as intersections) [Oikonomidis et al. 11b], that are dissimilar to some prior motions in a motion database [Martin and Crowley 95], or that are inconsistent with a predicted pose from the temporal sequence [Gall et al. 09]. The term γ is a weighting factor trading off between the data term and the smoothness term. Vision cues, such as silhouette, texture, edges, features, and estimated depth can all be integrated in the objective function to guide the optimization. Depending on the method used to solve for $\mathbf{\Theta}$, the generative motion capture approaches can be categorized into methods based on local pose optimization and global pose optimization.

Local Pose Optimization

Local optimization methods aim at fast computation of the desired $\mathbf{\Theta}$ and can only guarantee local convergence, therefore require a good initialization. Given the estimated skeleton pose $\tilde{\mathbf{\Theta}}$ and the deformed mesh model

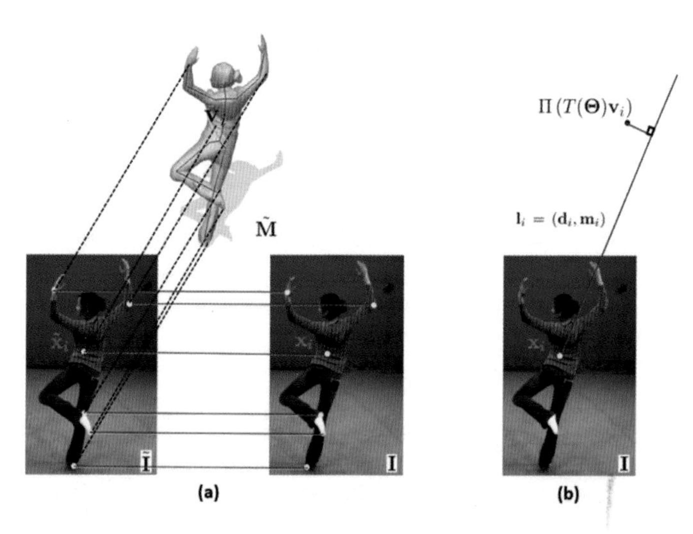

Figure 11.4: (a) 3D-to-2D alignment using SIFT matching. (b) The concept of a Plücker line.

$\tilde{\mathbf{M}}$ in a previous (multi-view video) frame, local pose optimization methods search for the best pose $\mathbf{\Theta}$ of the current frame in the vicinity of $\tilde{\mathbf{\Theta}}$. For instance, temporal cues can be used [Bregler et al. 04]. Such temporal cues are usually obtained by feature matching.

Specifically, for multi-view video input, the feature alignment is a 3D-to-2D alignment, namely, a correspondence set between the 3D model vertices $\mathbf{v}$ on the former tracked model $\tilde{\mathbf{M}}$ and 2D image pixels $\mathbf{x}$ in the current frame $\mathbf{I}$. Silhouette contour pixels and salient texture pixels are two commonly used features [Gall et al. 09]. In both cases, the 2D feature pixels $\mathbf{x}$ in the current frame are associated with a projected model vertex $\mathbf{v}$ yielding a 3D-to-2D correspondence $(\mathbf{v}, \mathbf{x})$. In the contour case, a contour vertex $\mathbf{v}_i$ on $\tilde{\mathbf{M}}$ is projected to the image plane forming a contour pixel $\tilde{\mathbf{x}}_i$, and the contour pixel $\mathbf{x}_i$ in image $\mathbf{I}$ which is closest to $\tilde{\mathbf{x}}_i$ is selected to form a correspondence $(\mathbf{v}_i, \mathbf{x}_i)$. In the texture case (Figure 11.4(a)), 2D-to-2D correspondences $(\tilde{\mathbf{x}}, \mathbf{x})$ between the former tracked frame $\tilde{\mathbf{I}}$ and the current frame $\mathbf{I}$ are first obtained by matching SIFT features [Lowe 04], and then model vertices $\mathbf{v}$ which are projected to $\tilde{\mathbf{x}}$ are associated with $\mathbf{x}$ to form the correspondence set.

Given the correspondence set $(\mathbf{v}, \mathbf{x})$, the projection of a deformed 3D point $T(\Theta)\mathbf{v}$ to the image coordinates should be as close to $\mathbf{x}$ as possible. To model this, the concept of a Plücker line [Stolfi 91] is used. A Plücker line $\mathbf{l} = (\mathbf{d}, \mathbf{m})$ is determined by a unit vector $\mathbf{d}$ and a moment $\mathbf{m}$, where $\mathbf{x} \times \mathbf{d} - \mathbf{m} = 0$ for all points $\mathbf{x}$ on the line. Since each 2D point $\mathbf{x}_i$ defines a projection ray that can be represented as Plücker line $\mathbf{l}_i = (\mathbf{d}_i, \mathbf{m}_i)$, the error of a pair $(T(\mathbf{\Theta})\mathbf{v}_i, \mathbf{x}_i)$ is given by the norm of the perpendicular vector between the line $\mathbf{l}_i$ and the transformed point $T(\Theta)\mathbf{v}_i$ (Figure 11.4(b)):

$$E_d(\mathbf{\Theta}) = \sum_i w_i \left\| \Pi \left(T(\mathbf{\Theta})\mathbf{v}_i \right) \times \mathbf{d}_i - \mathbf{m}_i \right\|_2^2, \tag{11.12}$$

where Π denotes the projection from homogeneous coordinates to non-homogeneous coordinates and w_i are the weights for the correspondences. These weights can be used to assign a confidence to each correspondence, for instance in order to give lower confidence to less trustworthy evidence. The objective can be locally minimized by an iterative Gauss–Newton scheme using the linear Eq.(11.8).

When depth cameras are used as opposed to video cameras for markerless motion capture, the feature alignment becomes a 3D-to-3D alignment between mesh model vertices $\mathbf{v}$ and 3D points (derived from the depth data) $\mathbf{p}$. The data term of Eq.(11.11) is then formulated as in Ref. [Ye et al. 12]

$$E_d(\mathbf{\Theta}) = \sum_i w_i \left\| T(\mathbf{\Theta})\mathbf{v}_i - \mathbf{p}_i \right\|_2^2. \tag{11.13}$$

Similar to Eq.(11.12), the energy function can be linearized and locally optimized using an iterative Gauss–Newton scheme. If both video data and depth data are available, Eqs.(11.12) and (11.13) can be combined to obtain a more informative energy term.

In order to stabilize the optimization, the linear system is regularized by the smoothness term in Eq.(11.11) defined as

$$E_S(\mathbf{\Theta}) = \|\mathbf{\Theta} - \tilde{\mathbf{\Theta}}\|_2^2. \tag{11.14}$$

This smoothness prior introduces a constraint by penalizing deviations from the previous pose $\tilde{\mathbf{\Theta}}$. In practice, feature alignment and energy optimization are usually iterated for about 10 to 20 times. For each iteration, the weights w_i can vary depending on the alignment confidence. The combined computation time for local pose optimization is low, and it takes less than one or two seconds to estimate the pose for a single time frame, even when using unoptimized code.

Global Pose Optimization

Compared with local pose optimization, global pose optimization defines the error measurement (11.11) between the input and the rendered images in a more direct and accurate way. The energy function is therefore usually non-linear, which demands time-consuming optimization like particle swarm optimization (PSO) [Oikonomidis et al. 11a] or interacting and annealing particle filters (ISA) [Gall et al. 10] for solving the optimal pose in a global manner.

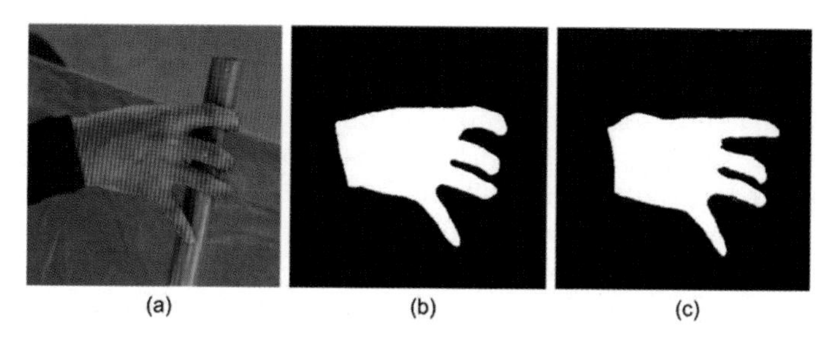

Figure 11.5: Silhouette term. (a) The original image; (b) the extracted silhouette map $\mathbf{S}_{i,v}$; (c) the synthesized silhouette map $\mathbf{S}_{r,v}$ automatically generated by the rendering process.

Since the energy function is not linearized, there is more flexibility in designing the objective function. For example, Eq.(11.11) can be defined as the sum of a silhouette term, a texture term, and a depth term, as

$$E_D(\mathbf{\Theta}) = \delta_{silh} E_{silh} + \delta_{tex} E_{tex} + \delta_{depth} E_{depth}. \tag{11.15}$$

Here, features are not matched to obtain correspondences, but each term in Eq.(11.15) measures by a more direct analysis-by-synthesis strategy the consistency of images data and current estimate [Gall et al. 09].

Silhouette Term This term measures the discrepancy between silhouette maps of the synthesized images and the silhouette maps extracted from the observed images (Figure 11.5). A silhouette image $\mathbf{S}$ is represented as an image whose foreground and background pixels are set to one and zero, respectively. The mismatch between the segmented silhouette $\mathbf{S}_i$ of the human in the foreground and the rendered silhouette $\mathbf{S}_r$ of its hypothesis model is measured pixel-wise for each camera view v by

$$E_{silh} = \sum_v \left(\frac{\sum \left(\mathbf{S}_{i,v} \cap \bar{\mathbf{S}}_{r,v}\right)}{\sum \mathbf{S}_{i,v}} + \frac{\sum \left(\mathbf{S}_{r,v} \cap \bar{\mathbf{S}}_{i,v}\right)}{\sum \mathbf{S}_{r,v}} \right). \tag{11.16}$$

The measurement is a bidirectional distance function minimizing the area of non-overlapping foreground regions between the segmented and synthesized silhouette images. Here, $\sum S$ sums the values of all binary pixels of S, and $\bar{\mathbf{S}} = 1 - \mathbf{S}$ is the inverse of the binary silhouette image.

Texture Term The texture differences between the synthesized image data and the observed image data are the most direct measurement, and can be defined as

$$E_{tex} = \sum_v \left(\sum_i D\left(\mathbf{W}_i\left(\mathbf{R}_v\right), \mathbf{W}_i\left(\mathbf{I}_v\right)\right) \right). \tag{11.17}$$

Specifically, for each pixel i in the rendered image $\mathbf{R}_v$, the difference D between a window $\mathbf{W}$ centered around pixel i in $\mathbf{R}_v$ and a window with the same position in the captured image $\mathbf{I}_v$ is computed. One of the measurement metrics for D is the zero-mean normalized cross-correlation (ZNCC) [Martin and Crowley 95].

The texture of the human model can be obtained in the first frame of a video and slowly updated over time. Without updates the texture remains constant over time and does not handle appearance changes or invisible parts in the first frame. If the texture is updated too frequently, tracking errors may occur due to accumulation of small errors that lead to drift.

Depth Term A depth term can be incorporated when a depth camera or a binocular video camera for stereo matching is used. The term measures the consistency of the observed 3D data and the current 3D model by pixel-wise comparison for each projected surface point of M:

$$E_{depth} = \sum_{v} \left(\sum_{i \in S_{r,v}} \| M(i) - P(i) \|_2 \right). \tag{11.18}$$

$\mathbf{S}_{r,v}$ is the projected silhouette of M.

A smoothness prior, as defined in Eq.(11.14), can also be added. The objective function can be optimized by interacting simulated annealing (ISA) [Gall et al. 09]. Although the optimization can be hastened by GPU computing [Shaheen et al. 09], it is still considerably slower than the previously described local optimization algorithm. The reason for this is that particle-based optimizers, such as ISA, require the evaluation of the pose energy function for a very high number of particles, where each particle is one pose hypothesis, i.e., one possible set of skeletal pose parameters. Local optimization, however, frequently gets stuck in local minima if the tracker is not initialized with a pose hypothesis sufficiently close to the true optimum. Therefore, an approach that combines local and global pose optimization is proposed in [Gall et al. 09] to achieve efficient and high-quality skeletal motion capture. Briefly, local optimization is first operated for each processing frame, followed by global optimization only for body parts with a high residual error. Figure 11.6 shows the results of using this joint optimization strategy.

11.4 Multi-Person Motion Capture

Many types of human motion can only be observed during close interaction between two or more humans. Examples are found in many types of sport, for instance wrestling, judo, or ballroom dancing. In these motions,

Figure 11.6: Markerless motion capture results with joint local and global optimization as proposed in [Gall et al. 09].

humans often move very close to each other, and there may be physical contact between them. From the perspective of markerless motion capture, this complicates the situation since the degree of pose ambiguity increases significantly. Commonly used features to determine the correct body pose, like silhouettes, color, edges, or interest points cannot be uniquely assigned to a person anymore. Due to frequent occlusions, these ambiguities become even more challenging when people interact closely. For these reasons, it is infeasible to directly apply single-person pose optimization algorithms that use such features, such as the one described earlier in this chapter, to the multi-person case. Attempting to use them and jointly optimize for the pose parameters of two body models will fail very quickly, since the model-to-data association is undetermined.

To successfully track multiple people in close interaction, [Liu et al. 11] and [Liu et al. 13] propose a method that employs a combination of tracking and robust segmentation of input multi-view video frames. The segmentation allows the method to generate separate silhouette contours and image features for each person. This way, a local pose optimization method, such as the one described earlier, can be applied in the multi-person case since model-to-data-association is known from segmentation. After each image pixel has been assigned to one of the observed persons or the background, pose estimation can be performed for each person independently (Section 11.3).

The core operation of the integrated tracking and segmentation algorithm is as follows. Before pose estimation on a new frame of multi-view

Figure 11.7: Multiview image segmentation results for a sequence with three subjects. The class labels for the three subjects are indicated by the colors red, green, and blue.

video commences, a label is assigned to each pixel in the foreground of the scene, denoting to which person the pixel belongs. This pixel assignment is computed by solving a discrete optimization problem which finds the most likely assignment of pixel labels based on a Markov random field (MRF) energy. The MRF energy formulation uses color information and a local regularization. But in addition to these appearance cues, it also uses shape cues of the scene that are derived from the reconstructed poses of all actors in the previous time step. The optimization generates high quality segmentation results even under serious occlusions and ambiguous appearance (Figure 11.7). In consequence, even under such challenging conditions, and even with three people in the scene, high-fidelity motion capture results can be obtained (Figure 11.8).

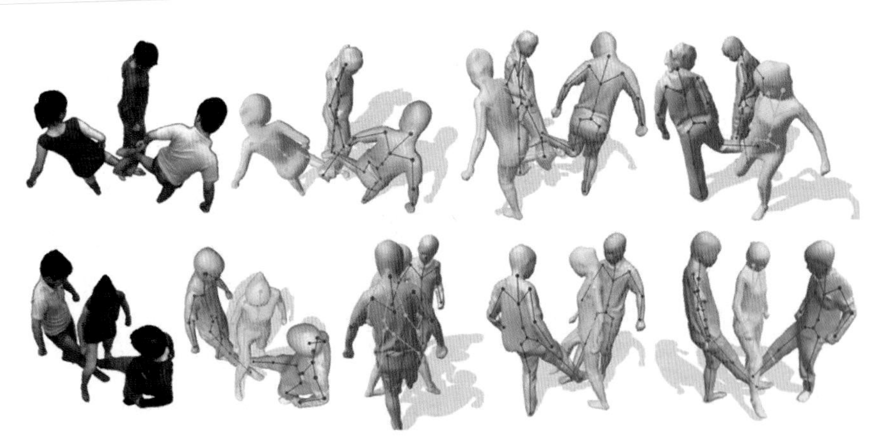

Figure 11.8: Multi-Person motion tracking results on two multi-view video sequences.

11.5 Capturing Hand Motion

Capturing the motion of hands is another important problem with many important applications in computer graphics, human computer interaction, and robotics. In principle, one could argue that hand motion can be captured by the same type of algorithms used for full body motion capture. In the end, the hand can be described by a kinematic skeleton itself that almost reaches the complexity of the full body. Markerless hand motion capture, however, is more challenging than full body motion capture. There are several reasons for this, for instance the fact that the hand has a very uniform appearance, and many fingers look very similar, which makes marker-free pose estimation challenging and ambiguous. Also, self-occlusions happen in many hand postures. Capturing the motion of one or two hands that are interacting with an object is an even more challenging task, because it requires not only the estimation of fine-grained hand articulation and object movement, but also subtle interactions and contact phenomena between the hand and the object. Two categories of approaches have been proposed recently to address this problem. One set of methods uses depth cameras as input sensors [Oikonomidis et al. 11b]; other algorithms resort to multi-view video input [Ballan et al. 12, Wang et al. 13b].

This section briefly describes the approach proposed in [Wang et al. 13b] for acquiring physically realistic hand motion from multi-view video while the hand is grasping and manipulating an object. The key idea of this approach is to use a composite physically based motion controller to simultaneously model hand articulation, object movement, and subtle interactions between the hand and the object. The motion controller of both hand and object motion is integrated in an analysis-by-synthesis framework, similar to the one described in Section 11.3.

Specifically, the global optimization model in Eq.(11.15) is parameterized by a motion controller

$$\operatorname*{argmin}_{\mathcal{M}} E(\Theta(\mathcal{M})), \tag{11.19}$$

where $\mathcal{M}$ are the parameters of the controller. In this way, the pose of the hand is not directly parameterized in kinematic joint parameters (11.9), such as described earlier for the full body case, but indirectly through the control input of a physics-based dynamical motion model. Using such a physics-based dynamical model is advantageous in the case of hand motion capture, as it naturally models the subtle interactions between the hand and an object. In consequence, they can help to eliminate physically implausible motions or contact phenomena between the hand and the surfaces it interacts with. Another benefit is that the recovered motion controller can be utilized to conveniently adapt the captured motion to a new object with different geometry, which is an important requirement in many

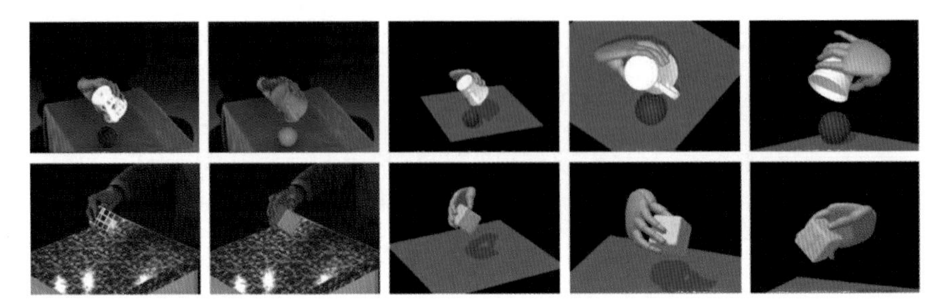

Figure 11.9: Video-based motion capture of hands that manipulate objects. Both rows show: input images, the reconstructed poses superimposed on the input images, the reconstructed poses from the same viewpoint, and the reconstructed poses from two different viewpoints (from left to right).

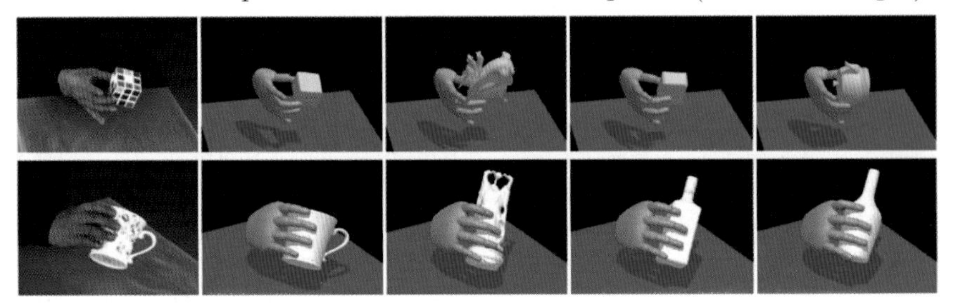

Figure 11.10: The captured motion controller is applied to animate the hand interacting with a new object with different geometry. Both rows from left to right show: the original image data, the captured pose of the hand and the object, and the controller applied to three different objects.

graphics and animation applications. Figures 11.9 and 11.10 show several results of hand motion capture and animation.

11.6 Summary

In this chapter the foundations of algorithms for markerless capture of human motion from multi-view video or depth data have been described. The focus was on generative approaches to the problem, and as examples, two methods have been discussed in more detail. Ongoing research in the field tries to further relax the constraints on situations under which the proposed methods are applicable. For instance, new methods for capturing the motion of multiple subjects or interactions with objects were developed. This imposes additional challenges and makes it necessary to combine generative approaches with physical scene models as well as discriminative approaches. There are several challenges to be addressed in the future. The pose estimation accuracy of current generative and discriminative approaches is still not quite as high as the one achieved with intrusive marker-based approaches. Reducing the error to a comparable range of a few millimeters is necessary for applications that require accurate measurements. Although there are several approaches for outdoor motion capture, motion capture under any general lighting condition is still not possible at the moment. While discriminative approaches generalize well to other subjects, generative approaches often require each tracked subject to strike a special initial pose, such that a personalized template model can be fitted to the subject. Further improving the initialization methodology is thus an active area of research. Finally, real-time performance with low latency is required for interactive applications. This chapter has shown that several approaches already satisfy some of these requirements. However, an approach that satisfies all of them and is applicable in general real-world scenes is still not in sight.

12 Dynamic Geometry Reconstruction

Edmond Boyer, Adrian Hilton, and Céline Loscos

12.1 Introduction

Recent progress in shape modeling allows the recovery of precise shape models from visual data, such as color and depth images. These models can come in different representations, usually points, surfaces, or volumes, depending on the strategy employed for their estimation, e.g., stereo vision or sensor fusion, which were discussed in preceding chapters. When considering visual information over time, as in videos, temporal sequences of models can be obtained. Yet, whatever the representation, static reconstructions performed independently over time do not provide clues on the motion and the deformation of a shape. However, natural scenes are usually dynamic and composed of shapes that evolve over time, for instance humans and objects (Figure 12.1). Several applications are based on the capture of shape evolution over time: in digital content production, for instance, for realistic animation of virtual characters; or in motion analysis for sport or medical applications. The next step in modeling reality is therefore concerned with the ability to recover or capture motion and deformation of shapes of unknown types by using visual information sampled over time.

Image observations over time are discrete by construction and the capture of shape motion, also referred to as shape tracking, consists in the recovery of temporal correspondences between successive time instants. To this aim, a large variety of directions can be employed depending on the representation chosen for objects and on the prior information assumed for shapes and their deformations. For example, when scenes are composed of humans only, articulated motions can be assumed and represented by the poses of a skeleton (Chapter 11). For more general scenarios with less prior information, motion models must be less constrained. At the end of the spectrum, when no prior information on the observed objects is available,

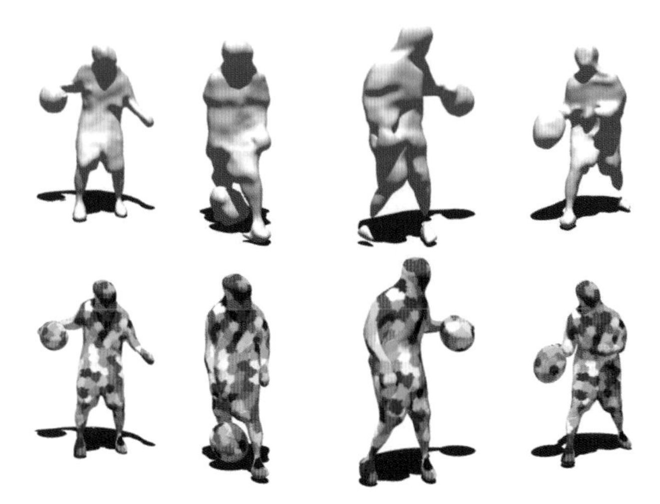

Figure 12.1: Example of captured dynamic scene: (top) independent reconstructed geometries; (bottom) registered shape model.

tracking reduces to finding the motion field between two successive time instants, i.e., matching, and global consistency over temporal shape models can hardly be enforced, and thus trajectory drifts along time sequences are often introduced. This is why numerous recent approaches assume a known surface model, such as a mesh, to start with. This allows the recovery of time consistent 3D models, also called 4D models, that encode both shape and motion information. 4D models include human models such as the templates presented in Chapter 11, but the concepts generalize to generic dynamic scenes.

In the following the strategies that exist in the literature to solve the tracking problem are explained. Different methods are exemplarily illustrated, starting with a mesh tracking approach that follows a popular sequential strategy and estimates the deformations of a mesh model between two time instants. Subsequently, a global strategy is described that can track a mesh model over several time sequences in an optimal and thus non-sequential way. Further on, an approach is presented that assumes little about the allowed deformation and that estimates motion fields. The chapter finishes with a discussion of current limitations of the existing approaches, as well as future research directions in the field.

12.2 Strategies

Existing strategies for shape tracking can be divided with respect to the amount of prior knowledge they assume and, consequently, to the constraints they impose on the observed shape. Probably the most widespread strategy consists in fitting a given deformable shape template to the observations at each time instant, which gives a 4D model (i.e., a time consistent 3D model). While providing robust approaches this strategy requires a deformation model that inherently limits the admissible observations, e.g., articulated movements only with a skeleton based model (Chapter 11). However, such limitation appears to be not critical in many applications and this strategy has received much attention in recent years as a result of the growing interest in human motion analysis. Another strategy worth mentioning consists in directly mapping observations from one frame to the next, which in turn yields motion trajectories over time. This approach requires less prior knowledge but, as a consequence, suffers more easily from drift over long sequences. In both cases the problem can be formulated as the estimation of motion parameters and the strategies can be classified with respect to the elements that feed into this estimation. To reduce the accumulation of drift errors for sequential tracking over long-sequences, and to allow alignment of tracked models across multiple sequences, non-sequential alignment approaches have been proposed. Non-sequential approaches align similar frames across the entire sequence to reduce the path length of sequential tracking.

Problem Formulation

Given observations $\mathcal{O}$, e.g., a point cloud or a mesh, at a given time instant, the problem is therefore to fit a given template, or to find a mapping from the previous time instant, to these observations. Existing methods usually formulate the problem as a maximum *a posteriori* (MAP) estimation of the deformation or mapping parameters $\hat{\Theta}$ that maximizes the posterior distribution $\mathcal{P}(\Theta|\mathcal{O})$ of the parameters Θ given the observations $\mathcal{O}$:

$$\hat{\Theta} = \arg\max_{\Theta} \; P(\Theta|\mathcal{O}) \simeq \arg\max_{\Theta} \; P(\mathcal{O}|\Theta) \; P(\Theta), \tag{12.1}$$

where $P(\mathcal{O}|\Theta)$ is the likelihood of the observations given the motion parameters and $\mathcal{P}(\Theta)$ is the prior information on these parameters. Taking the log of the above expression yields the following optimization problem:

$$\hat{\Theta} = \arg\max_{\Theta} \; E_{dt}(\mathcal{O}, \Theta) + E_r(\Theta), \tag{12.2}$$

where the data term E_{dt} corresponds to the log-likelihood and the regularization term E_r to the log-prior. Approaches then differ in the observations

$\mathcal{O}$ they consider, the parameterization Θ they use for the deformation, and the regularization E_r they impose over these parameters. Alternatives for these terms are discussed in the following.

Observations

Observations $\mathcal{O}$ are used, in the data term, in combination with a distance function to measure how good a motion prediction is. For instance Markers are used in traditional *mocap* systems to recover the pose of known skeletons. However, they only partially describe how shapes are moving, as they only describe the motion of sparse locations in space. To fully track shapes more densely, observations that vary from 2D image information to 3D geometric information can be considered.

Image Silhouettes Several model-based approaches deform a reference model so that the contour of the projected shape matches the contour of the observed silhouettes [Vlasic et al. 08, de Aguiar et al. 08b, Gall et al. 09, Straka et al. 12]. In these works, the silhouette overlap error is in general considered as a measure of the prediction error. This assumes that image silhouettes are accurate, which is not necessarily true, especially in unconstrained environments where foreground background discrimination is often difficult. In addition, due to the projections, large discrepancies along viewing lines can be missed. Nevertheless silhouettes are robust image primitives widely used for shape tracking.

Point Clouds Point clouds can, for instance, be obtained from range scanners or multi-view reconstructions methods, e.g., [Franco and Boyer 09, Furukawa and Ponce 10]. They are used as 3D observations in numerous approaches that typically proceed in two steps: (i) estimate correspondences between the model or the previous frame and the observed points; (ii) find the motion parameters that minimize distances between correspondents. The first step is either deterministic, as in the well known ICP method for rigid registration [Besl and McKay 92], or probabilistic, as in many recent approaches [Myronenko and Song 10, Horaud et al. 09, Cagniart et al. 10]. In the latter methods, observations are assumed to be drawn from Gaussian mixture distributions that are estimated with an EM strategy [Dempster et al. 77]. This allows the methods to explicitly handle outliers in the observations.

In the case where points are mesh vertices, shape matching methods that exploit local shape properties, such as [Anguelov et al. 04b, Starck and Hilton 05, Bronstein et al. 07, Varanasi et al. 08], could also be considered to find correspondences that would yet be deterministic. Besides, photometric information, as available in images, can be considered as well when searching for correspondences [Starck and Hilton 07a, Gall et al. 09].

Motion Parameterization

Parameters Θ in the estimation (12.2) are used to parameterize the mapping from a template model, or from the previous frame, to the current observations. They come from the knowledge that is available on the motion and explicitly or hardly constrain the motion to belong to a specific category as opposed to the regularization term where prior assumptions on the motion parameters, such as smoothness, are softly enforced. As a simple example of this principle, mocap systems parameterize the poses of a known skeleton with joint angles that are optimized in order to minimize the distances between markers and joints at each time instant. In the general situation, and with shape models or with point clouds, motion can be assumed to be:

Unconstrained In which case motion parameters are directly the unconstrained vector field over all vertices or points. This field being initialized using tracked features, e.g., [Varanasi et al. 08, de Aguiar et al. 08b] or using normal flow constraints, e.g., [Petit et al. 11, Blache et al. 14] (Chapter 8).

Rigid In this category, the shape is assumed to move rigidly and the optimized motion parameters are those of a 3D rotation and a 3D translation. This is the case with the well known ICP approach [Besl and McKay 92] that alternates between: (i) associating observations to the shape model, i.e., defining distances between the observations and the shape; (ii) optimizing for the best rigid motion parameters that minimize these distances.

Articulated As in [Mundermann et al. 07, Vlasic et al. 08, Horaud et al. 09, Gall et al. 09] where the parameters of a skeleton are optimized (e.g., the joint angles). Since the skeleton is not observed, a surface model that surrounds it must be defined in order to verify whether the estimation satisfies the shape observations. This surface is, for example, composed of ellipsoids in [Horaud et al. 09] or a mesh whose vertex positions are obtained via blend skinning [Vlasic et al. 08, Gall et al. 09] (Chapter 11).

Locally Rigid As in [Cagniart et al. 10, Huang et al. 13] where vertices of a template mesh are grouped into patches, each of which is assumed to move rigidly. This motion parameterization is more flexible than the articulated model and allows therefore for more general shape motions such as cloth deformations. On the other hand, this local model requires additional global constraints (e.g., regularization constraints) to ensure consistency between locally rigid motions.

Learned In contrast to the previous generative motion models, a body of work follows a discriminative strategy that consists in recognizing the most likely pose of a shape within a pre-learned dataset and given the observations. This is the case, for example, in [Agarwal and Triggs 06, Anguelov et al. 05, Sigal et al. 12] or with the well known Kinect application [Shotton et al. 11].

Regularization

The data term in Eq.(12.2) is usually not sufficient to identify a unique set of motion parameters, especially with noisy observations, as is often the case in practical and real scenarios. In order to reduce the space of admissible solutions and disambiguate potential candidates, additional knowledge on the motion is required. This comes in the form of constraints on the motion parameters Θ such as:

Smoothness of Motion Field In the case of point clouds or shapes, an intuitive constraint is to enforce smoothness of the estimated motion field, as in [Myronenko and Song 10, Jian and Vemuri 11] with non-rigid point registration.

Shape Preservation There is a good deal of work that assumes the preservation of local shape properties during motion. These properties can be the Laplacian coordinates [Sorkine et al. 04] as in [Varanasi et al. 08, de Aguiar et al. 08b, Gall et al. 09, Petit et al. 11, Straka et al. 12] or other local rigidity terms, e.g., [Cagniart et al. 10, Huang et al. 13]. This regularization is often stronger than the previous smoothness assumption. Consequently, it sometimes induces tracking failures when the observations do not satisfy the constraint and regularization adjustments may be necessary in that case.

Learned Models In a way similar to the parameterization of motion, a few works take benefit of known tracking examples to learn the space of possible motions beforehand. For example [Duveau et al. 12] proposes a supervised strategy that regularizes the estimated motion parameters based on their learned distribution in a low-dimensional latent parameter space (Gaussian Process Latent Variable Models). Also using a non-sequential strategy for tracking, [Klaudiny et al. 12, Budd et al. 13] learn a shape similarity tree in order to ease shape matching.

12.3 Sequential Shape Tracking

In this section, an approach is illustrated that sequentially tracks a mesh model over a time sequence [Cagniart et al. 10]. The observations $\mathcal{O}$

Figure 12.2: Illustration of sequential tracking: (top) input observations; (bottom) output tracked reference models.

over time are independent 3D meshes that are obtained with a multi-camera acquisition system and a silhouette-based reconstruction [Franco and Boyer 09]. The reference model is a triangle mesh that is usually one of the observed meshes. The approach adopts a locally rigid patch-based deformation model where vertices are grouped into N_p patches (Figure 12.2). While the approach is generic with respect to the shape category, it can be specialized through the prior (regularization) term to track a human surface model and its associated skeleton [Huang et al. 13]. In that case the triangle mesh model is combined with a skeleton model that is a tree structure of N_j nodes (3D joints) with the root set at the pelvis. It is rigged into the mesh using the `Pinocchio` software [Baran and Popović 07] that gives the associations between vertices and joints.

MAP Estimation

As explained earlier, the problem can be expressed as a MAP estimation of the motion parameters Θ given the observations $\mathcal{O}$. Here these parameters decomposed into mesh deformation parameters $\Theta_m = \{(\mathbf{R}_k, \mathbf{c}_k)\}_{k=1:N_p}$ that are the orientations and the locations of the k patches; and, when specialized to human shapes, skeleton pose parameters $\Theta_s = \{\mathbf{x}_j\}_{j=1:N_j}$

that are the 3D locations of the joints. Thus:

$$\hat{\Theta} = \underset{\Theta_m, \Theta_s}{\arg\max}\ P(\mathcal{O}|\Theta_m)\ P(\Theta_m, \Theta_s). \tag{12.3}$$

The skeleton is not observed in practice and the approach assumes that, given the mesh deformation parameters Θ_m, the joint locations Θ_s and the observations $\mathcal{O}$ are independent, i.e., $P(\mathcal{O}|\Theta_m, \Theta_s) = P(\mathcal{O}|\Theta_m)$. This means that the motion is parameterized with the patches only and that the skeleton is solely used as a regularization constraint. Taking the negative log of Eq.(12.3) the optimization becomes:

$$\hat{\Theta} = \underset{\Theta_m, \Theta_s}{\arg\min}\ E_{dt}(\mathcal{O}, \Theta_m) + E_r(\Theta_m, \Theta_s), \tag{12.4}$$

with $E_{dt}(\mathcal{O}, \Theta_m) = -\log P(\mathcal{O}|\Theta_m)$ and $E_r(\Theta_m, \Theta_s) = -\log P(\Theta_m, \Theta_s)$. The tracking is achieved by deforming the model on a frame-by-frame basis. That is, using $\hat{\Theta}^{t-1}$ as the initialization to solve Eq.(12.4) at frame t.

Data Term

The data term is not specialized and involves the patch parameterization only. Following [Cagniart et al. 10] the likelihood $P(\mathcal{O}|\Theta_m)$ is expressed using Gaussian mixture models (GMM) where an observed point p_i is associated to patches with respect to Gaussian distributions:

$$P(\mathcal{O}|\Theta_m) = \prod_i P(p_i|\Theta_m) = \prod_i \sum_{k=1}^{N_p} \Pi_k P(p_i|z_i = k, \Theta_m). \tag{12.5}$$

Here, z_i is the latent variable for each p_i: $z_i = k$ means that p_i is generated by the mixture component associated with the patch k. $\Pi_k = P(z_i = k|y_i, \Theta_m)$ are the mixture weights that are estimated alternatively with Θ_m using the expectation-maximization method [Dempster et al. 77]. The likelihood that p_i is generated by the k-th component is modeled as a multivariate Gaussian with mean located at the compatible vertex v_i^k (i.e., a vertex with a normal similar to p_i) on patch k closest to p_i and isometric covariances σ; and as a negligible value ϵ if there is no such compatible vertex:

$$P(p_i|z_i = k, \Theta_m) = \begin{cases} \mathcal{N}(\mathbf{p}_i|v_i^k, \sigma) & \text{if } v_i^k \text{ exists} \\ \epsilon & \text{otherwise.} \end{cases}$$

Interestingly, this model allows the algorithm to explicitly handle outliers through an additional virtual $N_p + 1$ component equipped with a uniform distribution $P(p_i|z_i = N_p + 1, \Theta_m) = cst$ [Cagniart et al. 10]. In this case, each observation has a constant probability to be generated from the virtual outlier patch which favors the association of distant observations to this outlier patch.

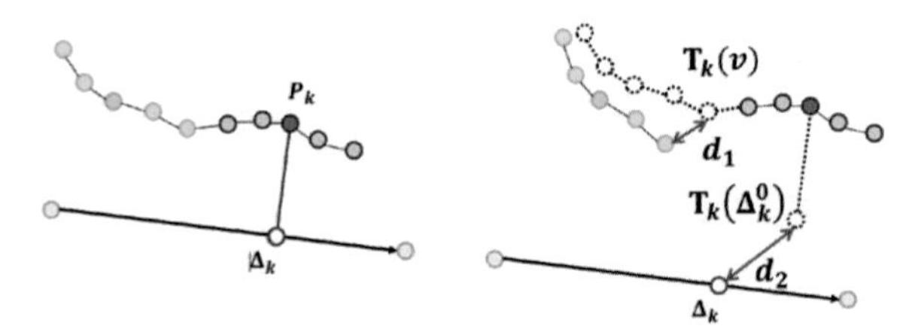

Figure 12.3: Illustration of the regularization: (left) original configuration; (right) distances being minimized in the optimization. d_1 impacts the mesh rigidity term E_m while d_2 impacts the bone-binding energy E_s.

Regularization Term

$P(\Theta_m, \Theta_s)$ decomposes into $P(\Theta_s|\Theta_m)P(\Theta_m)$. The regularization energy $E_r = -\ln P(\Theta_m, \Theta_s)$ is therefore a combination of a generic term $E_m = -\ln P(\Theta_m)$ that enforces rigidity between neighboring patches, and of an optional specialized term for humans $E_s(\Theta_s, \Theta_m) = -\ln P(\Theta_s|\Theta_m)$ that enforces the patches and the skeleton to satisfy binding constraints.

Mesh Rigidity E_m applies to the rigid transformation parameters of the patches and is defined over the set $\{(P_i, P_j)\}$ of neighboring patches [Cagniart et al. 10]:

$$E_m(\Theta_m) = \sum_{\{(P_i,P_j)\}} \sum_{v \in P_i \cup P_j} w_{ij} \| T_{\Theta_m(i)}(v) - T_{\Theta_m(j)}(v) \|,$$

where $T_{\Theta_m(k)}(v)$ is the predicted position of vertex v as transformed with patch k and w_{ij} weights the contribution of v with respect to its distances to the patch centers of P_i and P_j. This term favors similar rigid transformations between neighboring patches (Figure 12.3). Combined with the previous data term it allows to track arbitrary mesh model under the locally rigid assumption only (Figures 12.1 and 12.2). Additional prior knowledge about the scene, e.g., a human skeleton, can be exploited, when available, to further constraint the model evolution as explained in what follows.

Skeleton Binding Assuming a known skeleton for the tracked shape, a bone-binding energy $E_s(\Theta_s, \Theta_m) = -\ln P(\Theta_s|\Theta_m)$ is defined that builds on the approach introduced in [Straka et al. 12]. For each patch k the orthogonal projection of its center onto its associated bone is denoted Δ_k

Figure 12.4: Illustration of the sequential approach on standard datasets: example frames of the input videos and the overlaid tracked models that include skeleton poses here.

(Figure 12.3). Let Δ_k^0 be the location of Δ_k on the reference model, then:

$$E_s(\Theta_s, \Theta_m) = \sum_{k=1}^{N_p} w_k \| T_{\Theta_m(k)}(\Delta_k^0) - \Delta_k \|,$$

where $T_{\Theta_m(k)}(\Delta_k^0)$ is the predicted position of Δ_k^0 with the transformation of patch k and Δ_k is the location of the patch center projection onto the skeleton with parameters Θ_s. The term w_k weighs the patch contribution such that patches close to joint locations have less influence. This term involves both the transformation parameters Θ_m and the skeleton joint locations Θ_s. It favors rigid motions of all the patches associated to the same bone in the skeleton. The approach involving the skeleton binding term is illustrated in Figure 12.4 for various standard datasets.

12.4 Non-Sequential Mesh Tracking

Sequential frame-to-frame mesh tracking, as presented in the previous section, is prone to drift over time, especially when large displacements occur. Consequently, non-sequential alignment strategies have been introduced to limit the accumulation of errors associated with sequential frame-to-frame

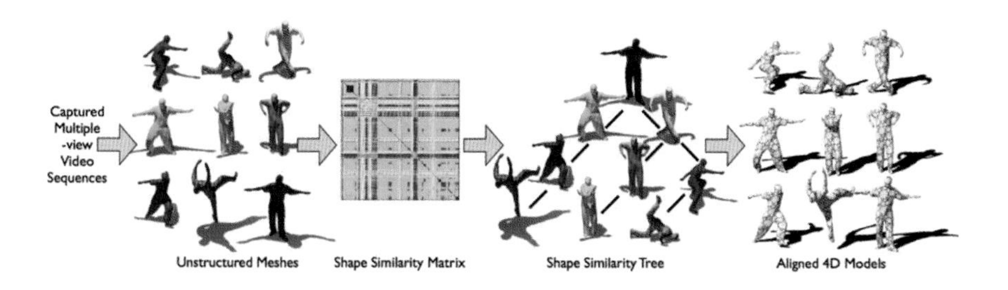

Figure 12.5: Overview of non-sequential global temporal mesh sequence alignment [Budd et al. 13].

mesh tracking by exploiting the similarity of frames across the entire sequence [Klaudiny et al. 12, Budd et al. 13, Beeler et al. 11]. Interestingly, this also allows the alignment across different sequences to produce 4D models from performance capture of different motions. Non-sequential approaches were first proposed in the context of reconstruction from video using structure-from-motion to reduce drift in reconstruction. In this case reconstruction is performed over sub-sequences of the video and fused into a single scene reconstruction using a hierarchical tree structure. This non-sequential reconstruction improves efficiency and reduces accumulation of reconstruction errors across the sequence.

For non-rigid mesh tracking Beeler et al. [Beeler et al. 11] presented an approach based on manually selected anchor frames to reduce drift for alignment of reconstructed non-rigid face sequences. The approach assumes that anchor frames are similar to a manually selected reference expression and are distributed across the sequence. Pairwise alignment of anchor frames with the reference reduces accumulation of errors by sequential alignment on shorter subsequences. This approach implicitly uses a non-sequential alignment based on a tree with branches from the reference pose to each of the anchor frame.

Shape similarity trees [Budd et al. 13] link frames with similar shape and motion providing a general representation for non-sequential alignment over one or more sequences. Non-sequential alignment changes the order in which frames are aligned to maximize the similarity of adjacent frames allowing existing pairwise sequential alignment algorithms to be applied while reducing drift and alignment failure. The branches of the shape similarity tree define the shortest path in similarity space between frames. This defines a non-sequential ordering of frames for alignment. Figure 12.5 presents an overview of the alignment process from raw unstructured mesh sequences to a single temporally coherent representation. Non-sequential alignment of a database of reconstructed mesh sequences based on the

Figure 12.6: Example frames for non-sequential alignment across multiple sequences of a StreetDancer dataset comprising 1800 frames.

shape similarity tree is performed in three stages:

1. **Shape Similarity Evaluation:** Evaluate the shape and motion similarity between all pairs of frames across all sequences.
2. **Shape Similarity Tree Construction:** Construct the minimum spanning tree given by the shortest path in shape similarity space for all frames.
3. **Global Non-rigid Alignment:** All frames for all sequences are aligned and re-meshed to have a single connectivity based on the shortest paths defined by the shape similarity tree and using, for instance, the patch based method presented in the previous section.

Figure 12.6 presents an example of non-sequential alignment to produce a 4D model from six sequences of a street dancer performing fast motion with loose clothing. The non-sequential alignment reduces drift and provides robustness to rapid motion by identifying optimal paths for tracking of mesh sequences. The approach links similar frames across the database rather than purely temporal adjacency. A potential drawback of non-sequential alignment is the independent accumulation for sequential alignment across different tree branches. The tree structure can be further optimized by clustering similar frames [Klaudiny et al. 12] to avoid short branches reducing alignment jitter due to fragmentation.

12.5 Motion Fields

As mentioned earlier, some approaches do not constrain motion through parameterization and, instead, estimate full motion fields over primitives, as in [de Aguiar et al. 07, Varanasi et al. 08, Petit et al. 11], and more recently in Blache et al. [Blache et al. 14]. In the latter approach the input is a temporal sequence of multi-view silhouettes transformed into a set of binary digital volumes using a visual hull reconstruction algorithm [Lucas et al. 13]. The visual hull is defined as the 3D intersection of the viewing cones associated to 2D silhouettes [Laurentini 94, Franco and Boyer 09].

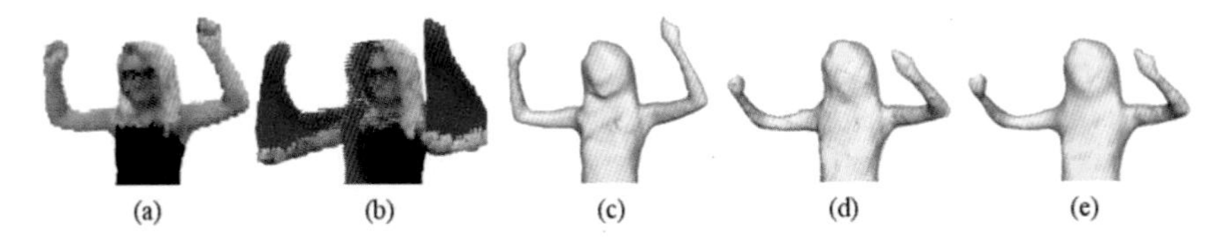

Figure 12.7: Starting from a colored volume sequence *(a)*, the motion flow is estimated *(b)*. The mesh of the first frame is used as a template *(c)*. After the deformation of this template by the motion flow *(d)*, the mesh is smoothed by a regularization algorithm *(e)* which matches the reconstruction with visual hull of the next pose (shown in yellow).

These binary digital volumes represent the observed shape at each video frame (Figure 12.7(a)). The method starts by computing a 3D motion flow between two consecutive frames, that extends the idea of optical flow described in Chapter 8 to 3D. This motion field is used in a subsequent step to deform a mesh model. Such a model can easily be obtained from any frame in the sequence by applying a surface reconstruction algorithm (such as marching cubes, Chapter 10) on the estimated volume at that frame. The tracking is then performed by deforming the mesh model at each time frame using the estimated flows.

Given two consecutive volumes V^t and V^{t+1} in the sequence, a motion field between the surface voxels $v_i^t \in V^t$ and $v_j^{t+1} \in V^{t+1}$ is determined. This 3D vector field is obtained by minimizing a distance function based on three criteria: a **proximity criterion** that accounts for the Euclidean distance between matching voxels, an **orientation criterion** that measures the difference between the normal vectors of matching voxels, and a **colorimetric criterion** that compares the photometric information between matching voxels. These criteria allow to match the voxels that correspond to the same part of the observed surface based on their proximity, orientation and appearance on the shape surface.

This estimation is performed backward and forward in time and each pair of matched voxels defines a 3D vector. The resulting 3D vector field describes the volumetric motion of the shape between t and $t+1$. However, the obtained motion field is corrupted by several inconsistent matches that remain. In order to reduce their effect, a Gaussian filtering step is applied on the 3D vectors which helps to obtain a coherent motion flow. This filtering is equivalent to a regularization of the motion field based on a smoothness assumption. In the final displacement field, each surface voxel is associated to a single motion vector (Figure 12.7(b)). In the second step of the approach, the template mesh is displaced along the motion field and each vertex is moved with respect to the closest vector. Already in previous sections, it was discussed that drift could lead to a deterioration of the result which yields an irregular mesh (Figure 12.7(d)). Therefore, a regularization algorithm is also applied here to obtain a proper mesh that fits the reconstructed volume at the next frame (Figure 12.7(e)). The mesh is considered to be a deforming mechanical system and each vertex is subject to a set of forces as follows:

- A **spring force** where each incident edge applies a force on the vertex to enforce similar edge lengths. This tends to regularize the vertex distribution.
- A **smoothing force**, based on a Laplacian operator, that tends to smooth the mesh surface.
- A **matching force** that uses the Euclidean distance field to keep each vertex close to the object's surface.

When compared to more traditional 3D optical flow approaches, this strategy proves to be more robust over time, failing only when strong topological changes occur. Using a 3D uniform grid to represent volumes facilitates the motion flow estimation. Such estimation would be tedious with 3D meshes reconstructed independently over time, and therefore inconsistent (i.e., with different number of vertices and faces). The evaluation of the results demonstrated that introducing the photometry in the matching criteria significantly increases the accuracy. This algorithm fails when topological changes occur (such as body parts in contact) and several reference meshes should be considered in those cases. Having topological changes in 4D reconstruction is one of the big open research questions.

12.6 Summary

The chapter discusses methods to capture the motions and deformations of general shapes recorded by multi-view video systems. Strategies that exist differ with respect to three main criteria: (i) the observations taken into account, (ii) the motion parameterization, and (iii) the regularization over these parameters. One of the most popular related strategies that is illustrated in the chapter consists in tracking a known reference template model of the scene, usually a mesh obtained with a reconstruction method from one of the video frames. The success of this tracking strategy, either the sequential or the non-sequential variants that are discussed, is most certainly due to its ability to produce time consistent shape models, known as 4D models. However, while good and robust results have been obtained, many issues are still unresolved. This includes, for instance, the capture of dynamic scenes with evolving topologies or complex scenes with many interacting objects.

Part III

Modeling Reality

13
Rigging Captured Meshes

Kiran Varanasi and Edilson de Aguiar

13.1 Introduction

Performance capture methods discussed in the earlier chapters reconstruct the geometry of real-world objects in motion. For realizing any computer graphics application, practical user controls for editing and modifying the scene content are essential. Thus, a second step of developing a *control rig* for the captured geometry is necessary, through which the 3D geometry can be easily posed and manipulated.

In practice, computer animation artists commonly use three different types of rigs depending on the motion they wish to control. Non-rigid deformations over an approximately rigid bone-structure such as facial expressions are manipulated using a set of example *blendshapes* that are provided by the modeling artist. New facial expressions can then be generated through blending between different examples. Highly fluid deformations dependent on external forces such as cloth folds are generated using physics-based simulation. These simulations can be controlled by the artist through positional or force-based constraints. Finally, if the captured geometry is that of an articulated character model such as a human actor, an intuitive means for editing the pose of that character is through a skeleton rig. In principle, performance capture methods that reconstruct the 3D geometry of real-world deformations can be applied to each of these settings. In this chapter, the focus is on rigged articulated models such as human characters.

The skeleton rig for controlling the pose corresponds only loosely to the physiological skeleton of the bone joints. Instead, it is an abstraction for driving the surface deformation through a set of geometric transformations. Essentially, the skeleton is a hierarchy of joints around which a 3D rotation can be performed with respect to a pre-defined axis of rotation. The movements of a parent joint are cascaded down, affecting all the joints below in the hierarchy. For example, the movements of the shoulder affect the position of the elbow, the arm, and so on. The overall motion of the character is thus a superposition of the various joint movements. In character animation, forward kinematics refers to how the motion encoded by a set of 3D rotations around joints in the skeleton rig is decoded into the overall motion of surface vertices of the character. Equivalently, inverse kinematics refers to how the 3D rotations around the joints in the skeleton rig can be inferred from the surface motion of vertices. To perform either of these steps, knowledge of the exact topology (i.e., hierarchy) of the character skeleton and of the relative positions of the joints in a neutral *rest* pose is required, as well as the mechanism through which the various surface vertices are affected by the skeleton joints. Conventionally in the character animation community, the first step is referred to as *rigging* and the second step as *skinning*. In this chapter, details about these steps are described to process 3D meshes captured from the real world.

Traditionally, for 3D meshes hand-sculpted by 3D artists, the rigging step involves bringing the input mesh into a neutral *rest* pose and carefully positioning the skeleton joints inside the interior volume of the 3D mesh, such that articulations around these joints generate plausible surface deformation results for the character. The skinning step involves partitioning the surface of the character into a set of body parts that correspond to *bones* linking joints in the skeletal hierarchy, as well as painting of *skinning weights* for each surface vertex that blend the effect of different joint transformations on that vertex. Typically, a surface vertex is affected by not more than 2–3 bones. A skeleton rig offers an artistically intuitive and computationally efficient means for controlling the character motion. Despite its simplicity, plausible results for character animation can be achieved, although with immense artistic skill required in both the rigging and the skinning steps. Performance capture methods can greatly simplify these steps.

In this chapter, assuming that a performance capture system is deployed to produce a sequence of temporally aligned triangle meshes with the same number of vertices and mesh topology, three methods are described for building a rigged character out of such meshes.

1. Fitting a skeleton into a static mesh such that the bone joints fall along the medial axis of the 3D shape and are aligned with plausible articulation points (Section 13.3).

2. Converting a mesh animation sequence into a skeletal animation sequence by optimizing for the joint transformations and skinning weights (Section 13.4).

3. Building a deformable model by learning the residual deformations of each surface vertex observed in the captured meshes, which are beyond the skinned deformations due to rigid transformations of the bone joints (Section 13.5).

In the following, the problem context of editing 3D surface deformations is introduced and viable solutions to the three problems mentioned above are presented.

13.2 Overview

Directly editing mesh animations It is possible to edit mesh animations without constructing a rig, by directly editing vertex positions. Kircher and Garland [Kircher and Garland 06] propose various such edits that align selected point trajectories to user-given locations. They parameterize the surface through dihedral angles between surface triangles, and deform the rest of the mesh surface such that local surface curvature and details are preserved. Sumner and Popovic [Sumner and Popović 04] propose an alternative motion parameterization through deformation gradients of mesh triangles. They are able to transfer the deformations across two different meshes with certain shape similarity. The deformation gradients can also be embedded into an automatically computed deformation graph structure [Sumner et al. 07] that can be edited. Hildebrandt et al. [Hildebrandt et al. 12] propose certain space-time editing operations that respect the modal energies of the surface. However, directly editing a mesh animation typically requires the user to place constraints on several vertices and thus needs a lot of edit-time. The range of edits that can be performed is also typically limited. In practical scenarios, an additional control rig that adapts to the motion of the object, and not just to its shape, is highly useful.

Surface deformation by rigs Skeleton rigs are a popular control structure for deforming articulated 3D objects [Lewis et al. 00]. Each vertex on the object's surface is associated to one or more skeletal bones and its motion is interpolated from the motion of the bones. The most common method for interpolation is known as *linear blend skinning*, where the vertex motion is computed as a weighted linear combination of rigid transformations of the bones [Magnenat-Thalmann et al. 88]. James and Twigg [James and Twigg 05] propose to use affine transformations on the bones to represent arbitrary surface motion, such as cloth deformation. Mean shift clustering is used to aggregate the vertex deformations into a set of bones. Kavan et al. [Kavan et al. 10] decompose the mesh animation into the relatively simpler linear blend skinning framework, and make use of the sparsity of the vertex weights to generate a fast animation. They use alternating least squares to optimize for the bone positions and skinning weights. Instead of a set of interior skeletal joints, the constraints for deforming a mesh can be provided through the vertices of an enclosing cage. The mesh is then deformed according to how the volume enclosed by the cage compresses or expands. The mesh vertices are typically represented through generalized barycentric coordinates inside the polyhedra of the cage. Cages are sometimes more intuitive to edit for artists than interior skeletal handles, e.g., to show subtle non-rigid deformations on an object. For smooth deformation, the weighting functions that relate the mesh vertices to the cage should vary continuously and smoothly. Cauchy-green coordinates and bounded biharmonic weights are proposed as options for such weighting functions [Weber et al. 09]. Jacobson et al. [Jacobson et al. 12] propose a general framework for mesh deformation through a combination of control rigs: skeletons, cages, or point control handles. The user specifies only a subset of the deformation space and their method automatically infers the remaining degrees of freedom and estimates geometrically appropriate skinning weights that connect the surface vertices to the control rig.

Surface deformation by blendshapes Artists have traditionally handcrafted target meshes, known as blendshape targets, for representing various facial expressions. Novel facial expressions are then synthesized on the virtual face by interpolating between these blendshape targets or by mapping direct surface manipulation to blendshape interpolation [Lewis and Anjyo 10]. New facial dialogue is synthesized by interpolating between specific blendshape targets, known as visemes, that correspond to a selection of speech phonemes. Facial rigs are typically limited to synthesizing rigid motions such as the rotation of the head and the jaw movement, while the rest of the facial motion is synthesized exclusively through the skill of the artist who sculpts facial expressions at vertex-level detail. This has been necessary because human visual perception is highly attuned to artefacts in facial expressions. Indeed, the virtual face is deemed less likeable by human observers even as the fidelity of its facial geometry and expressions increases, in a widely understood phenomenon known as the *uncanny valley*. However, with increased resolution of facial scans and high-fidelity performance capture, this barrier is being overcome. Vlasic et al. [Vlasic et al. 05] propose a multi-linear model for varying facial expressions across the axes of identity, emotion, and visemes. They align the data from a set of static 3D scans of human faces to a common geometric template. It is now also possible to capture dynamic facial geometry at high resolution on a single mesh topology. Beeler et al. [Beeler et al. 11] propose a high-resolution passive facial performance capture system from multi-camera recordings. Data-driven deformable models can exploit this data to synthesize virtual facial expressions. Neumann et al. [Neumann et al. 13b] automatically decompose an input mesh animation into a set of sparse localized deformation components, or *splocs*, that loosely correspond to blendshape targets that can be edited by artists for novel expression synthesis. Capturing and editing facial deformations for real-world visual effects is discussed in detail in Chapter 20. This chapter focuses on building rigged 3D models for articulated characters.

13.3 Fitting a Skeleton into a Static Mesh

Given a static 3D mesh as input, a skeleton structure can be extracted by successively thinning the shape [Gagvani and Silver 99]. For the purpose of character animation, the skeleton is a support structure of rigid line segments in the interior of the shape around which articulated movements can be executed. Teichmann and Teller [Teichmann and Teller 98] propose a method for semi-automatic recovery of the animation skeleton by simplifying the network structure of the Voronoi centers of the 3D shape using some user assistance. Since the precise motion semantics are not necessarily apparent from a static 3D mesh, automatically recovering an animation skeleton is an ill-posed problem. However, with certain assumptions, plausible animation results can be achieved. An informative geometric structure that matches our intuition about an underlying shape skeleton is the *medial axis* of the shape, which is the set of 3D points that are centers to medial balls which are equidistant to at least two different points on the shape's surface. The medial axis of a given shape is unique and completely informative about the shape. Several methods have been proposed for fast and robust computation of the medial axis given an input 3D shape [Chazal and Lieuter 05, Giesen et al. 09]. Although the medial axis is continuous for 2D shapes, it can be a set of disconnected sheets for 3D shapes. To remedy this problem, the curvilinear skeleton of the shape is proposed as a 3D curved line that connects a set of interior points [Sharf et al. 07], which can be built from approximations of the medial axis. Wade [Wade 00] uses the discontinuities in the distance field from the surface to approximate the medial surface and extracts the skeleton from this. Katz and Tal [Katz and Tal 03] propose a shape partitioning algorithm which can be used for skeleton extraction. Other methods for 3D surface segmentation can be similarly adapted for this purpose [Lien et al. 06]. Liu et al. [Liu et al. 03] use a different approximation by repulsive force-fields to estimate the skeleton. Tagliasacchi et al. [Tagliasacchi et al. 09] use cutting planes of the 3D shape to recover the set of interior points, and estimate a curve skeleton for incomplete point clouds. Among the various alternatives for shape approximation, methods based on sphere packing such as the medial axis have remained popular due to their simplicity, and current methods can robustly deal with surface noise.

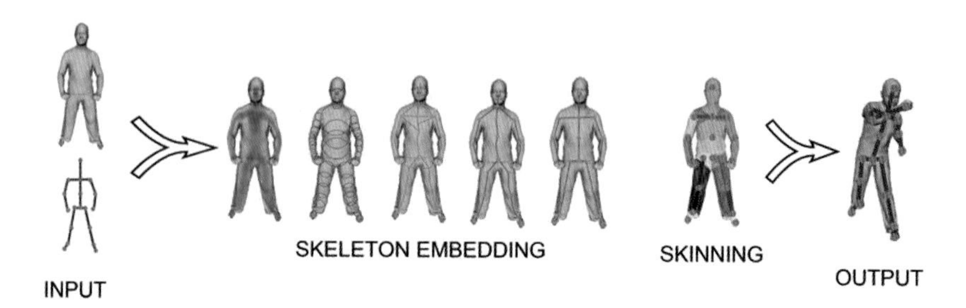

Figure 13.1: The *Pinocchio* system receives as input a character mesh and a skeleton. In the first step, the skeleton is embedded into the character mesh as follows: first, the surface medial axis distance is calculated, packed spheres are found, and the graph is constructed. Thereafter, the optimal embedding of the skeleton is found via discrete optimization and the resulting skeleton is refined. The second step computes skinning weights based on the embedded bones. The result is an automatic rig of the input character.

Instead of relying exclusively on 3D shape geometry, when one has prior information on the type of motion one expects from the input 3D shape, this can be used for the estimation of the skeleton rig. For example, a 3D shape resembling a humanoid can be expected to move in a human-like fashion with similar skeletal dynamics. When a skeleton rig is created with this prior, human motion, such as that captured from an actor's performance, can be easily retargeted to the input shape. Given a static surface mesh, biologically plausible animation skeleton can be extracted using constraints from prior knowledge based on the anatomy of humans or animals [Aujay et al. 07, Schaefer and Yuksel 07]. Baran and Popović [Baran and Popović 07] propose the *Pinocchio* system to embed a human skeletal rig into a static surface mesh of a character using a variety of priors such as the relative lengths of bones, the relative position of legs, etc. Their method also estimates automatically the skinning weights of the vertices from the bones. In comparison with skeleton extraction methods, skeleton embedding is more suitable for automatically animating a character. Extracted skeletons may have different topologies for similar character meshes, which may hinder its use with similar motion data. In addition, embedding a given skeleton provides information about the expected structure of the character, which can be difficult to obtain from just the character geometry.

As shown in Figure 13.1, the *Pinocchio* system [Baran and Popović 07] consists of two main steps: rigging (skeleton embedding) and skinning (skin attachment). The first component, skeleton embedding, computes the joint positions of the input generic skeleton inside the character. Afterwards, the skin attachment is performed by assigning bone weights to the vertices of the character model.

Intuitively, the joint positions need to be computed such that the embedded skeleton fits inside the character correctly and looks like the given skeleton as much as possible. This is achieved by first embedding the skeleton into a discretization of the character's interior and then by refining this embedding using continuous optimization. In summary, the skeleton embedding algorithm works as follows:

- Adaptive distance field is used to compute a sample of points approximately on the medial surface of the input character.
- A graph is constructed where the vertices represent potential joint positions and edges are potential bone segments. The graph is constructed by packing spheres centered on the approximate medial surface into the character and by connecting sphere centers with graph edges.
- The system finds the optimal embedding of the skeleton into this graph with respect to a discrete penalty function considering a variety of penalties like short bones, improper orientation between joints, length differences in bones marked symmetric, bone chains sharing vertices, etc.
- The resulting skeleton found by discrete optimization usually has the general character shape, but typically, it does not fit correctly inside the character. Therefore, embedding refinement is used to correct the embedded skeleton by minimizing a new continuous function that penalizes bones that do not fit inside the surface, bones that are too short, and bones that are oriented differently from the given skeleton.

As a result, the skeleton embedding step resizes and positions the input skeleton to fit inside the character. However, the character and the embedded skeleton are disconnected until the deformations of the skeleton to the character mesh are specified.

Deforming the mesh surface (or skin) by an underlying skeleton structure [Magnenat-Thalmann et al. 88] is the second step of the system: skin attachment or skinning. Briefly, the general standard linear blend skinning (LBS) method [Lewis et al. 00] works as follows: if $\mathbf{v}_i$ is the position of the vertex i, θ_j is the transformation of the j-th bone, and w_{ij} is the weight of the j-th bone for vertex i, the standard linear blend skinning (LBS) approach gives the position of the transformed vertex i as $\mathbf{v}'_i = \sum_j w_{ij}\theta_j(\mathbf{v}_i)$. Therefore, the skin attachment problem is to find bone weights $\mathbf{w}$ for all vertices, indicating how each bone affects each vertex.

Assigning bone weights purely based on proximity to bones will often fail because they ignore the character's geometry. An alternative approach is to use the analogy to heat equilibrium to find the weights. If the character volume is treated as an insulated heat-conducting body and the temperature of bone j is forced to be 1 while keeping the temperature of all of the other bones at 0, the equilibrium temperature at each vertex on the surface can be considered the weight of bone j at that vertex.

For simplicity, the *Pinocchio* system solves the equilibrium equation over the surface. The equilibrium over the surface for bone j can be written as $-\Delta \mathbf{w}_j + \mathbf{H}\mathbf{w}_j = \mathbf{H}\mathbf{p}_j$, where Δ is the discrete surface Laplacian, calculated with the cotangent formula [Meyer et al. 02], $\mathbf{p}_j$ is a vector with $p_{ji} = 1$ if the nearest bone to vertex i is j and $p_{ji} = 0$ otherwise, and $\mathbf{H}$ is the diagonal matrix with H_{ji} being the heat contribution weight of the nearest bone to vertex i.

The performance of *Pinocchio* has been demonstrated in a variety of papers and applications. The system can be used to automatically rig a character. It allows a user to go from a static mesh to an animated character automatically. As a result, users can animate many different characters using a generic skeleton with quality compared to modern video games.

13.4 Converting a Mesh Animation into a Skeletal Animation

It is popular to represent animations not by means of a classical skeleton-based model, but in the form of deforming mesh sequences, e.g., as a result of performance capture methods. The reason for this is that novel mesh deformation methods as well as surface-based scene capture techniques offer a great level of flexibility during animation creation. Unfortunately, the resulting scene representation is less compact than skeletal ones and there is a limited number of tools available which enables easy post-processing and modification of mesh animations. Several methods have been described to automatically rig a model using an animated mesh sequence [Schaefer and Yuksel 07, de Aguiar et al. 08a, Hasler et al. 10b]. The key idea of these methods is to first perform a motion driven clustering step to extract rigid bone transformations, then estimate the joint locations and bone lengths, and finally optimize the bone transformations and skinning weights.

In this section, a variant of this general algorithm will be described. A method to automatically extract a plausible kinematic skeleton, skeletal motion parameters, as well as surface skinning weights from arbitrary mesh animations will be presented [de Aguiar et al. 08a]. By this means, deforming mesh sequences can be fully automatically transformed into fully rigged virtual subjects.

An overview of the approach is shown in Figure 13.2. The input is an animated mesh sequence comprising N frames. An animated mesh sequence

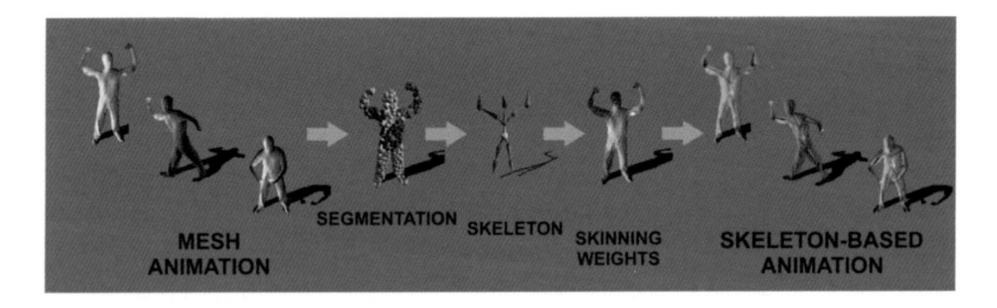

Figure 13.2: Using an animated mesh as input, the approach segments the model into plausible approximately rigid surface patches, estimates the kinematic skeleton and its motion parameters, and calculates the skinning weights connecting the skeleton to the mesh. The output is a skeleton-based version of the input mesh animation.

is represented by a mesh model M consisting of vertices V and triangulation T with positional data $\mathbf{p}_t(\mathbf{v}_i) = (x_i, y_i, z_i)_t$ for each vertex $\mathbf{v}_i \in V$ at all time steps t. In the first step of the algorithm, spectral clustering is used to group seed vertices on the mesh into approximately rigid segments. By using the clustered seed vertices it is possible to segment the moving mesh into kinematically meaningful approximately rigid patches. Thereafter, adjacent body parts are determined and the topology of the kinematic structure of the mesh is found. Using the estimated topology, joint positions between interconnecting segments are calculated over time. In the last step, appropriate skinning weights are calculated to attach the learned skeleton to the surface.

The first step of the algorithm segments the animated input mesh (given by M and $\mathbf{p}_t$) into spatially coherent patches that undergo approximately the same rigid transformations over time. The approach is initialized by selecting a subset of l seed vertices that are distributed evenly over the mesh M using a curvature-based segmentation method [Yamauchi et al. 05]. The motion trajectories of the seed vertices throughout the whole sequence form the input to a spectral clustering approach [Ng et al. 02] which automatically groups the l seeds into k approximately rigidly moving groups. The idea behind this clustering is to capitalize on the invariant that mutual distances between points on the same rigid part should only exhibit a small variance while the mesh is moving. Using the k optimal vertex clusters, the triangle clusters are created. The resulting clusters divide the mesh into k approximately rigid surface patches.

Given the list of body segments, their associated seed vertices and triangle patches, the kinematic skeleton structure is found by first finding its kinematic topology (i.e., find which body parts are adjacent). To determine which body segments are adjacent, the triangles at the boundaries of the triangle patches are analyzed. Body parts A and B are adjacent if they have mutually adjacent triangles in their respective patch boundaries. Unfortunately, in practice a patch may be adjacent to more than one other patch. Taking a heuristic approach, the method considers only those patches to be adjacent that share the longest common boundary (in terms of the number of adjacent boundary triangles). Note that the system assumes that the body part in the center of gravity of the mesh is the root of the hierarchy. A good estimate for the correct sequence of joint positions is the sequence of locations that minimizes the variance in joint-to-vertex distance for all seed vertices of the adjacent parts at all frames. Therefore, after aligning the segment poses to a reference time step, the algorithm proposed in [Anguelov et al. 04a] is used to calculate all joint positions. Additionally, the bone lengths are enforced to be constant over time.

In order to infer joint motion parameters, i.e., a rotational transformation for all joints at all times, a cyclic-coordinate-descent (CCD)-like algorithm [Luenberger 73, Badler et al. 87] is employed to calculate all motion parameters using Euler angle parameterization. The translation of the root is stored as an additional parameter for each frame. In the last step, the approach described in the previous (Section 13.3) is used to determine the skinning weight distribution for each bone considering the entire mesh sequence and the reconstructed skeleton poses.

The performance of the described system has been validated on a large variety of mesh animations. The fully-automatic approach is able to extract a kinematic skeleton, joint motion parameters, and surface skinning weights from a mesh animation. The original input can then be quickly rendered based on the new compact bone and skin representation, or it can be easily modified using the full repertoire of already existing animation tools.

13.5 Building a Deformable Model

A disadvantage of using a static motion template such as a skeleton or an enclosing cage is that the deformation priors induced by such a template may not correspond to the actual motion characteristics of a real-world object. Simulating the exact physiology and elastic properties of muscles and tendons is possible [Lee et al. 09], but this simulation is costly and time-consuming. Alternatively, real-world datasets of 3D motion can be used to build data-driven models for synthesizing new virtual deformations. These datasets can be acquired either from motion capture systems, or from artist-given example deformations in order to achieve an artistic effect or animation style. Kry et al. [Kry et al. 02] propose the *Eigenskin* system that synthesizes new deformations from examples. Weber et al. [Weber et al. 07] propose context-aware skeletal shape deformation that uses artist-given examples to enhance the realism of articulated deformations.

Motion capture systems, either based on optical markers or on visual feature tracks over high-resolution multi-view photography, can estimate the general 3D non-rigid motion of an object at a high detail. Such spatio-temporal data can be used to build dynamic 3D deformation models that mimic real-world motion. Park and Hodgins [Park and Hodgins 06] placed a set of 300 optical markers on an actor's body and recorded various muscle deformations. They used this data to build a deformation model for synthesizing physically realistic muscle deformations. They represent the 3D motion of the optical markers on the surface as residual angular transformations from an underlying skeleton. Another related research effort is to parameterize variation of human body proportions from a database of real-world 3D scans of people [Allen et al. 02, Allen et al. 06, Anguelov et al. 05, Hasler et al. 09c] (Chapter 14 gives a more thorough review). As the resolution of these 3D scans increases, surface deformation is parameterized independently from the underlying rigs, such as through deformation gradients [Sumner and Popović 04]. The surface deformation can then be represented in layers: that given by a control rig for editing pose, and that representing residual deformations either due to shape variation across people, or due to pose-specific variation.

The dynamic muscle deformation model of Park and Hodgins [Park and Hodgins 08] can be considered as a demonstrative example. This model is built as an extension of earlier work [Park and Hodgins 06] and able to synthesize dynamic effects such as the jiggling of muscles and body fat. A set of $400 - 450$ reflective markers are placed on an actor and his motion is captured in a studio with 16 near infrared Vicon MX-40 cameras running at 120 frames per second. Thus, an actor-specific database of roughly 10,000 frames of different motions are captured: flexing, twisting, running, jumping, punching, etc. Each motion is captured at different speeds to study the effect of dynamics. The motion capture data is then cleaned and smoothed to remove noise artifacts and to fill holes. The actor's body is segmented into a set of 17 near-rigid parts and each marker is placed in a local coordinate frame. The rigid motion of the body part is then removed and the residual displacement of the marker in the local co-ordinate frame is analyzed by decomposing it into static and dynamic components.

$$\mathbf{d} = \mathbf{d_s} + \mathbf{d_d} \tag{13.1}$$

where $\mathbf{d}, \mathbf{d_s}, \mathbf{d_d} \in \mathbb{R}^{3N}$ with N being the number of markers.

The static component is due to pose-specific muscle bulging, and is estimated using a locally weighted linear regression model.

$$\mathbf{d_s} = A\Theta \tag{13.2}$$

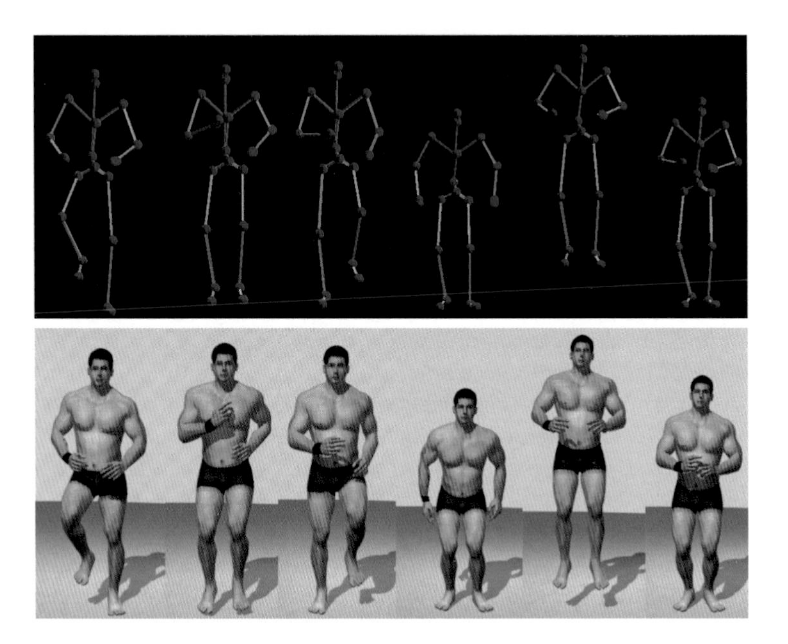

Figure 13.3: Surface deformation results from the data-driven deformation model of [Park and Hodgins 08]. The top row shows skeletal pose input; the bottom row shows detailed surface deformation. The data-driven model is built from dynamic motion capture data of 400 – 450 reflective markers placed on the actor's body.

where $\Theta = [\theta_1^\top \theta_2^\top 1] \in \mathbb{R}^7$ is a vector denoting the pose of the part (θ_1, θ_2 denote the angular displacement vectors between the body part and its inboard and outboard neighbors from the skeletal data). The regression matrix A is estimated by weighted linear least squares from ground-truth residual displacements. This regression model is not yet able to capture dynamic effects such as jiggling of the muscle and fat. These dynamics are captured by $\mathbf{d_d}$ using a second-order differential model.

$$m_k \ddot{d}_k + C_k \dot{d}_k + K_k d_k = f_k \tag{13.3}$$

where d_k is the k-th component of $\mathbf{d_d} \in \mathbb{R}^{3N}$. C_k and K_k are damping and stiffness coefficients, f_k is the component of the net external force $\mathbf{f}$ contributing to the dynamic local deformation component d_k, and m_k is an imaginary mass to provide inertia effects. By approximating $\ddot{d}_k, \dot{d}_k$ with numerical derivatives, the above equation can be linearized and solved using least-squares (the regression model can be made smoother and more robust by dimensionality reduction using principal component analysis (PCA) on $\mathbf{d_d}$ and solving for the projected components in a lower dimensional space).

Given a new body pose, the final skin deformation is synthesized by estimating the static and dynamic residual deformations for each vertex, and adding them to the rigid motion of the body part. High-quality animations of muscle and skin deformations are thus generated by exploiting the motion data (Figure 13.3).

One limitation of this work is the crude segmentation into rigid body parts, which was rectified in a later work by Hong et al. [Hong et al. 10], who proposed a method for automatically estimating the underlying skeleton rig for the shoulder-arm complex through a data-driven segmentation of the marker trajectories.

De Aguiar et al. [de Aguiar et al. 10] propose a similar dynamic model for learning stable spaces for cloth deformation (Chapter 15). De Aguiar and Ukita [de Aguiar and Ukita 14] proposed an approach to represent and manipulate a mesh-based character animation preserving its time-varying details by decomposing the input mesh animation into coarse and fine deformation components. A model for the coarse deformations is constructed by an underlying kinematic skeleton structure and blending skinning weights and a non-linear probabilistic model based on Gaussian processes is used to encode the fine time-varying details of the input animation.

Using multi-view camera recordings of real-world muscle exercises, Neumann et al. [Neumann et al. 13a] built a data-driven model for the shoulder-arm complex that synthesizes realistic deformation not only according to pose (Θ), but also with respect to the body shape parameters of the actor (β) as well as external forces acting on the arm (γ).

$$\mathbf{d}_s = \Psi(\Theta, \beta, \gamma) \tag{13.4}$$

They model only the static residual deformations $\mathbf{d_s}$ while the actors perform slow and natural movements, where dynamic effects can be neglected within the so-called *kinesostatic* assumption. They capture the surface deformations of the arm by applying a makeup of random dot patterns on human actors and imaging real-world muscle exercises at a high resolution through a multi-camera system. These muscle exercises are repeated with the subjects holding different weights in their arms. In this way, reactionary forces in the muscles and the resultant 3D surface bulging is captured from a passive multi-camera setup. Different arm muscles, such as biceps and triceps, get activated as a result of changes in the direction and magnitude of the external force on the hand. They propose a hybrid model, by fitting a linear regression model for variations due to body shape and external forces, and a non-linear regression model for more complex variations due to body pose. Figure 13.4 shows the variation of surface deformations produced by the model by varying the body shape parameters β and the addition of external force vector γ.

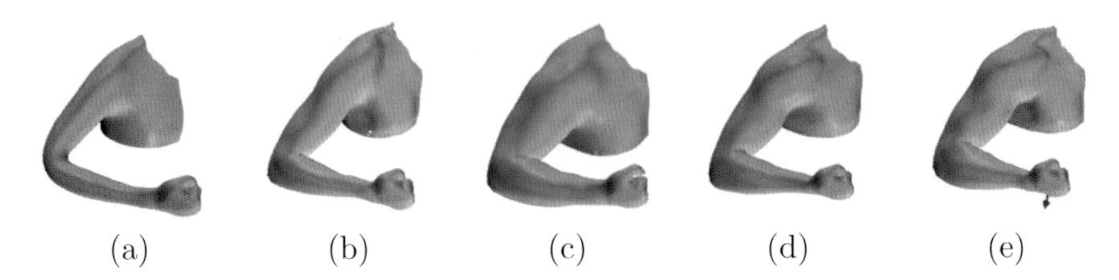

Figure 13.4: Statistical deformation model of the human arm. A simple pose-space deformation produces skinning artifacts (a) which are corrected by residual transformations learned by the model. The body shape parameters are varied between normal (b), high BMI (c), and high muscularity ratio (d). In (e), an external force vector is added to the muscular arm in (c) that sharply shows muscle-bulges. Figures adapted from [Neumann et al. 13a].

It is possible to extend such deformation models to account for specific muscle groups. In a later work [Neumann et al. 13b], Neumann et al. perform a sparse-matrix factorization on the meshes showing the second layer of residual transformations, after pose-specific deformation is subtracted. By concatenating these residual displacement vectors into an animation matrix $\mathbf{M}$, a basis of deformation components $\mathbf{C}$ is discovered by factoring the matrix $\mathbf{M}$.

$$\mathbf{M} = \mathbf{WC}. \tag{13.5}$$

It is desirable to obtain the deformation components $\mathbf{C}$ as corresponding to the actions of specific muscle groups that act in a localized manner on the surface. One option to achieve such an effect is to take the appropriate basis $\mathbf{C}$ from user-input. However, it is possible to achieve an approximate effect by giving appropriate priors on $\mathbf{C}$ and $\mathbf{W}$ for the decomposition. Specifically, the components $\mathbf{C}$ can be expected to be sparse and localized on specific surface regions. Neumann et al. [Neumann et al. 13b] use the sparsity inducing $L1$-norm as a regularizer on $\mathbf{C}$ in Eq.(13.5). Their paper discusses using other priors for locality and well-behaving weights. When run on a subset of the captured arm-muscle dataset of [Neumann et al. 13a] (bicep-curl movements of one subject), the decomposition components of their method loosely correspond to physical muscle-groups such as biceps, triceps, etc. This model can be used to enhance specific muscle effects (Figure 13.5).

In conclusion, data-driven deformation models learned from mesh sequences captured from the real world are a powerful artistic tool. They not only reduce the modeling time of the artists but also achieve more physically realistic rigs that can be manipulated easily.

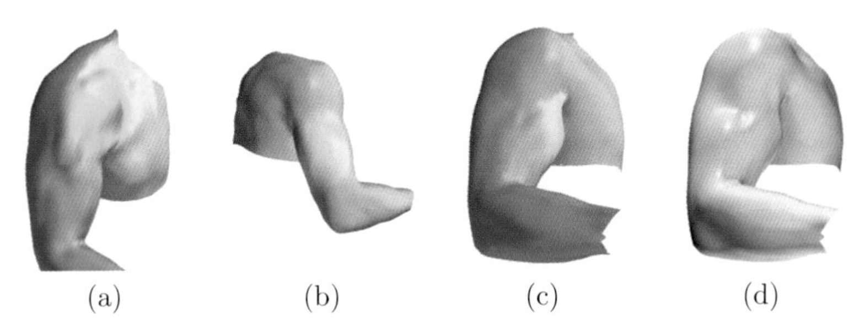

Figure 13.5: Sparse localized deformation components found on a subset of the muscle exercise dataset. (a,b,c) show localized effects on the deltoid, triceps, and biceps regions, respectively. In (d), the biceps bulge is exaggerated by amplifying the component in (c).

13.6 Summary

In this chapter, various methods for building virtual 3D characters from performance-captured mesh sequences were discussed, such that they can be controlled and edited by artists. With rising trends in modern 3D sensing technology and large-scale machine learning from big data, the deformation quality in the virtual characters can be expected to improve greatly. Historically, animation editing methods based on geometric templates and on statistical deformation models have evolved along different paths, taking inspiration from the mathematical disciplines of differential geometry and probabilistic modeling, respectively. In the future, these methods are likely to converge into a single unified discipline. Statistical deformation models built from real 3D motion datasets will also be used increasingly by professionals for real-world visual effects and computer games. These models will be built not only on traditional artistic toolkits such as bone skeletons, but also on more anatomically accurate skeletons and muscle models. In conclusion, capturing real-world 3D motion and building data-driven deformation models will help a professional 3D artist or computer programmer to create more realistic virtual characters.

14
Statistical Human Body Modeling

Stefanie Wuhrer, Leonid German, and Bodo Rosenhahn

14.1 Introduction

Human body models are required in many applications, such as creating realistic animations for gaming and movie productions, or creating accurate simulations for quality control in ergonomic design. In these application scenarios, the user wants to control the body shape and posture using a small number of intuitive parameters. To achieve this goal, application-dependent, manually generated control parameters can be designed. This is a common technique when rigging a character for a movie production (Chapter 13). However, generating these parameters manually is time-consuming and expensive.

A fully automatic alternative to this manual approach is to learn a small set of control parameters from a large database of human body scans using machine learning techniques. This method offers the advantage of learning application-dependent shape and posture variations without requiring extensive manual input.

However, performing statistics on a set of 3D scans is a challenging problem. To statistically analyze the shapes, a distance measure between pairs of shapes needs to be defined. This is challenging even in the case where the same subject is scanned multiple times in the same posture, because the scans are corrupted by noise and missing data, and because different scans contain different numbers of points. This problem is further complicated, because scans of different subjects scanned in different postures exhibit a large variation in body shape. Figure 14.1 depicts some noisy body scans that have very different shapes. This chapter outlines some methods to process these scans and to learn parametric models of body shape and posture.

To analyze 3D models of human bodies used in product design applications and entertainment, databases of 3D scans of different subjects are

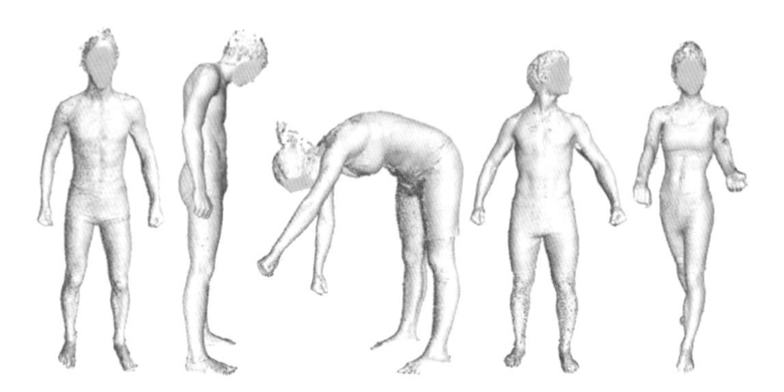

Figure 14.1: Human body scans of different subjects in different postures may contain large shape variation and be corrupted by noise and holes. Models from the MPI database [Hasler et al. 09c].

required. Therefore, several large databases have been collected. The first large 3D database of human models was the Civilian American and European Surface Anthropometry Resource (CAESAR) [Robinette et al. 99], whose goal was to provide accurate anthropometric data for various design applications. This commercially available database contains over 4500 subjects in three postures each and was collected at multiple locations across North America and Europe. Similar datasets have been subsequently compiled all over the world in order to find typical, representative shapes of a population to improve the sizing systems used for product design (e.g., the SizeGERMANY survey [SizeGermany 07]). All of these surveys have the goal to analyze variation in body shape only and, therefore, collect scans in very few standard postures. To use statistical models for animation, subjects need to be scanned in many different postures in order to allow for the automatic extraction of posture variations from the database. For these applications, Hasler et al. [Hasler et al. 09c] compiled a database of over 100 subjects in up to 35 postures each. Unlike the previously discussed databases, it is freely available for research purposes. Figure 14.1 shows some scans from this database.

14.2 Overview

This section gives a brief overview of the challenges related to human body modeling. The goal is to learn a small number of parameters to control the body shape and posture of a human body model from a large database of human body scans acquired in different postures.

Once a 3D model database of human body scans has been acquired, each model is represented by a set of points that are possibly connected by triangles. At this stage, the different body scans cannot be directly compared, because different models can have different numbers of vertices, and they are corrupted by noise and missing data. Furthermore, no information is available about which points or triangles correspond to which body part. To compare the models, intrinsic correspondence information needs to be established. Computing dense correspondences between 3D models is a challenging problem that continues to receive considerable attention in the computer vision and computer graphics research communities [van Kaick et al. 11, Tam et al. 13].

To compute correspondences between human models, *templates* are commonly used. Templates describe the rough shape and deformation behavior of a human model and are usually represented by a rigged triangle mesh (Chapter 13). This technique was first proposed by Allen et al. [Allen et al. 03] to compute correspondences between human models in a standard posture and subsequently extended in numerous works to allow for posture variation [Anguelov et al. 05, Hasler et al. 09c]. To initialize the template fitting, anatomical markers corresponding to joint positions are commonly used. Acquiring these positions is time-consuming and expensive. For instance, acquiring the CAESAR database which includes 73 anatomical marker positions per scan took four years and cost $6 million.[1] To avoid this cost, data-driven methods learn localized information about the geometry of the marker positions and use this learned information to predict marker positions on a newly available scan automatically [Azouz et al. 06, Wuhrer et al. 10]. These automatically predicted landmarks have been shown to be sufficiently accurate to allow for the computation of reliable correspondences for most scans [Wuhrer et al. 11]. Section 14.3 discusses methods that predict marker positions and methods that fit a template to a scan in more detail.

Once a database of human body models is in correspondence, shape variations across different subjects and postures can be analyzed using machine learning techniques. To this end, each shape is represented as a vector consisting of the x-, y-, and z-coordinates of its n vertices in an arbitrary but fixed order. This representation of each model in $\mathbb{R}^{3n}$ allows to compute distances between shapes as vector distances. Machine learning techniques aim to represent each human model in a low-dimensional *shape space*. Allen et al. [Allen et al. 03] analyze a database of processed human scans captured in a standard posture using principal component analysis (PCA), which finds a shape space as the orthogonal linear sub-space of $\mathbb{R}^{3n}$ that captures the largest proportion of the shape variability present in the database.

[1] http://www.sae.org/standardsdev/tsb/cooperative/caesumm.htm

To facilitate the simultaneous analysis of shape and posture variations, Anguelov et al. [Anguelov et al. 05] couple this shape space with an additional linear shape space that models the deformation of a single subject captured in multiple postures to allow for posture variation. The advantage of this method, which is called SCAPE, is that the shape and posture spaces can be modeled using linear mappings, and shape and posture can be analyzed independently. Section 14.4 discusses the commonly used SCAPE model that represents human shape and posture changes in more detail. The main disadvantage of this model is that the posture variation is learned from a single subject.

In order to extend the analysis of posture variations to a population of subjects, Allen et al. [Allen et al. 06] propose the use of multiple subjects captured in multiple postures to learn a correlated shape space capturing body shape and posture variations. A substantial drawback is that transforming a model from $\mathbb{R}^{3n}$ to this shape space requires a non-linear transform. This was addressed by Hasler et al. [Hasler et al. 09c] with the use of a rotation-invariant encoding of the models to find a linear shape space that encodes human body shape and posture in a correlated way. This method allows to model muscle deformations accurately using a linear shape space, which has applications in fitting human models to scans acquired with clothing [Hasler et al. 09b]. However, this model does not allow to control human shape and posture variations independently. To obtain more control, Hasler et al. [Hasler et al. 10a] propose the use of a bilinear shape space to estimate the 3D shape of a human model from a single image. Recently, Chen et al. [Chen et al. 13b] combined such a bilinear model with the SCAPE model to allow for intuitive parameters to control posture variation while achieving realistic shape deformations near joints.

In some applications, such as in ergonomic design, it is desirable to analyze the variations due to body shape differences independently of posture changes of the models. To achieve this goal, Wuhrer et al. [Wuhrer et al. 12] propose to perform PCA on a shape representation based on localized Laplace coordinates of the mesh.

Statistical models offer the advantage of allowing the control of realistic human models through few parameters. Applications include reconstructing 3D human models from possibly dynamic RGB or depth image data [Balan et al. 07, Balan and Black 08, Guan et al. 09, Hasler et al. 10a, Weiss et al. 11, Helten et al. 13], controlling human models using a small set of semantic parameters, such as one-dimensional measurements or body weight [Allen et al. 03, Wuhrer and Shu 13, Rupprecht et al. 13], and editing image and video data [Zhou et al. 10, Jain et al. 10].

14.3 Parameterization of Human Body Models

This section discusses how to process raw human body scans, such that they can subsequently be used for statistical analysis. This is a difficult problem, where the goal is to correspond two surfaces T and S containing $n^{(T)}$ and $n^{(S)}$ vertices, respectively. When trying all possible combinations, the search space for solving the correspondence problem has size $O\left(n^{(T)}n^{(S)}\right)$. In practice, this problem is further complicated by acquisition artifacts, such as noisy and incomplete data. Usually it is difficult to filter the noise and fill the holes in the acquired data, because the geometry of the data can be complex.

To remedy these problems, a template T consisting of a rigged triangle mesh that represents the shape and deformation behavior of a typical human body is often used to reduce the search space. The task of computing a correspondence between a population of human body shapes $S^{(1)}, \ldots, S^{(m)}$ represented by possibly noisy and incomplete point clouds can now be solved by deforming T to each shape $S^{(i)}$. After this step, there are m deformed versions of T (and hence, correspondences are known), and the i-th deformed version of T is close to $S^{(i)}$. This approach has the advantage of leading to m complete surfaces in correspondence even when the original acquisitions are incomplete, without solving the challenging task of filling holes in scan data explicitly.

This template fitting approach uses a non-rigid iterative closest point (ICP) [Besl and McKay 92] algorithm and was first proposed by Allen et al. [Allen et al. 03]. Non-rigid ICP approaches proceed by repeatedly computing the current correspondences between T and a scan S using nearest neighbors in $\mathbb{R}^3$ and use these correspondences to smoothly deform T to S. The methods stop when the alignment no longer changes significantly. Hence, these approaches need a good initialization to avoid getting trapped in local minima.

To find a good initial shape alignment, anthropometric landmarks are commonly used. In a first step, the landmarks are used to roughly align the initial shape with the data by using a skeleton model. Afterwards, the landmarks give a small set of corresponding points on T and on S which can be used to guide the deformation from T to S. To fit T to S, two fitting steps are commonly used. First, the landmarks are used to refine the posture of the skeleton of T, such that the deformed version of T is close to S. Second, the vertices of T are deformed to be close to S using non-rigid ICP.

Regarding the first step, to fit the skeleton of T such that the posture of T is close to the posture of S, the method takes advantage of the rigging information of T. That is, a skeleton model of T consisting of b bones along with a set of rigging weights describe how the vertices of T move with respect to the skeleton. Let $\mathbf{A}_k$ denote the 3×4 transformation matrix that encodes the rigid transformation of the k-th bone. Note that $\mathbf{A}_k$ depends on few parameters since most joints only allow a rotation. Let $\mathbf{l}_j^{(T)}$ and $\mathbf{l}_j^{(S)}$ denote the corresponding landmarks of T and S, respectively. Furthermore, in the following, let $\tilde{\mathbf{p}}$ denote the homogeneous coordinates of the 3D point $\mathbf{p}$. Landmark $\mathbf{l}_j^{(T)}$ can be deformed to $\sum_{k=0}^{b-1} w_{jk}\mathbf{A}_k\left(\tilde{\mathbf{l}}_j^{(T)}\right)$, where w_{jk} is the rigging weight for the k-th bone and the j-th landmark of T. A commonly used approach to fit the posture of T to the posture of S is to use a variational approach to solve for the bone transformations $\mathbf{A}_k$, such that

$$E_{\text{lnd}}(\mathbf{A}_k) = \sum_j \left(\sum_{k=0}^{b-1} w_{jk}\mathbf{A}_k\tilde{\mathbf{l}}_j^{(T)} - \tilde{\mathbf{l}}_j^{(S)}\right)^2 \tag{14.1}$$

is minimized. This is also referred to as *skeleton transfer*.

After the posture of T has been initialized to be close to the posture of S, in the second fitting step, the vertices of T are deformed to be close to the surface of S using non-rigid ICP. This step is also called *shape alignment*, and it deforms each vertex $\mathbf{p}_j$ of T using a 3×4 transformation matrix $\mathbf{A}_j$.

Different methods allow for different classes of transformation matrices $\mathbf{A}_j$. To allow for elastic deformations, affine transformation matrices are often allowed. Alternatively, to restrict the deformations to be piecewise rigid, rigid transformations are allowed per vertex. The goal is to fit T to S while preserving the overall shape of the template surface. This is achieved by minimizing the energy

$$E_{\text{fit}} = \omega_{\text{data}}E_{\text{data}} + \omega_{\text{smooth}}E_{\text{smooth}}, \tag{14.2}$$

with respect to the deformations $\mathbf{A}_j$. The term

$$E_{\text{data}} = \sum_j \left(NN^{(S)}\left(\mathbf{A}_j\tilde{\mathbf{p}}_j\right) - \mathbf{A}_j\tilde{\mathbf{p}}_j\right)^2 \tag{14.3}$$

is an energy that pulls the vertices of T toward the sensor data S, where $NN^{(S)}\left(\mathbf{A}_j\tilde{\mathbf{p}}_j\right)$ is the nearest neighbor of the transformed vertex $\mathbf{A}_j\tilde{\mathbf{p}}_j$ on S. The term E_{smooth} encourages close-by vertices to have similar deformations, e.g., rotation around a common axis as opposed to translation by different distances. Thereby transformations which incur low E_{smooth} penalties are more likely to preserve the overall shape of the template. $E_{\text{smooth}} = \sum_j \sum_{(j,k)\in E} |\mathbf{A}_j - \mathbf{A}_k|_F^2$ is commonly used, where E is the set of edges of T, and where $|\cdot|_F^2$ denotes the squared Frobenius norm.

It is noteworthy that the location of the minimum is influenced only by the ratio of the weighting factors, e.g.,

$$E_{\text{fit}} = \omega_{\text{data}}E_{\text{data}} + \omega_{\text{smooth}}E_{\text{smooth}} \propto E_{\text{data}} + \frac{\omega_{\text{smooth}}}{\omega_{\text{data}}}E_{\text{smooth}}. \tag{14.4}$$

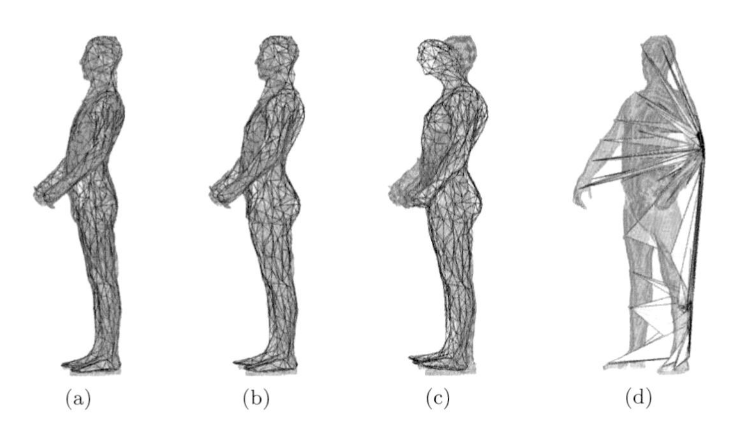

Figure 14.2: The scan is shown in gray and the fitting result as wireframe. (a) Successful fit with normal values for ω_{data} and ω_{smooth}. (b) A fit with normal smoothness, data not considered. (c) A fit with very high ω_{smooth}. (d) A fit with very high ω_{data}.

The influence of the two fitting terms E_{data} and E_{smooth} is demonstrated in Figure 14.2. Figure 14.2(a) shows a successful template fitting result; ω_{data} and ω_{smooth} were well-balanced in this case. Figure 14.2(b) shows the result if E_{data} is not considered during the fitting. The mesh as a whole was affinely transformed using landmarks, and the pose adjusted to the scan, but the mesh was not fitted to the scan surface. Figure 14.2(c) shows a fitting with an exaggerated value for ω_{smooth}. A translation and centering occurred, but the pose of the mesh is unchanged, as that would increase the smoothness term which is penalized strongly in this case. Minimizing the distance between landmarks on the scan and the template, the mesh was shrunk slightly and transformed in other ways that do not increase E_{smooth}, leaving it a little warped. Figure 14.2(d) shows the result if a disproportionately large value is chosen for ω_{data}. The mesh has completely lost its form, as most vertices are simply attracted to the closest points on the scan. Certain vertices, however, are still attracted to landmarks; their adjacent faces are seen overlapping the scan surface.

One remaining question is how to find the anthropometric landmarks $\mathbf{l}_j^{(S)}$ on a scan S automatically. This can be achieved by taking advantage of the observation that most anthropometric landmarks represent skeletal features and correlate with geometric surface features. This allows the use of machine learning techniques to learn statistical properties about the geometry of these landmark locations from a training database of human models with annotated landmark positions.

To this end, the landmarks $\mathbf{l}_j$ are considered as nodes V of a graph and are linked by edges E in a skeleton-like structure. For each landmark, a node descriptor is computed, and a probability distribution is then fitted to the descriptor values of the landmark location over all shapes of a training set. Commonly used node descriptors include surface curvatures, spin images [Johnson and Hebert 97], and measures related to the lengths of or the areas enclosed by geodesic isolines at fixed distances from the landmark [Sun and Abidi 01]. Similarly, for each edge, an edge descriptor is computed, and a probability distribution is then fitted to the descriptor values of the edges over all shapes of a training set. Commonly used edge descriptors include the length or orientation of the edge measured in a pre-aligned pose of the shape.

This learned information is then used to model a Markov random field (MRF) [Duda et al. 01], where a random variable $\mathbf{x}$ associated with node $\mathbf{l}_j$ is considered conditionally independent of all other variables given the variables associated with the neighbors of $\mathbf{l}_j$. A MRF models the joint probability of a set of random variables $\mathbf{x}_j$ corresponding to potential locations of $\mathbf{l}_j$ as

$$p(X) = \frac{1}{Z} \prod_{\mathbf{l}_j \in V} \phi_j(\mathbf{x}_j) \prod_{(\mathbf{l}_j, \mathbf{l}_k) \in E} \psi_{j,k}(\mathbf{x}_j, \mathbf{x}_k), \tag{14.5}$$

where $\phi_j(\mathbf{x}_j)$ is the likelihood that vertex $\mathbf{x}_j$ corresponds to landmark $\mathbf{l}_j$ (computed using the previously learned probability distribution over node descriptors), $\psi_{j,k}(\mathbf{x}_j, \mathbf{x}_k)$ is the joint likelihood that vertices $\mathbf{x}_j$ and $\mathbf{x}_k$ correspond to landmarks $\mathbf{l}_j$ and $\mathbf{l}_k$ (computed using the previously learned probability distribution over edge descriptors), and Z is a normalizing factor. This model allows to find the vertices $\mathbf{x}_j$ on a new scan that maximize the joint probability $p(X)$. These are the most likely positions of the landmarks $\mathbf{l}_j$ on this scan according to the geometric descriptors used to model the nodes and edges.

Figure 14.3 shows some manually picked landmarks in red and automatically predicted landmarks in blue. The algorithm used to compute these landmarks used 200 models of the MPI database in 35 different postures for training and is described in more detail in Wuhrer et al. [Wuhrer et al. 10].

14.4 Statistical Analysis of Human Body Models

This section discusses methods to analyze shape variations of a database of human scans that are in correspondence. Analyzing shape variations allows to find a small number of parameters that can be used to control human body shape and posture changes.

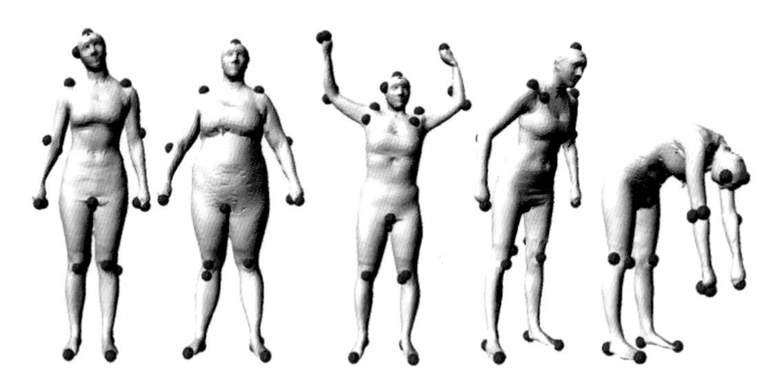

Figure 14.3: Manually picked (red) and automatically predicted (blue) landmarks on models from the MPI database. For correctly predicted vertices, only the blue landmark is visible.

Let $S^{(1)}, \ldots, S^{(m)}$ denote a database of human body shapes consisting of n vertices that are each in correspondence and that have been aligned rigidly to be located in the same coordinate system. To align a population of shapes rigidly, generalized Procrustes analysis (described in more detail in Dryden and Mardia [Dryden and Mardia 02]) can be used. The shapes $S^{(i)}$ can be represented as vectors $\mathbf{s}^{(i)} = \left[x_1^{(i)}, y_1^{(i)}, z_1^{(i)}, \ldots, x_n^{(i)}, y_n^{(i)}, z_n^{(i)}\right]^T$ in $\mathbb{R}^{3n}$, where $(x_j^{(i)}, y_j^{(i)}, z_j^{(i)})$ are the coordinates of the j-th vertex of $S^{(i)}$. It is straightforward to convert between the representations $S^{(i)}$ and $\mathbf{s}^{(i)}$, and the new representation $\mathbf{s}^{(i)}$ allows to compute distances between pairs of shapes as Euclidean distances in $\mathbb{R}^{3n}$.

To analyze variations in body shape only, PCA can be used. PCA is an approach that aims to reduce the dimensionality of a dataset given as vectors in $\mathbb{R}^{3n}$ to $\mathbb{R}^d$ with $d < \min(m, 3n)$ using a linear transformation, while aligning the reduced coordinate system along the main modes of variation of the data. More specifically, the aim is to express the data vectors $\mathbf{s}^{(i)}$ as

$$\mathbf{s}^{(i)} = \left(\sum_{j=1}^{d} a_{ij}\mathbf{e}_j\right) + \bar{\mathbf{s}}, \tag{14.6}$$

where $\mathbf{e}_j$ are a set of basis vectors, a_{ij} are scalar weights, and $\bar{\mathbf{s}} = \frac{1}{m}\sum_{i=1}^{m}\mathbf{s}^{(i)}$ is the mean body shape. This way of reducing the dimensionality can be expressed in matrix form as

$$\mathbf{S}^{(\text{cent})} = \mathbf{AE}, \tag{14.7}$$

where $\mathbf{S}^{(\text{cent})}$ is a $(m \times 3n)$ matrix that contains the centered data $\mathbf{s}^{(i)} - \bar{\mathbf{s}}$ as its rows, $\mathbf{A}$ is a $(m \times d)$ matrix with entries a_{ij}, and $\mathbf{E}$ is a $(d \times 3n)$ matrix that contains the basis vectors $\mathbf{e}_j$ as its rows.

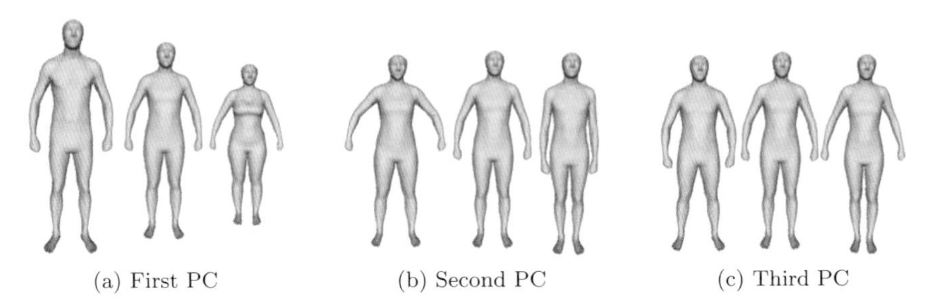

Figure 14.4: Shape variations along the first three principal components for models of the MPI database.

It remains to compute the basis vectors $\mathbf{e}_j$ and the weight matrix $\mathbf{A}$. PCA computes the basis vectors with the help of an eigendecomposition of the $3n \times 3n$ sample covariance matrix

$$\boldsymbol{\Sigma} = \frac{1}{m-1}\sum_{i=1}^{m}(\mathbf{s}^{(i)} - \bar{\mathbf{s}})(\mathbf{s}^{(i)} - \bar{\mathbf{s}})^T. \tag{14.8}$$

Let $\lambda_1, \ldots, \lambda_d$ denote the largest eigenvalues of Σ in non-increasing order. The eigenvectors corresponding to λ_j are used as new basis vectors $\mathbf{e}_j$. These basis vectors are orthogonal and have the property that $\mathbf{e}_j$ corresponds to the j-th largest direction of data variability. The coordinate axes $\mathbf{e}_j$ are called the principal components (PC) of the data. Given $\mathbf{S}^{(\mathrm{cent})}$ and $\mathbf{E}$, $\mathbf{A}$ can be computed. Dryden and Mardia [Dryden and Mardia 02] and Duda et al. [Duda et al. 01] give more extensive discussions of PCA and its properties.

PCA can be used to visualize the main variations of a population of shapes by sampling points, denoted by $\mathbf{a}$, along the coordinate axes in the reduced space $\mathbb{R}^d$ and by computing and visualizing the shape vectors $\mathbf{s}$ as $\mathbf{s} = \mathbf{aE}+\bar{\mathbf{s}}$. Figure 14.4 shows the shape variations computed along the first three principal components for all models of the MPI database captured in standard standing posture. This approach was first applied to analyze the shape variations of 3D human body shapes by Allen et al. [Allen et al. 03].

PCA is a global method, that is, each principal component influences each vertex coordinate. This is not always desirable. For instance, to analyze both body shape and posture variations, it is better to segment the body into several parts and to perform shape analysis separately on different parts.

Anguelov et al. [Anguelov et al. 05] proposed SCAPE, which is an extension of PCA that allows to analyze both shape and posture variations independently. Shape variations are analyzed by performing PCA over a dataset of different human subjects captured in a standard posture. Posture variations are analyzed using a dataset $P^{(1)}, \ldots, P^{(k)}$ of a single human subject captured in k different postures. The two types of variations are then linearly combined to allow for both body shape and posture changes.

To analyze posture variations, SCAPE decomposes the human body into different body parts. Let $r_j^{(i)}$ denote the j-th body part of $P^{(i)}$, and let $P^{(1)}$ denote the model acquired in the same posture that was used to analyze body shape variations. SCAPE computes for each training posture and each body part a rigid transformation that aligns $r_j^{(1)}$ to $r_j^{(i)}$. This rigid alignment is not sufficient to deform $r_j^{(1)}$ to $r_j^{(i)}$, as posture changes cause non-rigid deformations, such as muscle bulging. To model these effects, an affine transformation is computed for each triangle in $r_j^{(1)}$ that deforms this triangle in $r_j^{(1)}$ to its corresponding triangle in $r_j^{(i)}$. It remains to link the body shape differences that occur due to posture changes to a small set of skeleton parameters. This is achieved by computing a linear regression between a small set of skeleton parameters and the computed rigid and affine transformations.

PCA does not necessarily give semantically meaningful variations. The variations are merely the most significant ones in terms of data variability. To remedy this, a regression between the PCA space and semantically meaningful parameters, such as body measurements, is commonly learned [Allen et al. 03]. This regression can then be used to generate synthetic human body models with desired measurements provided as input.

In the following, some typical results are shown that can be obtained by applying such a regression method within the learned PCA-space. Here, the intuitive input parameters are body height and weight. All results are obtained using an iterative regression with the mean shape as starting position.

Figure 14.5 shows the regression results of synthesizing a typical male body shape with input parameters height 180 cm and weight 80 kg, and a typical female body shape with input parameters height 170 cm and weight 50 kg. Note that realistic body shapes are generated.

Figure 14.6 shows the influence of changing the height and weight parameters separately. The left side shows the influence of changing the height for male body shapes. Results are synthesized for heights of 160 cm and 180 cm when keeping the weight fixed at 70 kg. The right side shows the influence of changing the weight parameter for female body shapes. Results are synthesized for weights of 40 kg and 80 kg when keeping the height fixed at 170 cm. Note how these input parameters lead to realistic changes in body proportions for all synthesized body shapes.

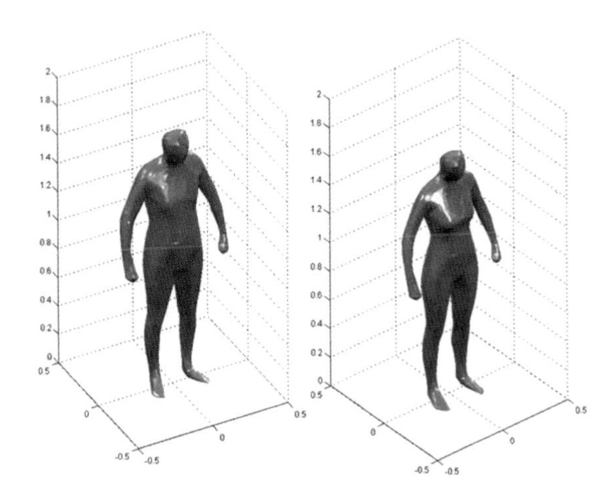

Figure 14.5: Results of regression based on body height and weight. From left to right: Male with height/weight of (180 cm, 80 kg) and female with height/weight of (170, 50 kg).

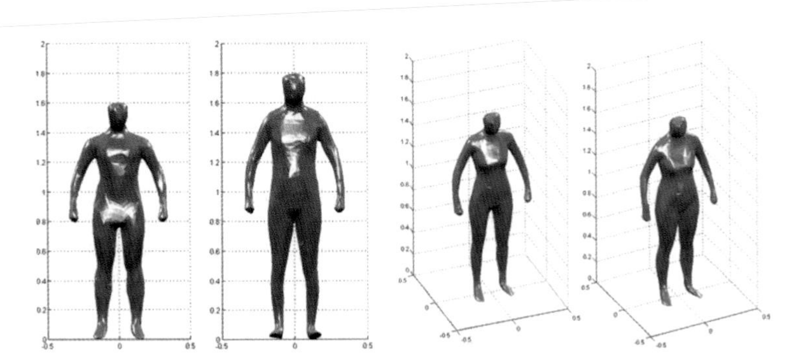

Figure 14.6: Influence of the height and weight parameters on regression results. Left: Regression of a male body shape with input parameters height 160 cm and 180 cm, respectively, and weight 70 kg. Right: Regression of a female body shape with input parameters height 170 cm and weight 40 kg and 80 kg, respectively.

14.5 Summary

This chapter presented commonly used methods to parameterize and statistically analyze human body models. To parameterize the body models, a template model can be deformed to the data using a variational technique that jointly optimizes a data fitting energy and a deformation smoothness energy. To use a template fitting method, the template needs to be aligned to the data. This can be achieved with the help of a sparse set of anthropometric landmarks, which can be predicted automatically with the help of machine learning techniques.

PCA, which is a commonly used statistical method to analyze a population of human body models, was discussed. It finds the main modes of variation of the population. The result of this analysis can be used in various applications, such as the presented application of predicting synthetic body models from a sparse set of intuitive measurements.

An interesting direction for further research is to statistically analyze human body shape and posture changes as the subjects perform different types of motion, such as walking or running.

15
Cloth Modeling

Anna Hilsmann, Michael Stengel, and Lorenz Rogge

15.1 Introduction

Realistic looking garments and clothes are an important component when it comes to modeling reality and producing photorealistic images including people. In modern movies, many scenes contain completely virtual actors driven by motion capture data, or real actors are visually augmented with virtual clothes in a post-process. Thus, realistic virtual clothing and proper models for fabric behavior and appearance are very important. Cloth modeling and simulation is a challenging task, because cloth deformation and drapery exhibit many degrees of freedom, and wrinkles produce complex deformations and shading. Furthermore, due to the complex structure of fibers, modeling realistic reflection properties of cloth is very difficult. Yet, these complex details are essential for realistic appearance of the virtual clothes. Different approaches exist for modeling and rendering realistic looking pieces of clothing with correct or plausible behavior and appearance, ranging from sophisticated and very complex methods, accurately simulating all fine details with physics-based methods, to real-time methods that focus on producing visually plausible rather than accurate results.

The field of cloth modeling and simulation has a very long tradition in computer graphics and there exist a number of surveys and state-of-the-art reports [Gibson and Mirtich 97, House and Breen 00, Magnenat-Thalmann et al. 04, Nealen et al. 05]. This chapter presents an overview of cloth modeling and simulation techniques as well as realistic appearance modeling of clothing, focusing on recent trends, especially on methods that are directly inspired by the *real world* to properly model physical characteristics, deformation behavior as well as appearance of clothing. The chapter follows the pipeline of modeling and rendering photorealistic pieces of cloth (Figure 15.1). Section 15.2 starts with recent advances in modeling the geometry as well as mechanical parameters from real-world cloth samples, followed by Section 15.3 describing cloth simulation and deformation modeling approaches. Section 15.4 focuses on modeling appearance and reflection properties of cloth.

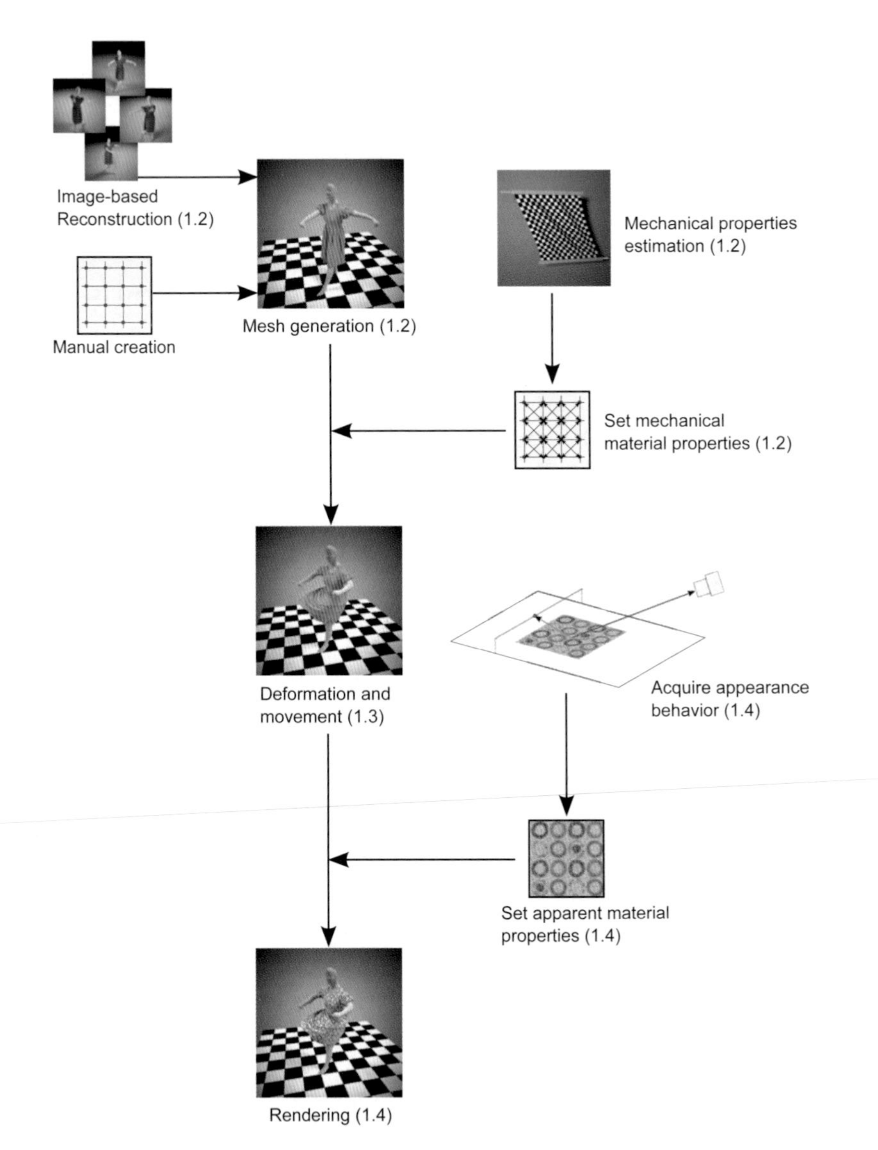

Figure 15.1: Data-driven cloth simulation pipeline.

15.2 Cloth Geometry and Mechanics Modeling

The main goal of cloth modeling is to simulate a piece of cloth as realistically as possible, ideally indistinguishable from real cloth. Usually, modeling clothes relies on a 3D model, e.g., a polygonal mesh, and folds as well as dynamics of clothes are synthesized. Traditionally, the parameters for the synthesis process are calculated with physics-based models based on a character's pose, motion, external forces as well as material properties (Section 15.3). In these models, the mechanical properties of a specific material are modeled by a number of parameters, describing its resistance to bending, stretching, and shearing. To fully exploit the capabilities of these methods, and to realistically simulate the specific characteristics of different cloth materials, the parameters must be tuned with great care, which is a tedious process if done manually by a professional artist. To achieve a realistic modeling of cloth, several *real-world*-driven (or data-driven) methods have been proposed as an alternative to pure simulation. These methods are directly inspired by the real world to extract the geometry of cloth as well as intrinsic material parameters from visual and/or other sensors. This section gives an overview on modeling cloth geometry and physical properties from real-world samples.

Geometry Modeling

Early approaches focused on capturing and modeling the geometry of moving garments and clothing at high detail from multi-view image or video data (Section 8.4). These approaches generate a detailed and temporally consistent animated mesh model from the input data. For this purpose, reliable correspondences must be established between the different camera views and time steps. One approach to establish these correspondences is to use markers on the piece of clothing [Guskov et al. 03, Scholz et al. 05, White et al. 07] (Figure 15.2). A similar approach is presented in Section 13.5 to reconstruct the shape of a human actor. The nature of the marker layout allows a unique identification of surface patches across the different camera views and timesteps. In order to remove the constraints of a markered cloth sample, which needs to be manufactured specifically for the capturing process, other researchers rely on natural image features, e.g., SIFT or SURF features (Section 7.2), to track surface points over different views and time in multi-view video setups [Pritchard and Heidrich 03, Hasler et al. 06]. However, relying on image features still puts constraints on the captured piece of clothing, as these features require a strongly and preferably anisotropically textured material. This limits the applicability of feature-based approaches to the capturing of highly textured pieces of clothing. Geometry modeling from untextured *off-the-shelf* clothing has been addressed by Bradley et al. [Bradley et al. 08]. By combining imperfectly reconstructed mesh models of every frame in a video sequence with a template mesh, a consistent and complete mesh animation of the captured clothing can be modeled.

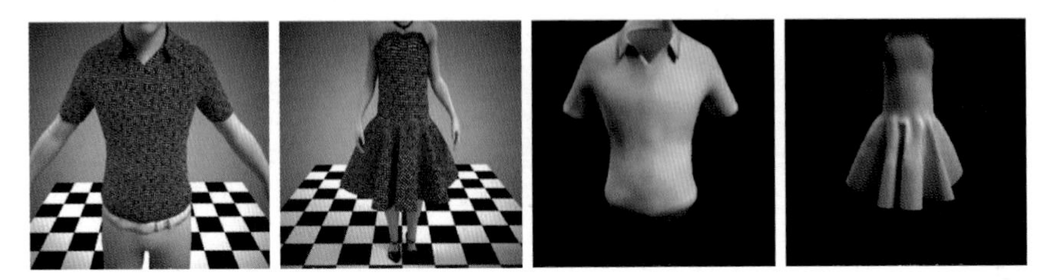

Figure 15.2: Marker-based garment reconstruction according to [Scholz et al. 05]. Left: Input images of one camera in a multi-view setup. Right: Reconstructed 3D models.

Modeling Mechanical Properties

Besides geometry, other approaches try to explicitly extract cloth material parameters describing the physical properties of a piece of cloth directly from visual data [Bhat et al. 03, Kunitomo et al. 10, Bouman et al. 13]. For example, Bhat et al. [Bhat et al. 03] captured video data of a person wearing different pieces of clothing to estimate the location of cloth folds from gradient vector fields in the video frames. The retrieved information is used to determine stiffness parameters of a mass-spring model [Baraff and Witkin 98] (Section 15.3). Similarly, Bouman et al. [Bouman et al. 13] proposed to automatically analyze videos of fabrics moving under various unknown wind forces to estimate stiffness and area weight parameters.

While purely vision-based approaches appeal through a simple setup, it is generally very difficult to measure the true material properties accurately without physical contact and control over external forces. Moreover, it is difficult to separate internal from external parameters. Therefore, recently, data-driven methods have emerged in the literature that measure cloth parameters by applying physical forces to a sample piece of cloth in very controlled conditions [Wang et al. 11a, Miguel et al. 12, Miguel et al. 13]. The external forces are usually measured by force sensors, and deformation is captured by computer vision systems. The information on deformation under controlled forces is then used to fit the parameters of a cloth deformation model to the data. Since these models rely on real-world test data, they can mimic the fabric behavior very realistically during simulation. For example, Wang et al. [Wang et al. 11a] presented a data-driven and piecewise linear model with 39 material parameters to approximate the non-linear, anisotropic deformation behavior of cloth in a continuum-based deformation model [Baraff and Witkin 98] (Section 15.3). By applying external forces to the cloth samples and capturing the planar deformations from camera images, they estimate stretching and bending parameters for ten different samples of cloth strongly varying

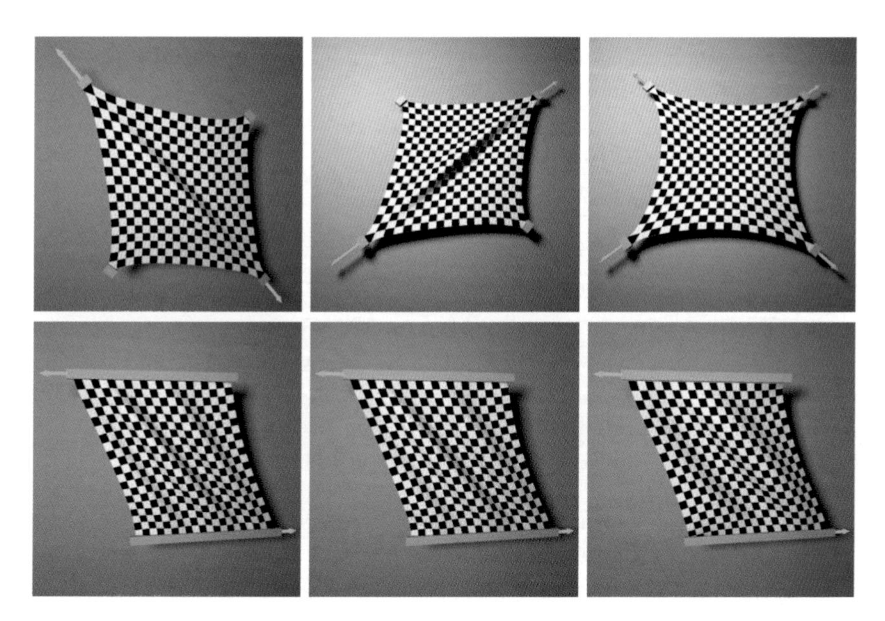

Figure 15.3: Different stretching and bending states of a cloth sample according to the measurement setup of Miguel et al. [Miguel et al. 13]. Top row: Different diagonal forces applied to a cotton sample. Lower row: Horizontal forces applied to cotton, leather and polyester (left to right).

in their stretching and bending behavior. Similarly, Miguel et al. [Miguel et al. 12] developed a system for the estimation of non-linear material parameters of different coefficients of cloth. In contrast to Wang et al., they reconstructed 3D deformation instead of 2D planar deformation using a stereo vision system (Chapter 2, and Figure 15.3). They later extended their work by developing a vision-based technique to estimate internal friction parameters from cloth samples [Miguel et al. 13].

15.3 Cloth Deformation Modeling and Simulation

Having a specific cloth model at hand, a physically correct simulation of garments requires sophisticated simulation techniques to model the deformation behavior. Physics-based methods aim at an accurate and physically correct modeling of deformation as well as wrinkling behavior. However, these methods usually need substantial computation time because highly detailed cloth simulation requires a model with a large number of degrees of freedom. In contrast to that, hybrid and example-based techniques focus on producing visually plausible and realistic deformation and wrinkling with lower computational complexity.

Physics-Based Deformation Models

A number of different physics-based cloth models have been developed, which estimate the motion and drapery of cloth based on its intrinsic material parameters, as well as external forces, e.g., induced by a character's pose, motion, gravity, collisions, etc. [Terzopoulos et al. 87, Breen et al. 94, Eberhardt et al. 96, Baraff and Witkin 98, Choi and Ko 02]. These models have been widely studied in the past decades, and detailed surveys have been presented in [House and Breen 00, Magnenat-Thalmann et al. 04, Magnenat-Thalmann and Volino 05, Nealen et al. 05, Choi and Ko 05]. The most widely used models are discrete particle models. These models represent a piece of cloth as a discrete set of particles (e.g., vertices in a polygonal mesh) forming the shape of its surface [Breen et al. 94, Eberhardt et al. 96, Baraff and Witkin 98, Bridson et al. 03, English and Bridson 08]. The particle positions are determined by forces applied to their topological neighborhood. One of the most popular particle-based models is the mass-spring model [Provot 95, Choi and Ko 02], which models the interaction between the particles as linear massless springs. The state of the system, i.e., the positions of all particles, is defined by their current positions $\mathbf{x}_i$, their masses m_i as well as their velocities $\mathbf{v}_i$. The forces $\mathbf{f}_i$ on each particle are computed based on external as well as internal forces, defined by the mesh topology. The motion of each particle is then driven by Newton's second law $\mathbf{f}_i = m_i\ddot{\mathbf{x}}_i$. Hence, calculating the position of all particles results in solving a system of ordinary differential equations $\mathbf{M}\ddot{\mathbf{x}}_i = \mathbf{f}(\mathbf{x}, \mathbf{v})$. The matrix $\mathbf{M}$ is a $3n \times 3n$ diagonal mass matrix, where n denotes the number of particles in the system. The solution has to be found by advanced numerical integration methods, which determine the computation performance. The various methods differ in the way the forces are calculated and the time integration is performed [Magnenat-Thalmann and Volino 05, Nealen et al. 05]. While particle systems are an intuitive and simple model for deformation, they are not necessarily accurate and need a large number of particles to model fine-scale deformations. Moreover, these models are known to suffer from *post-buckling instability* when wrinkles are shaped [Choi and Ko 02]. An assumption that compressing forces lead to buckling rather than compression (*immediate buckling assumption*) can solve this problem and lead to improved stability [Choi and Ko 02].

An alternative to particle models are continuum models, which are built on elasticity theory. Very popular models in this class are finite element models [Etzmuss et al. 03, Thomaszewski et al. 09, Volino et al. 09]. These models are based on energy-functions defined on a continuous model, which are approximated by polynomial patches. Compared to particle-based models, continuum-based models have the advantage that they are parameterization independent, i.e., they can theoretically be applied to arbitrary shaped unstructured meshes, and can reproduce the anisotropic deformation behavior of cloth more accurately. However, these advantages come at the cost of substantially more numerical operations than particle systems.

Hybrid and Data-Driven Deformation Models

With physics-based approaches, modeling and simulating very fine detailed folds and wrinkles requires a very large number of triangles and substantial computation times for both simulation and rendering [Choi and Ko 05]. Therefore, with these methods, usually there is trade-off between speed and quality. Thus, the challenge is to reduce the amount of numerical operations while maintaining visual quality. Exploiting the fact that fine wrinkles and buckles usually appear in local regions on the surface, adaptive remeshing techniques can be used to dynamically refine and coarsen polygonal meshes so that they automatically conform to the geometric and dynamic detail of the simulated cloth [Narain et al. 12]. This can reduce computation time to a fraction of the original time while keeping fine-scale wrinkles.

Various other techniques have been developed to improve the computational speed, giving up the mechanical and physical accuracy of the model and rather focusing on visual realism. Hybrid approaches tackle the trade-off between computation time and simulation quality by combining a low-resolution physical model with geometric detail synthesis. These approaches use a coarse deformable model with reduced detail to quickly generate a plausible simulated animation. To improve the final result, the animated model is augmented with synthetic geometric details, such as folds and wrinkles, which are typically very difficult to simulate in real-time. The position and shape of these synthetic details can for example be based on a deformation analysis of the low-resolution mesh [Hadap et al. 99, Decaudin et al. 06, Rohmer et al. 10, Müller and Chentanez 10], or directly on the image data, e.g., by extracting prominent folding edges [Popa et al. 09].

Recently, data-driven deformation models have become very popular to learn wrinkling and deformation models [Cordier and Magnenat-Thalmann 05, Kim and Vendrovsky 08, Wang et al. 10, Feng et al. 10, Guan et al. 12, Zurdo et al. 13]. The idea of these methods is to establish a mapping between low-resolution and highly detailed cloth models, in such a way that during rendering, a low-resolution mesh can be augmented with predicted fine details that have been learned a priori from highly detailed examples. For this purpose, an appropriately chosen parameterization needs to be defined, describing characteristic wrinkling behavior of the piece of clothing. For example, Kim and Vendrovsky [Kim and Vendrovsky 08] as well as Wang et al. [Wang et al. 10] focus on tight-fitting clothing for which they assume that wrinkling mainly depends on the articulated pose of a character. Hence, they model the shape of the clothing mesh as a function of pose, i.e., as a function of skeleton or animation parameters (Chapter 11). Similarly, Guan et al. [Guan et al. 12] addressed not only pose- but also body shape-dependency of wrinkling and drapery of clothing. Their approach learns a clothing model from highly detailed physics-based models of clothing on bodies of different shapes and in different poses. During rendering, based on a parameterized model of the human body (Chapter 14), describing its shape and pose, a piece of clothing is synthesized from the learned database. Using more abstract embeddings and directly establishing a relationship between low-resolution and high-resolution cloth deformations instead of assuming pose- or shape-dependence allowed several researchers to develop learning-based models, which are valid also for very loose pieces of clothing, e.g., dresses or skirts, or other types of cloth like flags [Feng et al. 10, Kavan et al. 11, Zurdo et al. 13]. Since temporally consistent cloth models at very high detail together with the required additional information (e.g., pose, body shape, external forces) are difficult to acquire from real cloth samples, these methods rely on highly detailed simulations, giving a known alignment between the test data. However, some efforts have been made to synthesize dynamic wrinkles based on real captured data [Ryan White 07].

Cloth Modeling at the Yarn Level

Common methods approximate cloth as an elastic sheet. In contrast, recent developments model cloth at high detail directly at the yarn level [Kaldor et al. 08, Kaldor et al. 10, Yuksel et al. 12]. Driven by the observation that sheet-based deformation models cannot realistically simulate the complex interactions of yarn loops in real woven material, these methods directly model the material as a complex structure of interwoven yarn. The yarns in the fabric are for example represented as inextensible but flexible spline curves. This allows modeling virtual clothing using common garment layout, exchanging the types of fibers and stitching techniques used in real-world production processes. Single parts of a cloth model can be virtually stitched together, forming a complex piece of cloth that behaves like real woven clothing.

15.4 Cloth Appearance Modeling

Besides modeling geometric properties and behavior, realistic visual photometric characteristics of the cloth material are important. However,

Table 15.1: Characteristics of different appearance models [Schröder et al. 12].

Model	Surface-based	Volumetric	Explicit
Translucency	✓	✓	✓
Silhouettes		✓	✓
Light Diffusion		✓	✓
Real-time	✓		
Scalability	✓	✓	
Viewing Distance	far/medium	medium	close

modeling the appearance of cloth in high visual quality is a non-trivial task, since the look of a fabric is determined by complex light interaction inside fine surface structures, and often, highly anisotropic single and multiple scattering and self-shadowing effects dominate the appearance. For example, the complex reflection properties of finely structured fabrics like silk or velvet are very difficult to model. This part of the chapter presents approaches which have been inspired by the real world to realistically model the appearance of clothes by reproducing the reflection properties of cloth as closely as possible.

In the last years, different methods for appearance modeling of clothes have been developed, which differ in terms of visual plausibility but also in runtime complexity during rendering. A good overview of appearance models has been presented in [Schröder et al. 12]. The applicability of each method strongly depends on the requirements of the visual quality, runtime and on the viewing distance. Fast methods for high-quality real-time applications or rendering from a far viewing distance may use precomputed bidirectional reflectance distribution functions (BRDFs) or bidirectional texture functions (BTFs). However, with these methods, a specific BRDF or BTF needs to be built for each individual type of fabric.Volumetric models enable a wider field of fabrics and closer viewing distances, but usually need more time and memory to render. For very close viewing distances, explicit or so-called comprehensive models can be used, which allow a visualization of lighting effects on the detailed geometry of the textile on yarn level. This chapter gives an overview of these appearance modeling techniques. The properties of the different models are summarized in Table 15.1.

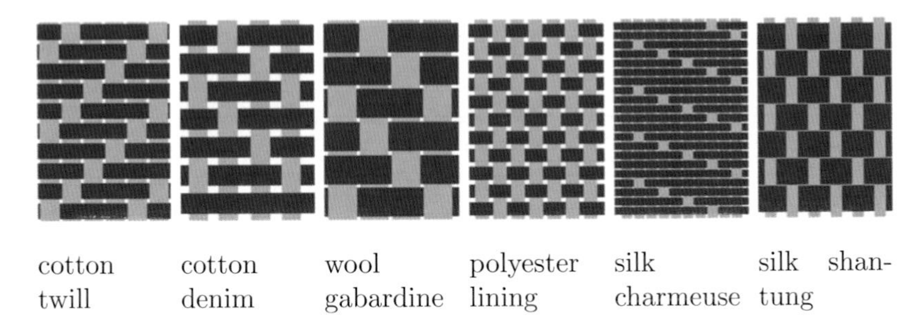

Figure 15.4: Examples of weaving patterns as used by Irawan [Irawan 08].

Surface-Based Models

Much work for modeling reflection properties of cloth has focused on statistical surface-based models such as bidirectional reflectance distribution functions (BRDFs) and bidirectional texture functions (BTFs) [Müller et al. 04] (Chapter 20). These functions model the reflection properties of a material as a function of viewing and illumination direction and offer a simplified representation of a material. BTF- or BRDF-based methods can be heuristic or data-driven. Among the heuristics models, microfacet models describe the surface as a number of flat micromirrors (facets) which reflect light only in one direction. The BRDF is then determined by the number of visible microfacets at the respective orientation reflecting light to the camera. A generation process for 2D microfacet orientation distributions, for example to model satin and velvet, is described by Ashikmin et al. [Ashikmin et al. 00]. Also, weaving information available from textile computer aided design data can be used to generate specific BRDFs for each fabric material [Adabala et al. 03]. Another heuristic approach is to derive BTFs and BRDFs at the yarn level, as it is done for woven fabrics by Irawan [Irawan 08]. The basis of this approach is a complex empirical model for light interacting with threads of fabric (Figure 15.4). This complex and precise model achieves plausible results for a variety of materials. However, it needs substantial computing time, making it more appropriate for offline ray-tracing methods than for real-time visualization techniques.

Data-driven approaches measure the BRDF or BTF directly from real cloth samples, e.g., by capturing a dataset of images of a surface under different viewing and illumination directions [Sattler et al. 03, Wang et al. 08]. Illumination effects like occlusion, global illumination, self-shadowing, and parallax effects at the yarn level are captured by the images and thereby implicitely incorporated in the measurement data without explicitly modeling them (Figure 15.5). This drastically simplifies the underlying model but requires a defined BTF for every kind of fabric. A publicly available BTF

Figure 15.5: Example images of the BTF Database Bonn for three different kinds of cloth under 5 different illumination directions [Sattler et al. 03]. For each sample, 81 views and 81 illumination directions are captured in total. Note the differences in the images regarding self-shadowing and interreflections.

dataset has been presented by Sattler et al. [Sattler et al. 03] (Figure 15.5). They captured planar patches of different materials under 81 viewing and 81 illumination directions. In their BTF representation, all images are registered such that a complete set of reflection values is assigned to one 2D image coordinate. To reduce the size of the data, they performed a principal component analysis (PCA) for each of the viewing directions. In a single view approach, Wang et al. [Wang et al. 08] acquired reflectance data using a setup of a single fixed camera and a linear light source. The light source moves above a sample of cloth, and the camera captures the intensity of the reflected light (Figure 15.6). A microfacet-based BRDF is derived by interpolating between the scattered data. The basic assumption for this single view approach is that for any surface point, there is another surface point with similar but rotated microstructure and that, therefore, a single view contains different slices of the BRDF at surface points with similar reflectance. Since this process is related to example-based texture synthesis [Efros and Freeman 01, Kwatra et al. 03], the method is also called example-based microfacet synthesis.

BRDF- or BTF-based appearance modeling can be extremely efficient for simple lighting setups, and for real-time graphics applications, BTFs are

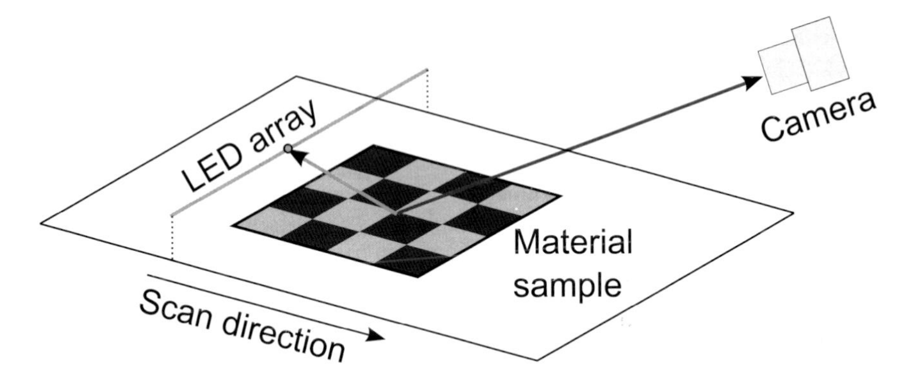

Figure 15.6: The setup of lighting unit, camera, and material sample used for example-based microfacet synthesis [Wang et al. 08].

currently among the most popular techniques. Limitations of BTFs include light diffusion at shadow boundaries, which cannot be reproduced properly. Also, silhouette effects are commonly ignored and handling of transparency is difficult.

Volumetric Models

Volumetric models overcome some of the above-mentioned limitations by additionally capturing the thickness of the fabric. This improves the visual quality when viewing a piece of cloth from a medium or close distance. In the micro-flake model, scattering events in the volume are computed by introducing a directional flake distribution that describes the interaction between particles [Schröder et al. 11, Zhao et al. 11]. This distribution represents the orientation of idealized mirror flakes at any point in the volume. These models can also be driven by real-world measurements of real cloth samples. For example, Zhao et al. [Zhao et al. 11] used CT scans of fabric samples for volumetric rendering. Their approach is able to generate renderings in high quality. However, very sophisticated hardware is needed to record the data for the model.

Explicit Models

In contrast to surface- or volumetric models, explicit models describe the reflection properties by precisely computing light intersections with actual fiber geometry. For example, Irawan and Marschner [Irawan and Marschner 12] presented a comprehensive model to reproduce the look of woven cloth in both small scale and large scale. Their model is based on measurements of the light interaction with fiber threads and can reproduce a wide range of cloth appearances, e.g., black cotton twill, denim, silk, or polyester. Since the numerical fit to the measured data within their approach is very time-consuming, the number of possible thread directions is reduced. However, this also reduces the number of specular highlights. Sadeghi et al. [Sadeghi et al. 13] extend this model to render more complex highlights. They use a microcylinder appearance model to render anisotropic highlights and color shifts by combining BRDF measurements with scattering profiles. The scattering is represented by a scattered radiance distribution function (BSDF) measured by a spherical gantry. By providing the measured parameters as well as weave definitions for a variety of individual threads of fabrics, e.g., linen, silk, polyester, and velvet, realistic visualization is possible for these types of cloth.

Image-Based Models

An alternative to following the complete pipeline from physically modeling mechanical material parameters over deformation simulation to reflection modeling is to model all these characteristics purely by appearance in image-based representations. These representations use a large database of images of a piece of clothing, e.g., captured from different viewpoints and in different body poses. Similar to image-based BTFs, where the images implicitly capture all reflection properties at fine scale, these images capture geometric properties, like wrinkling behavior and deformation, as well as appearance, e.g., reflection and shading properties, at a larger scale. Hence, a database of images showing a piece of clothing can be used as a database of examples to model a piece of clothing. Early methods that followed this idea focused on retexturing. These methods extract 2D texture deformation and shading from a single image or video frame to render a new piece of clothing under the same deformation and shading conditions [Scholz and Magnor 06, White and Forsyth 06, Hilsmann et al. 10, Hilsmann et al. 11]. While these methods are restricted to the conditions present in the original image (regarding deformation and shading), recent methods learn a mapping between the pose of a human body and appearance of clothing captured by a large number of images [Zhou et al. 12, Hauswiesner et al. 11, Hauswiesner et al. 13, Hilsmann et al. 13]. This way, deformation and appearance of clothes is modeled in a pose-dependent way, and images for new pose configurations can be synthesized and merged from the database. The direct use of real images of clothing allows a photorealistic modeling and rendering with very fine details without the need of computationally demanding simulation (Figure 15.7). However, a dataset must be captured for each individual piece of clothing. Also, as these methods interpolate images from precaptured data, the variety of possible animations and simulations strongly depends on the number and variety of examples in the database.

Figure 15.7: Modeling geometric as well as photometric characteristics of clothing in appearance [Hilsmann et al. 13]. Images of clothes are synthesized based on pose information from a set of pre-recorded database images (below). Different body parts, e.g. the left and the right parts are synthesized from different example images and the final result is merged via image blending.

15.5 Summary

This chapter gives an overview on recent work in cloth modeling and simulation. The ultimate goals of cloth modeling are (i) to produce realistic looking and behaving cloth models that are indistinguishable from real cloth and (ii) to achieve a simulation at an acceptable frame rate, ideally in real-time. Modeling realistic cloth comprises modeling of cloth geometry and mechanical properties, deformation behavior as well as appearance. For each of these steps, different approaches exist, which differ in accuracy and complexity, and for each specific application, different methods are suitable. This chapter focused on recent approaches that try to infer information on cloth properties and appearance from the real world in data-driven and example-based methods. These approaches determine realistic geometry and mechanical properties as well as reflection and appearance properties, to model a piece of cloth as realistically as possible.

16
Video-Based Character Animation

Dan Casas, Peng Huang, and Adrian Hilton

16.1 Introduction

Current interactive character authoring pipelines commonly consist of two steps: modeling and rigging of the character model which may be based on photographic reference or high-resolution 3D laser scans (Chapters 13 and 14); and move-trees based on a database of skeletal motion capture together with inverse dynamic and kinematic solvers for secondary motion. Motion graphs [Kovar et al. 02] and parameterized skeletal motion spaces [Heck and Gleicher 07] enable representation and real-time interactive control of character movement from motion capture data. Authoring interactive characters requires a high-level of manual editing to achieve acceptable realism of appearance and movement.

Chapters 11 and 12 introduced recent advances in performance capture [Starck and Hilton 07b, Gall et al. 09] that have demonstrated highly realistic reconstruction of motion using a temporally coherent mesh representation across sequences, referred to as 4D video. This allows replay of the captured motions with free-viewpoint rendering and compositing of performance in post-production while maintaining photo-realism. Captured sequences have been exploited for retargeting surface motion to other characters [Baran et al. 09] and analysis of cloth motion to simulate novel animations through manipulation of skeletal motion and simulation of secondary cloth movement [Stoll et al. 10]. However, these approaches do not enable authoring of interactive characters which allow continuous movement control and reproduction of secondary motion for clothing and hair.

This chapter presents a framework for authoring interactive characters based on actor performance capture. A Surface Motion Graph representation [Huang et al. 09] is presented to seamlessly link captured sequences, allowing authoring of novel animations from user specified space-time constraints. A 4D parametric motion graph representation [Casas et al. 13] is described for real-time interactive animation from a database of captured

4D video sequences. Finally, a rendering approach referred to as 4D video textures [Casas et al. 14] is introduced to synthesize realistic appearance for parametric characters.

16.2 Surface Motion Graphs

Multiple view reconstruction of human performance as a 3D video has advanced to the stage of capturing detailed non-rigid dynamic surface shape and appearance of the body, clothing, and hair during motion [Starck and Hilton 07b, de Aguiar et al. 08b, Vlasic et al. 08, Gall et al. 09] (Chapters 11 and 12). Full 3D video scene capture holds the potential to create truly realistic synthetic animated content by reproducing the dynamics of shape and appearance currently missing from marker-based skeletal motion capture. There is considerable interest in the reuse of captured 3D video sequences for animation production. For conventional skeletal motion capture (MoCap), *motion graph* techniques [Molina-Tanco and Hilton 00, Kovar et al. 02, Arikan and Forsyth 02, Lee et al. 02] are widely used in 3D character animation production for games and film. This section presents a framework that automatically constructs motion graphs for 3D video sequences, called *surface motion graphs* [Huang et al. 09], and synthesizes novel animations to best satisfy user specified constraints on movement, location, and timing. Figure 16.1 shows an example novel animation sequence produced using a surface motion graph with path optimization to satisfy the user constraints.

Character Animation Pipeline

A frame-to-frame similarity matrix is first computed between all frames across all 3D video motion sequences in the 3D video database. Potential transitions between motions are automatically identified by minimizing the total dissimilarity of transition frames. The idea behind this is to minimize the discontinuity that may be introduced by transitions when transferring

Figure 16.1: An example of a synthesized 3D character animation (10 transitions).

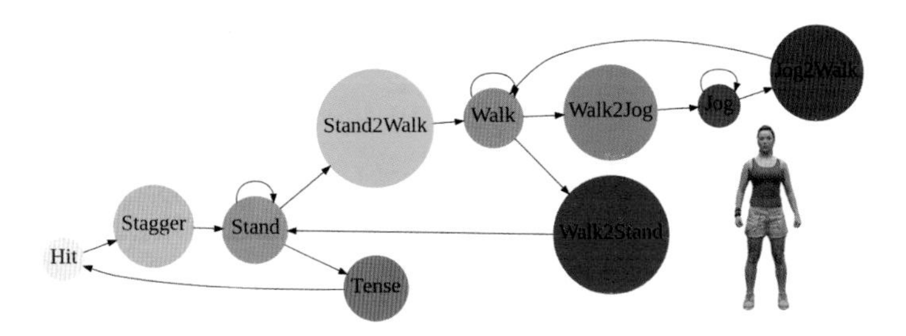

Figure 16.2: An example of a surface motion graph for a game character.

within or across different motions. A surface motion graph is then constructed using these transitions. Once the graph structure is computed, a path on the graph is optimized according to user input such as key-frames, global timing, and distance constraints. Finally, concatenative motion synthesis and rendering is performed to produce video-realistic character animation.

Graph Representation

A surface motion graph represents possible inter- and intra-sequence transitions for 3D video sequences, analogous to motion graphs [Kovar et al. 02] for skeletal motion capture sequences. It is defined as a directed graph: each node denotes a 3D video sequence; each edge denotes one or multiple possible transitions. A path on the graph then provides a possible motion synthesis. Figure 16.2 shows an example of surface motion graph constructed from multiple 3D video motion sequences for the actor shown. The method to automatically identify transitions is described in Section 16.2.

Graph Optimization

Graph optimization is performed to find the path through the surface motion graph which best satisfies the required animation constraints. Intermediate key-frames selected by the user provide hard constraints defining the desired movement. Start and end key-frame locations specify the target traverse distance d_V and the target traverse time t_V chosen by the user. Both target traverse distance and time are used as soft constraints, which define global constraints on the animation. The cost function for graph path optimization to satisfy the constraints is described as follows:

Combined Cost. The cost function for a path P through the surface motion graph between a pair of key frames is formulated as the combination of three costs, C_{tran} representing cost of transition between motions, soft constraints on distance C_{dist} and time C_{time},

$$C(P) = C_{tran}(P) + w_{dist} \cdot C_{dist}(P) + w_{time} \cdot C_{time}(P). \tag{16.1}$$

where w_{dist} and w_{time} are weights for distance and time constraints, respectively. Throughout this chapter, the parameters are set to $w_{dist} = 1/0.3$ and $w_{time} = 1/10$, which equates the penalty for an error of 30 cm in distance with an error of 10 frames in time [Arikan and Forsyth 02].

Distance Cost. $C_{dist}(P)$ for a path P with N_f frames on the surface motion graph is computed as the absolute difference between the user-specified target distance d_V and the total travelled distance $dist(P)$, given the 3D frames on the path of P is $\{M(t_f)\}, f = [0, N_f - 1]$,

$$C_{dist}(P) = |dist(P) - d_V|. \tag{16.2}$$

$$dist(P) = \sum_{f=0}^{N_f - 2} |center(M(t_{f+1})) - center(M(t_f))|. \tag{16.3}$$

where function $center()$ computes the projection of the centroid of the mesh onto the ground.

Timing Cost. $C_{time}(P)$ for a path P with N_f frames is evaluated as the absolute difference between the user-specified target time t_V and the total travelled time $time(P)$,

$$C_{time}(P) = |time(P) - t_V|. \tag{16.4}$$

$$time(P) = N_f \cdot \Delta t. \tag{16.5}$$

where Δt denotes the frame rate (e.g., 25 frames per second).

Transition Cost. $C_{tran}(P)$ for a path P is defined as the sum of distortion for all transitions between concatenated 3D video segments. If the index for concatenated 3D video segments is denoted $\{f_i\}$, $i = 0, ..., N_{f-1}$, the total transition cost $C_{tran}(P)$ is computed as

$$C_{tran}(P) = \sum_{i=0}^{N_P - 2} D(S_{f_i \to f_{i+1}}). \tag{16.6}$$

where N_P denotes the total number of transitions on path P and $D(S_{f_i \to f_{i+1}})$ from Eq. (16.9) is the distortion for transition from motion sequence S_{f_i} to motion sequence $S_{f_{i+1}}$.

Path Optimization. Finally, the optimal path P^{opt} for a given set of constraints is found to minimize the combined cost $C(P)$, as defined in Equation 16.1,

$$P^{opt} = \arg\min_{P} C(P). \qquad (16.7)$$

An efficient approach using interger programming to search for the optimal path that best satisfies the user-defined soft constraints can be found in [Huang et al. 09].

Transitions

A transition of the surface motion graphs $S_{i \to j}$ is defined as a seamless concatenation of frames from two 3D video sequences S_i and S_j. If m, n denotes the central indices for the overlap, the length of overlap as $2L + 1$, the blending weight for the kth transition frame is computed as $\alpha(k) = \frac{k+L}{2L}$, $k \in [-L, L]$. The kth transition frame $M_{i \to j}(t_k) = G(M_i(t_{m+k}), M_j(t_{n+k}), \alpha(k))$ will be generated by a non-linear 3D mesh blend [Tejera et al. 13]. The distortion measure of a transition frame $M_{i \to j}(t_k)$ is then defined as the weighted 3D shape dissimilarity (Equation 16.11) between frame $M_i(t)$ and $M_j(t)$,

$$d(M_{i \to j}(t_k)) = \alpha'(k) \cdot c(M_i(t_{m+k}), M_j(t_{n+k})). \qquad (16.8)$$

where $\alpha'(k) = \min(1 - \alpha(k), \alpha(k))$. The total distortion for a transition sequence $S_{i \to j}$ is then computed as the sum of the distortion of all transition frames,

$$D(S_{i \to j}) = \sum_{k=-L}^{L} d(M_{i \to j}(t_k)). \qquad (16.9)$$

The optimal transition $S_{i \to j}^{opt}$ is then defined to minimize the distortion cost,

$$S_{i \to j}^{opt} = \arg\min_{S_{i \to j}} D(S_{i \to j}). \qquad (16.10)$$

3D Shape Similarity. To measure the similarity of the 3D mesh geometry within each frame of the 3D video database, shape histograms are used. This has previously demonstrated to give good performance for measuring non-rigid deformable shape similarity for 3D video sequences of human performance [Huang et al. 10]. The volumetric shape histogram $H(M)$ represents the spatial occupancy of the bins for a given mesh M. A measure of shape dissimilarity between two meshes M_r and M_s can be defined by optimizing for the maximum overlap between their corresponding radial bins with respect to rotation about the vertical axis.

$$c_{shape}(M_r, M_s) = \min_{\phi} \|H(M_r) - H(M_s, \phi)\|. \qquad (16.11)$$

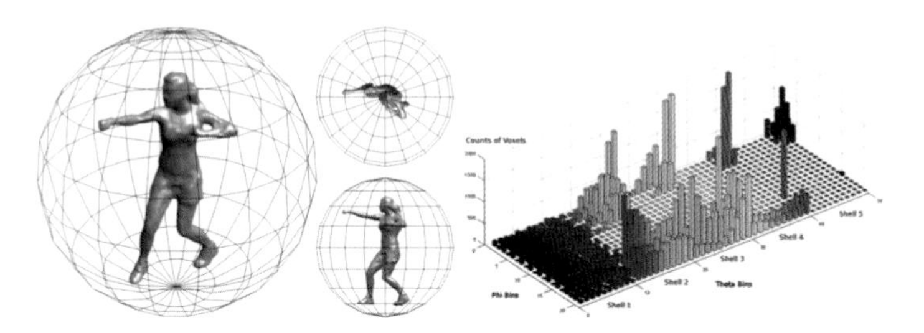

Figure 16.3: A 3D shape histogram of 5 shells, 10 bins for θ, and 20 bins for ϕ together with the space partitions on the 5th shell is illustrated.

The volumetric shape histogram $H(M)$ partitions the space which contains a 3D object into disjoint cells and counts the number of occupied voxels falling in each bin to construct a histogram as a signature for this 3D object. The space is represented in a spherical coordinate system (R, θ, ϕ) around the center of mass. An example of a 3D shape histogram is shown in Figure 16.3.

A frame-to-frame static similarity matrix $\mathbf{C} = [c_{r,s}]_{N \times N}$, where N denotes the total number of frames, between all frames across all 3D video sequences in the 3D video database is pre-computed according to Eq.16.11,

$$c_{r,s} = c_{shape}(M_r, M_s). \tag{16.12}$$

Adaptive Temporal Filtering. Each transition is determined by a tuple (m, n, L). The global optimization is then performed by testing all possible tuples (m, n, L) and so finding optimal arguments for the minimum,

$$(m^{opt}, n^{opt}, L^{opt}) = \arg \min_{m,n,L} \sum_{k=-L}^{L} \alpha'(k) \cdot c_{m+k,n+k}. \tag{16.13}$$

This equates to performing an adaptive temporal filtering with window size $2L + 1$ and weighting $\alpha'(k)$ on the pre-computed static similarity matrix $\mathbf{C}$. The process is computationally efficient. To obtain multiple transitions between a pair of sequences, top T best transitions are preserved. T is pre-determined by the user. An example of the frame-to-frame 3D shape similarity matrix and transition frames (marked in yellow) evaluated using a temporal window $L = 4$ and $T = 4$ are shown in Figure 16.4.

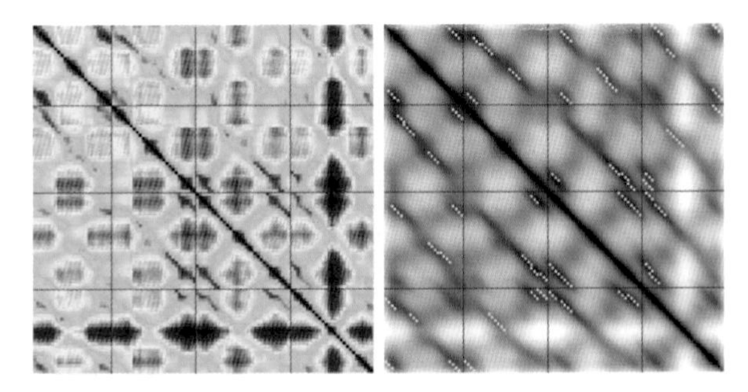

Figure 16.4: Shape similarity matrix and identified transitions for motions *Jog*, *Jog2Walk*, *Walk*, and *Walk2Jog* (left to right and top to bottom) .

16.3 4D Parametric Motion Graphs

Parametric Control of Mesh Sequences

Parametric animation requires the combination of multiple captured mesh sequences to allow continuous real-time control of movement with intuitive high-level parameters such as speed and direction for walking or height and distance for jumping. Methods for parameterization of skeletal motion capture have previously been introduced [Kovar et al. 02, Rose et al. 98] based on linear interpolation of joint angles. Analogously, as described by Casas et al. [Casas et al. 12b], parametric animation of mesh sequences can be achieved by interpolation between mesh sequences.

This requires temporal alignment of frames across multiple 3D video mesh sequences, such that all frames have a consistent mesh connectivity and vertices correspond to the same surface point over time. The set of temporally aligned 3D video sequences are referred to as 4D video [Budd et al. 13] (Chapter 11).

Given a set of N temporally aligned 4D mesh sequences $\mathbf{M} = \{M_i(t)\}_{i=1}^{N}$ of the same or similar motions (e.g., high jump and low jump), some sort of parametric control is sought by blending between multiple mesh sequences

$$M_B(t, \mathbf{w}) = b(\mathbf{M}, \mathbf{w}) \tag{16.14}$$

where $\mathbf{w} = \{w_i\}_{i=1}^{N}, w_i \in [0..1]$ is a vector of weights for each input motion and $b()$ is a mesh sequence blending function. This function must perform at online rates, ≥ 25 Hz, and the resulting mesh $M_B(t, \mathbf{w})$ to maintain the captured non-rigid dynamics of the source $\{M_i(t)\}_{i=1}^{N}$ meshes.

Three steps are required to achieve high-level parametric control from mesh sequences: time-warping to align the mesh sequences; non-linear mesh blending of the time-warped sequences; and mapping from low level blending weights to high-level parameters (speed, direction, etc.).

Sequence Time-Warping. Each 4D video sequence of related motions (e.g., walk and run) is likely to differ in length and location of corresponding events, for example foot-floor contact. Thus, the first step for mesh sequence blending is to establish the frame-to-frame correspondence between different sequences. Mesh sequences $M_i(t)$ are temporally aligned by a continuous time-warp function $t = f(t_u)$ [Witkin and Popovic 95] which aligns corresponding poses of related motions prior to blending such that $t \in [0, 1]$ for all sequences. Temporal sequence alignment helps in preventing undesirable artifacts such as foot skating in the final blended sequence.

Real-Time Mesh Blending. Previous research in 3D mesh deformation concluded that linear-methods for mesh blending, despite being computationally efficient, may result in unrealistic results [Lewis et al. 00]. Nonlinear methods based on differential surface representation [Botsch and Sorkine 08] overcome this limitation, achieving plausible surface deformation. However, the price paid is a significant increase in processing requirements, hindering their use for online applications. Piecewise linear blending methods [Casas et al. 13] have demonstrated successful results for online applications by precomputing a set of non-linear interpolated meshes. At run time, any requested parametric mesh is computed by linearly blending the relevant offline interpolated meshes. This results in an approximation to the robust non-linear blending approach with a computational cost similar to the linear approach.

High-Level Parametric Control. High-level parametric control is achieved by learning a mapping function $f(\mathbf{w})$ between the blend weights $\mathbf{w}$ and the user specified motion parameters $\mathbf{p}$. A mapping function $\mathbf{w} = f^{-1}(\mathbf{p})$ is learned from the user-specified parameter to the corresponding blend weights required to generate the desired motion because the blend weights $\mathbf{w}$ do not provide an intuitive parameterization of the motion. Motion parameters $\mathbf{p}$ are high-level user specified controls for a particular class of motions such as speed and direction for walk or run, and height and distance for a jump. As proposed by Ahmed et al. [Ahmed et al. 01], the inverse mapping function f^{-1} from parameters to weights can be constructed by a discrete sampling of the weight space $\mathbf{w}$ and evaluation of the corresponding motion parameters $\mathbf{p}$.

Figure 16.5 presents three examples of mesh sequence parameterization, enabling control of speed, jump length, and jump height, respectively. Change in color represents change in parameterization. In each example,

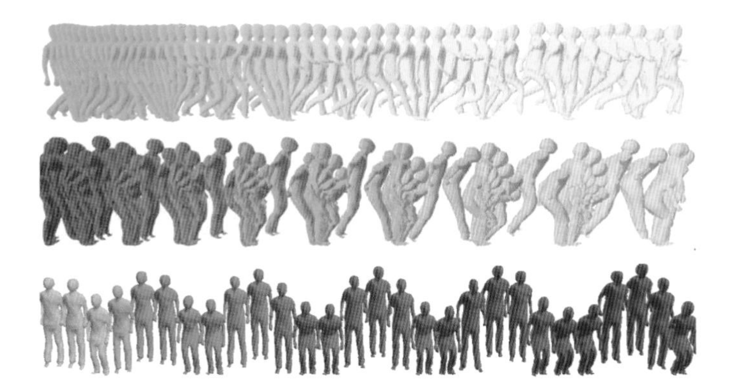

Figure 16.5: Mesh sequence parameterization results [Casas et al. 12b]. In each row, two input motions are interpolated, generating in-between parametric motion control. Top row, speed control; middle row, jump length control; bottom row, jump height control.

two motions are interpolated to generate in-between poses, enabling interactive parametric control of the motion.

4D Parametric Motion Graphs

Given a dataset of mesh sequences for different movements with a consistent mesh structure at every frame, referred to as 4D video, parametric control of the motion can be achieved by combining multiple sequences of related motions (Section 16.3). This gives a parameterized motion space controlled by high-level parameters (for example walk speed/direction or jump height/length). However, parametric motions can only be synthesized by combining semantically similar sequences (i.e., walk and run), whereas the interpolation of non-similar sequences, such as jump and walk, would fail in generating a human-realistic motion. Therefore, in order to fully exploit a dataset of 4D video sequences, methods for linking motions performing different actions are required.

An approach referred to as 4D parametric motion graph, 4DPMG, [Casas et al. 12a, Casas et al. 13] tackles this problem by employing a graph representation that encapsulates a number of independent mesh parametric spaces as well as links between them, to enable real-time parametric control of the motion. Figure 16.6 presents an illustration of a 4DPMG created using 10 different motions used to create 4 parametric spaces.

The nodes of a 4DPMG are created by the combination of similar motions. Each node can be considered as an independent parametric motion space created (Section 16.3). The problem is then how to find transitions

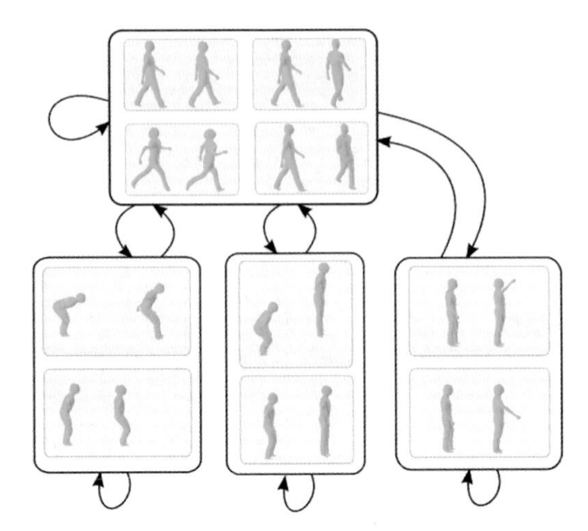

Figure 16.6: Illustration of a 4D parametric motion graph. Ten different motions are combined to create a 4-node 4DPMG, enabling speed, direction, jump, and reach parametric control. Each node represents an independent parametric space; edges represent links between them.

between parametric motions at run time. Natural transitions require a similar shape and non-rigid motion between the linked meshes, otherwise the resulting animation will not look realistic due to sudden change of the character's speed and pose. In the literature, parametric transitions for skeletal data have been approached by precomputing a discrete set of *good* transitions, evaluating the similarity in pose and motion between pairs of the linked motion spaces [Heck and Gleicher 07]. However, precomputation of a fixed set of transition points may result in a relatively high latency due to the delay between the current pose and the next pre-computed good-transition pose.

In a 4DPMG, to evaluate the best transition path P_{opt} between two parametric points, a cost function representing the trade-off between similarity in mesh shape and motion at transition, $E_S(P)$, and the latency, $E_L(P)$, or delay in transition for a path P, is optimized

$$P_{opt} = \arg\min_{P \in \Omega} (E_S(P) + \lambda E_L(P)) \quad (16.15)$$

where λ defines the trade-off between transition similarity and latency. The transition path P is optimized over a trellis of frames as illustrated in Figure 16.7, starting at the current frame $M^s(t^s, \mathbf{w}^s)$ in the source motion space and a trellis ending at the target frame $M^d(t^d, \mathbf{w}^d)$ in the target motion space, where M^s and M^t are interpolated meshes as defined by Eq. (16.14).

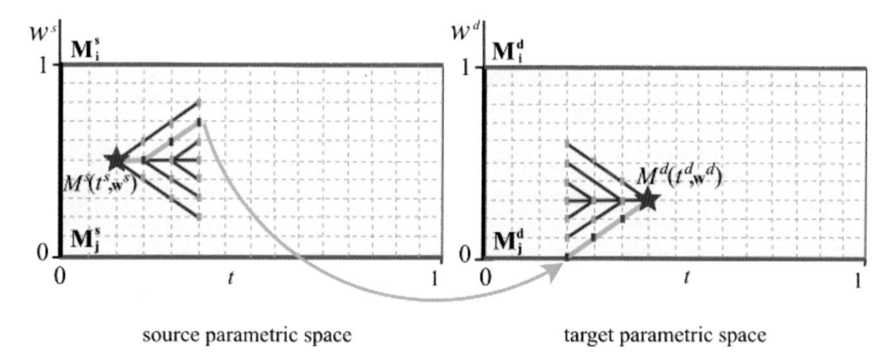

Figure 16.7: Diagram depicting the process of finding a transition between two parametric meshes, $M^s(t^s, \mathbf{w}^s)$ and $M^d(t^d, \mathbf{w}^d)$, here arbitrarily selected for illustration purposes, and marked with a purple star inside their corresponding parametric spaces. Two trellises, depicted in red in each parametric space, are built to find the sets of transition mesh candidates. Optimal transition path P_{opt}, also arbitrarily selected in this diagram, is highlighted in green over the trellis.

The trellis is sampled forward in time at discrete intervals in time Δt and parameters $\Delta \mathbf{w}$ up to a maximum depth l_{max} in the source space. Similarly from the target frame a trellis is constructed going backward in time. This defines a set of candidate paths $P \in \Omega$ with transition points between each possible pair of frames in the source and target trellis.

For a path P, the latency cost $E_L(P)$ is measured as the number of frames in the path P between the source and target frames. Transition similarity cost $E_S(P)$ is measured as the similarity in mesh shape and motion at the transition point between the source and target motion space for the path P. Online mesh similarity computation is prohibitively expensive for large sets of transition candidates. To overcome this, Casas et al. [Casas et al. 12a] proposed a method based on precomputing a set of similarities between the input data, and interpolated them at runtime to approximate any requested parametric pose similarity.

Figure 16.8 presents a parametric character interactively synthesized combining 10 different motions. Qualitative and quantitative results using 4DPMG have demonstrated [Casas et al. 12a] realistic transitions between parametric mesh sequences, enabling interactive parametric control of a 3D-mesh character.

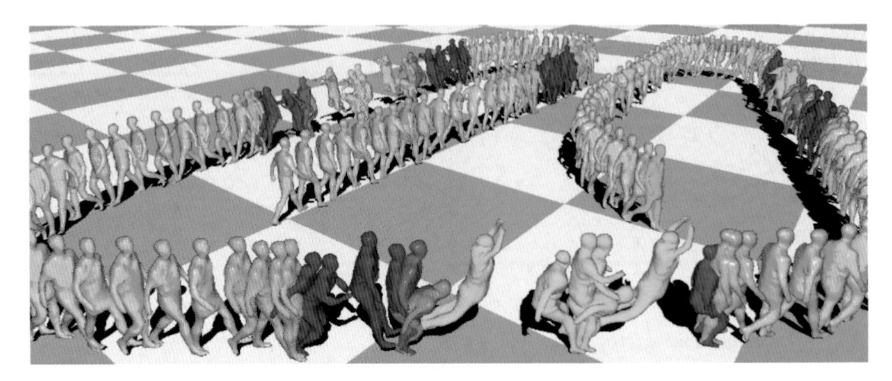

Figure 16.8: An interactively controlled character generated using 4DPMG, combining walk, jog, left turn, right turn, short jump, and long jump motions. Grey meshes indicate transitions between parametric spaces.

4D Video Textures

The 4D parametric motion graph allows real-time interactive control of a character's motion to produce novel animation sequences. However, this is missing the realistic appearance of the source video for free-viewpoint rendering. To achieve video-realistic rendering of the dynamic appearance for 4D parametrically controlled animation a representation referred to as 4D video textures (4DVT) is used [Casas et al. 14].

With 4DVT, appearance for an intermediate motion is produced by aligning and combining multiple-view video from the input examples to produce plausible video-realistic dynamic appearance corresponding to the modified movement. As the character motion changes, so does the dynamic appearance of the rendered view reflecting the change in motion.

A 4D video $F(t) = \{V(t), M(t)\}$ combines multiple view video sequences $V(t) = \{I_c(t)\}_{c=1}^{C}$ with C camera views with a 4D proxy of the dynamic surface shape represented as a mesh sequence $M(t)$, where vertices correspond to the same surface point over time. This form of representation has previously been employed for free-viewpoint video rendering of dynamic scenes [Carranza et al. 03, Starck and Hilton 07b, de Aguiar et al. 08b, Vlasic et al. 08]. Free-viewpoint video renders a novel video sequence $I(t, \mathbf{v})$ for a viewpoint $\mathbf{v}$ from the set of input videos $V(t)$ using the mesh sequence $M(t)$ as a geometric proxy [Buehler et al. 01]. The objective of free-viewpoint video is to maintain the visual realism of the source video while providing the flexibility to interactively control the viewpoint. However, free-viewpoint video is limited to the replay of the captured performance and does not allow for any change in scene motion.

Figure 16.9: A character animated using 4D Video Textures jumping over obstacles of varying length. Notice the texture detail in face and clothing.

4DVT overcomes this limitation [Casas et al. 14]. Given a set of motion control parameters $\mathbf{w}$ and viewpoint $\mathbf{v}$, the aim is to render a novel video $I(t, \mathbf{w}, \mathbf{v})$:

$$I(t, \mathbf{w}, \mathbf{v}) = h(F_1(t), ..., F_N(t), \mathbf{w}, \mathbf{v}), \tag{16.16}$$

where $h(.)$ is a function which combines the source 4D videos according to the specified motion parameters $\mathbf{w}$ and viewpoint v. The rendered video $I(t, \mathbf{w}, \mathbf{v})$ should preserve the visual quality of both the scene appearance and motion.

A two-stage approach is used to synthesize the final $I(t, \mathbf{w}, \mathbf{v})$ video. First, a 4D shape proxy $M(t, \mathbf{w})$ is computed by the combination of the input mesh sequences using the approach presented in Section 16.3. Finally, exploiting the known vertex-to-vertex correspondence across sequences, view-dependent rendering of the source videos $V_i(t)$ is performed using the same 4D shape proxy $I(t, \mathbf{w}, \mathbf{v})$. The output video $I(t, \mathbf{w}, \mathbf{v})$ is generated based on real-time alignment of the rendered images.

Qualitative and quantitative results have demonstrated that 4DVT succesfully enables the creation of video characters with interactive video and motion control and free-viewpoint rendering which maintain the visual quality and dynamic appearance of the source videos [Casas et al. 14]. Intermediate motions are rendered with plausible dynamic appearance for the whole-body, cloth wrinkles, and hair motion. Figure 16.9 presents an animation interactively created combining 4D video textures and 4D parametric motion graph; a character combines different styles of jumps and walks to avoid the obstacles, and finally performs a sharp left turn.

16.4 Summary

Recent research on multi-camera performance capture has enabled detailed reconstruction of motions as 3D mesh sequences with a temporally coherent geometry. This chapter has presented a set of methods to allow the reutilization of reconstructed motions, with the goal of authoring novel character animation while maintaining the realism of the captured data.

A representation referred to as surface motion graph has been introduced to synthesize novel animations that satisfy a set of user-defined constraints on movement, location, and timing. A graph optimization technique is used to find transitions between originally captured sequences that best satisfy the animation constraints. A 3D shape similarity descriptor is used to evaluate the transitions between different motions.

A parametric approach referred to as 4D parametric motion graph has been presented to synthesize novel interactive character animation by concatenating and blending a set of mesh sequences. This enables real-time control of a parametric character that preserves the realism of the captured geometry. Video-realistic appearance for novel parametric motions is generated at run-time using 4D video textures, a rendering technique that combines multi-camera captured footage to synthesize textures that changes with the motion while maintaining the realism of the captured video.

Part IV

Authentic Rendering, Display, and Perception

17 Image- and Video-Based Rendering

Christian Lipski, Anna Hilsmann, Carsten Dachsbacher, and Martin Eisemann

17.1 Introduction

The purpose of image- and video-based rendering (IVBR) is to be able to synthesize photo-realistic new, virtual views of real-world scenes and events from no more than a set of conventional photographs or videos of the scene. Purely image-based, or plenoptic, approaches densely sample scene appearance using a large number of images (Sections 17.2 and 5.3). New views are generated by simply re-sampling the captured image data. In contrast, geometry-assisted methods require much less input images. Here, 3D scene geometry is either a priori known, reconstructed from the captured imagery, or acquired separately by some other means, e.g., ranging imaging (Chapter 4). With (approximate) scene geometry available, views from arbitrary viewpoints are synthesized using the acquired images as geometry texture (Section 17.3). Over the years, various IVBR methods have been proposed that can all be categorized in-between these two limiting cases [Lengyel 98] (Figure 17.1). Their respective advantages and limitations have been discussed in several IVBR surveys [Shum and Kang 00, Smolic et al. 09, Linz 11, Germann 12].

17.2 Plenoptic Approaches

Light Fields and Lumigraphs

The idea underlying light field rendering is to represent the plenoptic function of a real-world scene, i.e., its appearance from any direction, as a four-dimensional lookup table [Levoy and Hanrahan 96] (Figure 17.2). By assuming the space around the scene to be transparent, each light ray can be parameterized by four scalar values (u, v, s, t). (u, v) are the intersection

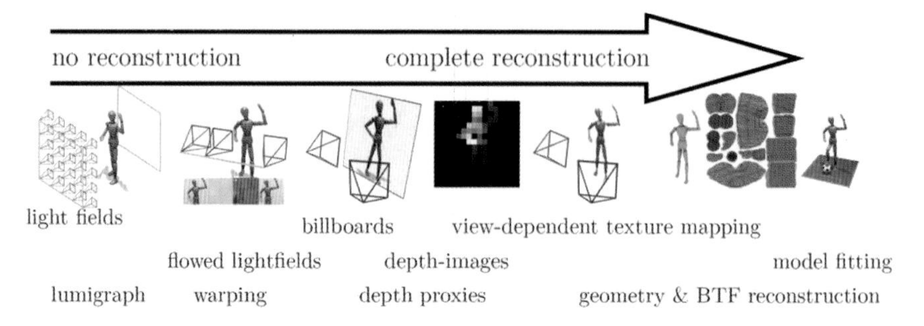

Figure 17.1: Overview of video and image-based rendering systems. While some approaches are based on densely sampling scene appearance with many images (far left), others rely on having available high-quality 3D scene geometry (far right). Numerous techniques can be located somewhere between these two limiting cases.

coordinates of the light ray with the camera plane, while (s, t) are the ray's intersection coordinates with the fronto-parallel image plane. Light field acquisition consists of sampling the uv plane by taking images from regularly spaced uv grid positions (Figure 17.3). For high-quality light field acquistion, a rectifying homography transformation is typically applied to align the image plane of all captured images with the st plane [Kim et al. 13].

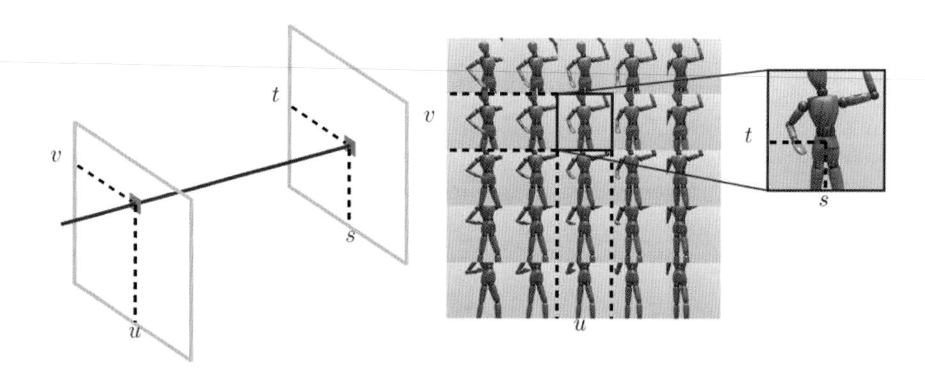

Figure 17.2: Light field rendering: views of the scene from arbitrary vantage points are obtained by tracing a view ray for each pixel and determining the intersection coordinates with the uv- and the st-planes (left). The $uvst$ coordinates are used to look up pixel color from the 4D light field dataset (right).

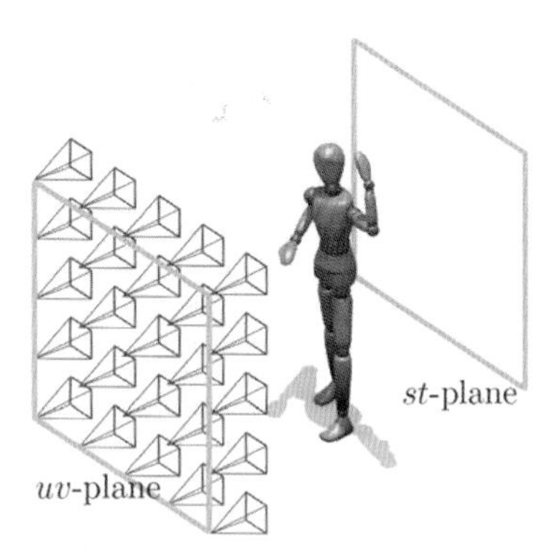

Figure 17.3: Light field acquisition: in the two-parallel-plane parameterization, the plenoptic function of a scene is regularly sampled by camera positions on the uv plane and coinciding image plane st. To acquire the light field of a static scene, a motorized camera gantry can be used (Section 5.2), while for dynamic scenes an array of video cameras is needed.

For static scenes, a mechanical gantry enables sequentially capturing many light field images with one camera. The capture hardware has to be precisely calibrated, however, and the capture process may take a long time. Alternatively, single-chip light field cameras have been propsed [Rodriguez-Ramos et al. 11, Lytro, Inc. 12, Wietzke 12] that employ lenslet arrays to acquire the entire light field simultaneously (Chapter 5). With single-chip systems, however, there is a trade-off between image resolution (st plane) and viewpoint range (uv plane).

Several modifications and extensions to light field rendering have been proposed. In Lumigraph rendering, capturing the scene is simplified by allowing for non-regular placement of cameras [Gortler et al. 96]. Before rendering, the image data is re-parameterized to the (u, v, s, t) representation in a rebinning step. Following a similar approach, light fields can also be captured using mobile phones [Davis et al. 12].

Light field rendering requires neither scene geometry nor image correspondence information. In theory, photo-realistic, high-quality rendering results can be obtained. For aliasing-free rendering, however, unrealistically high sampling rates are required [Chai et al. 00]. To reduce discretization artifacts when light field-rendering from undersampled data, filtering schemes can be employed [Stewart et al. 03, Eisemann et al. 07]. On the other hand, light field data is highly redundant which allows for efficient compression, storage, and streaming [Vaish and Adams 12].

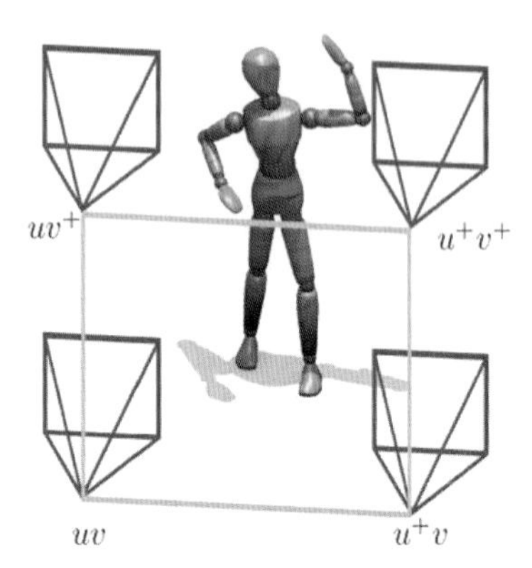

Figure 17.4: Sparse light field capture: with increasing distance between camera positions, uv plane sampling becomes less dense. To avoid ghosting artifacts during rendering, disparity between light field images must be compensated.

Flowed Light Field Rendering

To render from only sparsely sampled light fields, image warping can be employed to synthesize in-between camera positions on the uv-plane [Shade et al. 98, Heidrich et al. 99, Einarsson et al. 06] (Figure 17.4). Prior to rendering, dense image correspondences are established between adjacent light field images by estimating the optical flow fields. During rendering, each ray is intersected with the uv-plane, but instead of just looking up pixel color in the closest-by light field image or linear blending, backward warping is applied to correct for parallax. In backward warping, the pre-computed flow fields are first individually forward-warped to the desired location [Shade et al. 98]. For each ray intersection with the uv plane, the flow vectors to all input images are then looked up and the corresponding light field image pixels are weightedly blended (Figure 17.5).

Flowed light field rendering requires considerably fewer input images than standard light field rendering. For a full 360° surround capture of an actor, 3×30 images are sufficient to obtain convincing rendering results [Einarsson et al. 06]. In addition, the warping step can correct for small errors, e.g., misaligned images due to calibration errors or unsynchronized cameras. This way, even dynamic scenes may be light field-captured sequentially from different vantage points [Einarsson et al. 06].

The main limitation of flowed light field rendering is its dependency on correct, dense flow fields. With increasing distance between neighboring camera positions or for complex scenes, however, optical flow estimation algorithms tend to fail. Another limitation is that backward warping cannot

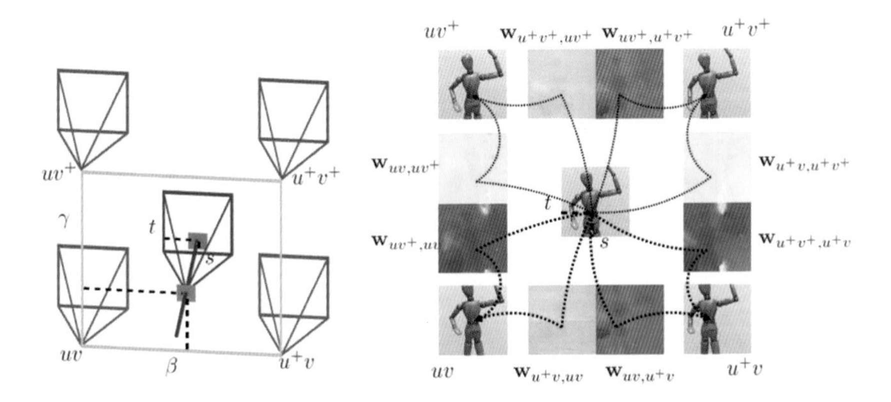

Figure 17.5: Flowed light field rendering. For each viewing ray that intersects the uv plane, the four surrounding images $u[^+]v[^+]$ are determined. In a pre-processing step, optical flow fields $\mathbf{w}_{u[+]v[+],u[+]v[+]}$ have been computed between adjacent light field images. During rendering, backward warping $\mathbf{w}$ is applied to obtain the ray's corresponding pixel color in each of the four images. The rendered pixel is then computed as a weighted sum from the four light field images.

cope with object occlusions, causing ghosting and other artifacts during rendering.

Warping-Based Approaches

In contrast to flowed light field rendering where for each viewing ray adjacent light field images are locally queried, in warping-based rendering, also known as image morphing or correspondence-based rendering, the acquired light field images are being warped completely using flow fields (Figure 17.6). Warping-based approaches are not restricted to light fields but have a long-standing tradition in view interpolation and the creation of smooth transitions between similar images [Stich et al. 08]. For example, warping has been used to create transitions between different actors who are performing an identical choreography [Beier and Neely 92]. An extension to more than two images has been proposed by Lee et al. [Lee et al. 98]. Chen and Williams [Chen and Williams 93] proposed to use forward image warping for viewpoint interpolation based on previously estimated flow fields. Image warping is applied to each input image, and the final result is obtained by weighted blending of the warped images. The combination of image warping and blending is also frequently referred to as image morphing.

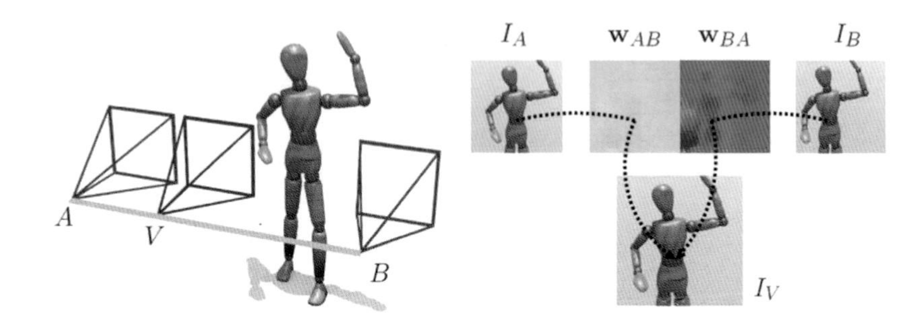

Figure 17.6: Warping-based rendering: to interpolate viewpoint I_V, each pixel in the input images I_A and I_B is forward-warped according to its flow vector $\mathbf{w}_{AB}$, $\mathbf{w}_{BA}$. The final image is weightedly blended from the two warped images.

For real-world images, perceptually convincing dense correspondence fields can be estimated either automatically [Stich et al. 11] or assisted by additional user input [Ruhl et al. 12a]. For multi-view video footage of dynamic scenes, additionally loop consistency among subsequent video frames from neighboring cameras can be exploited [Sellent et al. 12]. For synthetic scenes, the flow vector for a given pixel location $\mathbf{x}$ can be easily derived from per-pixel depth d and camera matrices $\mathbf{P}_A, \mathbf{P}_B$:

$$\mathbf{w}_{AB}(\mathbf{x}) = \mathbf{P}_B({\mathbf{P}_A}^{-1}(\mathbf{x}, \mathbf{d})) \tag{17.1}$$

where ${\mathbf{P}_A}^{-1}$ is the inverted projection matrix of camera A (Section 1.6). To ensure geometrically undistorted in-between views during warping-based image interpolation, rectifying homography transforms are applied to the input images prior to warping [Seitz and Dyer 96]. Warping enables convincing viewpoint interpolation of panoramas [McMillan and Bishop 95] (Chapter 3), dynamic light fields [Goldlücke et al. 02], uncalibrated images [Fusiello 07] as well as uncalibrated and unsynchronized multi-view video [Lipski et al. 10a].

Similarities exist to flowed light fields as well as to depth-based rendering (Section 17.3). If backward warping is used, i.e., if each target pixel of the view to be synthesized is queried for its location in the captured light field images, image warping is a special case of flowed light fields: here, the interpolated viewpoint is located on the manifold spanned by all camera positions, as opposed to flowed light fields where the target view does not have to lie on the capture manifold. On the other hand, if forward-warping is used, i.e., if each pixel of a captured light field image gets shifted to its new location in the target view, warping is akin to depth-based rendering:

For rectified input imagery, the flow vectors are scanline-aligned and reduce to one-dimensional *disparity* that is proportional to the inverse scene depth at a given pixel position.

Analogous to flowed light fields, warping-based rendering requires considerably fewer input images than pure light field rendering. The amount of image data can be further reduced if cameras do not have to span a 2D manifold but can be arranged in arc-like setups around the scene. Warping is able to compensate for calibration inaccuracies and works with unsynchronized multi-view video footage. It performs robustly even for complex outdoor scenes [Lipski et al. 10a] and lends itself to creating numerous space-time visual effects [Linz et al. 10b].

In comparison to traditional light field rendering, one limitation of warping-based IVBR is that the virtual viewpoint must lie on the manifold spanned by the capture positions of all input images/videos. The quality of the rendered output depends on the accuracy of the dense correspondence maps. For multi-view acquisition setups of up to about 10° between adjacent cameras, robust correspondence estimation algorithms exist [Lipski et al. 12] (Chapter 8). Still, occlusion effects cannot always be handled correctly by warping alone, motivating the use of geometry proxies in IVBR.

17.3 Geometry-Assisted Approaches

Geometry Proxies

Image-based occlusion detection is an active research area [Ince and Konrad 08, Herbst et al. 09]. Alternatively, actual depth information is needed to render occlusion effects robustly [Chen and Williams 93]. Fortunately, even if actual scene geometry is too complex for faithful reconstruction, simple geometry proxies often already suffice to achieve visually convincing rendering results.

The most basic scene geometry approximation consists of a single fronto-parallel plane located in the middle of the scene, often referred to as a billboard (Figure 17.7). By default, the billboard is always oriented perpendicular to the current viewing direction. For rendering, one or more captured scene images are projected onto the billboard and rendered as textures, cross-blending between projected images [Snavely et al. 06, Snavely et al. 08a]. If the scene consists of several objects, separate billboards may be used, one for each object. An individual object may also be represented by more than one billboard. In microfacet billboarding [Yamazaki et al. 02], the object is divided into many thousand small billboards. Further rendering improvements can be achieved by faithfully reconstructing the boundary colors of neighboring proxies [Germann et al. 10] by

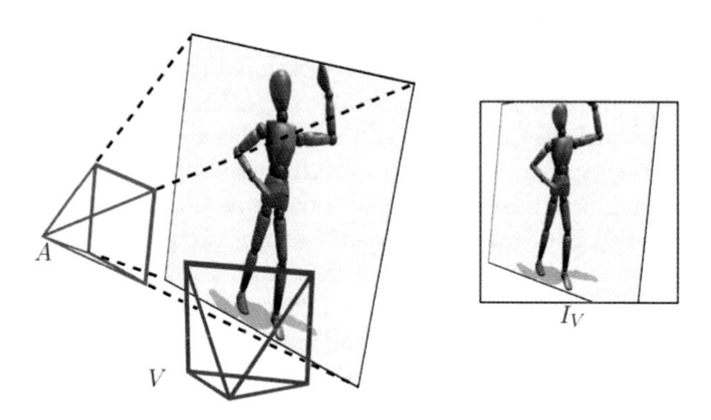

Figure 17.7: In geometry proxy-based rendering, only a coarse geometric representation of the scene may be required. In the depicted case, a single view-dependent, fronto-parallel billboard is used. During rendering, the source image is projected onto the proxy geometry.

merging them in image space [Hornung and Kobbelt 09], or by applying local displacements to billboards [Waschbüsch et al. 07]. The main advantage of billboarding is its modest computational requirements, enabling real-time, on-the-fly rendering from live-captured multi-video footage [Goldlücke and Magnor 03]. Billboard rendering has been successfully applied to free-viewpoint video of actors, sports broadcasts, and architectural scenes [Germann et al. 10, Schwartz et al. 10].

If for some image region no reliable billboard depth can be obtained, to avoid more annoying artifacts such regions may deliberately be visualized in a blurred manner (Section 17.4). In ambient point cloud rendering, for example, random depth values are assigned to pixels of unknown depth so that they form an amorphous, unobtrusive point cloud [Goesele et al. 10].

Geometry proxy-based approaches are able to achieve visually pleasing results without pixel-accurate depth information. Coarse scene representations such as billboards can be estimated very robustly, and computational as well as memory requirements are small due to the limited amount of estimated geometric information. Billboard rendering yields improved rendering results from undersampled light field data. Also, viewpoint position is not restricted to any kind of camera manifold. Still, rendering artifacts may be apparent (Section 17.4). Especially when using a simple billboard proxy, strong ghosting artifacts become visible when cross-blending from widely separated input images. The impact of such artifacts in image-based rendering on perceived image quality has been studied by Vangorp et al. [Vangorp

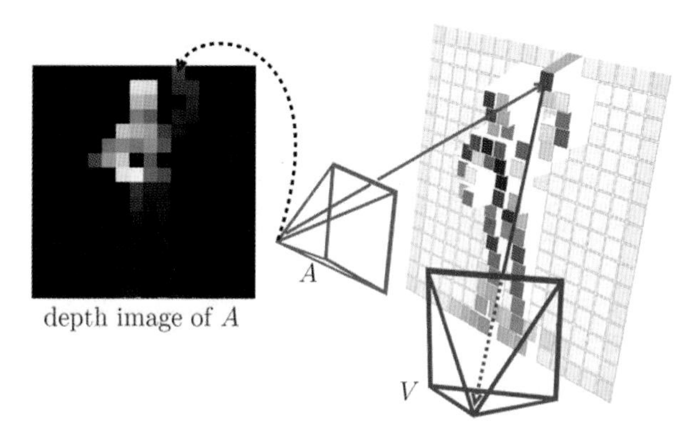

Figure 17.8: In depth-image-based rendering (DIBR), the depth of each pixel is known (visualized by grayscale depth map, left). According to pixel depth and camera matrix of camera image I_A, each pixel is projected to its world space position. Using the projection matrix of the virtual camera I_V, the image I_A is projected to the image plane of I_V.

et al. 11]. Another constraint is that although only coarse depth information is required, the input images must be calibrated.

Depth Image-Based Rendering (DIBR)

Depth image-based rendering relies on dense depth information for every pixel of all input images. To obtain reliable depth, dense per-pixel stereo matching algorithms are typically used (Chapter 8). For rendering, in general each pixel in reference image I_A is reprojected into the world space and re-rendered from the desired viewpoint I_V [Fehn 04] (Figure 17.8). In point cloud rendering [Zabih and Woodfill 94, Addison et al. 95] and point splatting [Hornung and Kobbelt 09], reprojection and re-rendering is accomplished by treating all pixels independently. Alternatively, the source images can be considered as a connected mesh [Zitnick et al. 04, Zheng et al. 09]. To avoid artifacts along object silhouettes, single quads of the mesh that feature large depth discontinuities must be locally discarded. Alpha matting is used to estimate local foreground color and alpha values, and an additional boundary layer is rendered to guarantee smooth transitions between different depth layers.

A very challenging problem is disocclusion handling. Both point splatting and mesh rendering methods may produce holes in the final image [Tauber et al. 07]. One solution is to use two or more source images for rendering that hopefully fill in the holes in the final image [Zitnick et al. 04, Zheng et al. 09]. Another possibility is to use a depth-layered representation of the scene [Shade et al. 98, Müller et al. 08]. Still, infilled image regions cannot be ruled out and may remain annoyingly visible as empty holes in the synthetized novel view. To remedy the problem, several inpainting techniques have been proposed to assign plausible color information to such unfilled regions [Criminisi et al. 03, Moreno-Noguer et al. 07, Debevec et al. 98]. Common inpainting techniques have also been surveyed and benchmarked for their applicability in image-based rendering [Schmeing and Jiang 11].

If dense and correct depth information can be obtained, depth image-based rendering can be very accurate. Similar to depth proxies, the location of the virtual viewpoint is arbitrary. Also, only few input views are needed to achieve useful rendering results. On the other side, accurate, dense depth reconstruction is notoriously difficult and error-prone for the general case. Acquisition cameras must be calibrated and synchronized accurately, and scene content may not change appearance by too much between different viewpoints, neither due to occlusion effects nor to non-Lambertian reflectance characteristics. In essence, input images may not be separated by more than about 10°, else even state-of-the-art reconstruction methods fail (Chapter 8).

To overcome the problems specific to depth-based and warping-based rendering, a hybrid approach has been proposed [Lipski et al. 14]. By exploiting both dense, pairwise image correspondences as well as depth information simultaneously, convincing rendering results can be obtained even from imprecisely reconstructed depth and inaccurately calibrated, asynchronously captured multi-view footage.

3D Geometry Reconstruction and View-Dependent Texture Mapping

If the scene is not too complex, instead of estimating local per-pixel depth it may be possible to reconstruct a complete, consistent 3D geometry model of the scene prior to rendering (Chapter 12). During rendering, the 3D mesh can then be projectively textured using a view-dependent selection of input images [Debevec et al. 98] (Figure 17.9). If the scene consists of a single object of interest, visual hulls are a common, conservative 3D geometry approximation [Baumgart 74, Potmesil 87, Matusik et al. 00]. While visual hull reconstruction and rendering is very fast and allows for real-time applications [Li et al. 03], due to its inherently limited geometric accuracy attainable rendering quality is limited. Instead, more elaborate, off-line 3D reconstruction schemes can be employed. One approach consists of performing structure-from-motion calibration [Snavely et al. 06] (Chapter 7), prior to applying quasi-dense multi-view reconstruction of surface patches [Furukawa and Ponce 10, Snavely 12, Lipski 12] (Chapter 8). Alternatively,

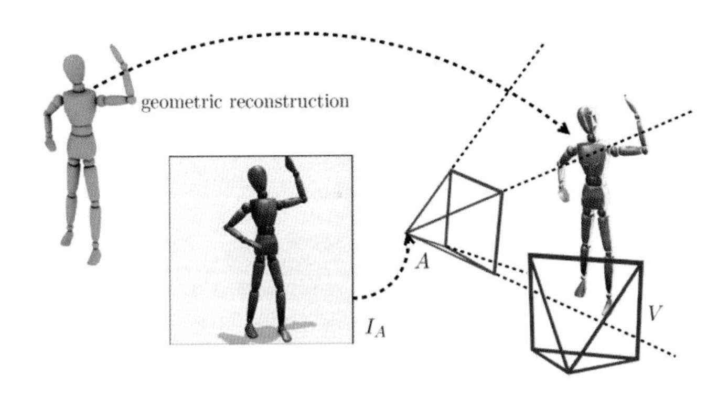

Figure 17.9: View-dependent texture mapping: if an accurate 3D geometry model of the scene is available, it can be projectively textured using only a few captured input images. Typically, more than one input image is used for projective texturing to cover all visible regions of the object. The selection of input images and their respective blending weights are assigned based on the position of the viewpoint.

depth cameras or 3D scanners may be used to obtain partial object geometry [Wood et al. 00] (Chapter 9). Finally, a watertight 3D model is obtained from the acquired point cloud using Poisson surface reconstruction [Kazhdan et al. 06] (Chapter 10). Solutions to estimate 3D geometry exist also for scenes that contain non-Lambertian objects [Vogiatzis et al. 06]. Many reconstruction algorithms rely on silhouette extraction, effectively limiting applicability to scenes consisting of a single, easily segmentable object of interest. If automatic reconstruction is infeasible, scene geometry reconstruction can also be user-guided, e.g., by specifying points and edges of the mesh to determine 3D-positions semi-automatically [van den Hengel et al. 07]. Because of their relevance for popular 3D map services, specific tools have been developed for architectural scenes that allow for user assistance and correction [Debevec et al. 96, google 12]. User-assisted methods are, however, labor-intensive.

For view-dependent texture mapping, only a small number of input views is needed to reproduce highly realistic scene appearance. Given an accurate 3D model, the scene can be rendered from any viewpoint, and the rendered viewpoint is not restricted to any particular area. Having a 3D geometry model available has additional advantages, e.g., it can receive and cast shadows in a virtual scene. On the other hand, exact alignment between projected images and 3D geometry is essential for authentic rendering results. If camera calibration is only slightly off, or if the geometry

Figure 17.10: Model-based IVBR: A priori information about scene content can be exploited to obtain robust modeling results. Instead of reconstructing 3D geometry from scratch, a parameterized 3D model may be fitted to the recorded footage.

model exhibits even small inaccuracies, annoying rendering artifacts occur [Eisemann et al. 08].

Model-Based IVBR

3D geometry can be reconstructed from multi-view imagery based on first principles (Chapter 8). However, scene reconstruction can be considerably improved if knowledge about scene content is exploited and a parameterized 3D model can be provided and a paraterized 3D model can be provided, Figure 17.10. By fitting an a priori 3D model to the recorded data, parameter space is greatly reduced and consistency enforced (Chapter 12). Prominent application areas are human pose estimation and motion capture (Chapter 11). For free-viewpoint video rendering of an actor, for example, joint angles and shape parameters of a 3D human body model as well as time-varying textures may be derived directly from sparse multi-video footage [Carranza et al. 03]. A statistical human body model enables modeling almost any person's physique [Hasler et al. 09a], even from a single video recording [Jain et al. 10]. Alternatively, laser scanning or other depth sensors can be employed to model an individual's 3D geometry precisely [de Aguiar et al. 08b, Ye et al. 11, Kuster et al. 11]. Of course, known 3D geometry and appearance of scene background or other parts of the scene can also be exploited. Many spectator sports, for example, take place on a well-defined playing field. In TV sports broadcasting applications (Chapter 21), this a priori knowledge is used for reliable background segmentation as well as scene augmentation [Hilton et al. 11, Germann et al. 10].

Dynamic Objects and Scenes

Many image-based approaches do not explicitly provide any special means to deal with dynamic scenes recorded with multiple video cameras simultaneously. Although it is always possible to apply image-based techniques independently to each consecutive frame of a multi-video sequence, this is no way to assure temporal coherence. Consequently, independent processing of consecutive time frames can result in flickering artifacts, severely impacting perceived rendering quality. In contrast, video-based rendering approaches are specifically designed for dynamic scene content by both ensuring and exploiting temporal coherence [Magnor 05].

If multi-video acquisition is synchronized across all cameras, an initial geometry model may be estimated for the first frame [Furukawa and Ponce 10] and tracked over successive frames [Furukawa and Ponce 08]. Occlusion of scene parts, however, can lead to holes in the surface model. Mesh completion may be employed to obtain a temporally coherent representation [Li et al. 12a]. For deformable objects like the human face, an initial mesh may be tracked using one or several reconstruction anchor frames [Bradley et al. 10, Beeler et al. 11].

Alternatively, instead of reconstructing a geometry model first and enforcing its temporal consistency in a second step, dynamic geometry may be reconstructed globally as a weighted minimal 3D hyper-surface in 4D space-time [Goldlücke and Magnor 04]. The hyper-surface is defined by the minimum of an energy functional which is given by an integral over the entire hypersurface and which is designed to optimize photo-consistency. A PDE-based evolution derived from the Euler–Lagrange equation maximizes consistency with all of the given video data simultaneously. The result is a globally photo-consistent, closed 3D model of the scene that varies smoothly over time [Goldlücke and Magnor 04].

Besides surface geometry, some IVBR applications also benefit from dense 3D scene motion. This so-called scene flow can be reconstructed from synchronized multi-video footage by estimating the optical flow per camera view and reprojecting the flow fields onto 3D scene geometry [Vedula et al. 05].

While synchronized multi-video recordings considerably simplify subsequent processing, mass-market cameras typically do not provide any technical means for inter-camera synchronization [Hasler et al. 09a]. Even with high-end camera equipment, temporally misaligned frames and complete frame drops can occur [Imre and Hilton 12, Imre et al. 12]. Only a few IVBR approaches explicitly allow for non-synchronized multi-view input imagery. One option is to synchronize dynamic light field recordings prior to rendering via temporal interpolation [Wang et al. 07]. By deliberately offsetting camera recording times, temporal resolution can even be improved [Li et al. 12b]. For the initial spatial reconstruction, however, still a subset of cameras has to record the dynamic scene in sync.

Besides the technical or practical inability to synchronize multiple video cameras, e.g., in the field, also multi-camera calibration can be a tedious, difficult, and error-prone procedure (Section 2.3). A warping-based approach still enables free-viewpoint video of complex outdoor scenes from completely unsynchronized, uncalibrated, sparse multi-video footage [Lipski et al. 10a]. The underlying idea is to simulate a virtual video camera by interpolating between recorded video frames across space and time. Prior to rendering, dense image correspondences must be estimated between consecutive video frames of each camera sequence as well as between adjacent cameras [Stich et al. 11]. View interpolation takes place in the spatio-temporal domain spanned by all recorded video frames [Stich et al. 08]. By subdividing the interpolation domain into tetrahedrons, with the recorded video frames as vertices and dense correspondence maps along the edges, free viewpoint navigation, slow motion, freeze-and-rotate shots, and many more special effects can be photo-realistically rendered [Linz et al. 10b].

17.4 Advanced Image-Based Methods and Extensions

The classification introduced in the previous sections gives an overview of fundamental IVBR approaches. When looking at some actual rendering systems it is apparent that many of them do not fit precisely into one single category. Some approaches have been proposed that combine different techniques. Unstructured lumigraphs [Buehler et al. 01], for example, generalize both lumigraphs and view-dependent texture maps. Depending on the level of detail of the proxy geometry, they behave like one of the extremes, or a mixture of them. The view-dependent texture mapping approach [Debevec et al. 96] also employs depth-based rendering at a fine level. For each reconstructed facade, the original textures can be projected onto the geometry, and local depth maps are computed to compensate local projection errors. View-dependent textured splatting [Yang et al. 06c], on the other hand, constitutes a mixture between view-dependent texture mapping and point splatting.

For many practical applications it proves to be beneficial to segment the scene into different regions (e.g., actors and background) and to treat them differently. In outdoor sports scenarios, for example, players are separated from the field at an early stage of the processing pipeline [Hilton et al. 11, Germann et al. 10]. While billboard representations or 3D surfaces are reconstructed for the individual players, it is often sufficient to represent the playing field by a single plane. Alternatively, user-supervised segmentation and billboard rendering is used for the foreground person while the background is reconstructed in high detail and rendered using view-dependent texture mapping [Ballan et al. 10].

Error-Concealed Rendering

As each image-based method has its own advantages, it also has its particular limitations and failure cases. In some scenarios, the user can assist the reconstruction process to obtain pleasing results. Several approaches have been suggested that require user input for geometry reconstruction [Debevec et al. 96, van den Hengel et al. 07]. Other approaches require sparse user input for scene segmentation [Ballan et al. 10, Guillemaut et al. 10] and view interpolation [Chaurasia et al. 11]. Floating textures provide an automatic correction mechanism at render time that does not require any manual intervention [Eisemann et al. 08]. Prior to the blending stage in image-based rendering, the different source image projections on the geometry proxy are locally aligned based on optical flow estimated in real-time. Alternatively, the 3D geometry may be aligned with the images using sparse feature matches, circumventing dense optical flow estimation [Germann et al. 12].

Comprehensive Reconstruction

For free-viewpoint video, i.e., rendering a dynamic, real-world scene from arbitrary vantage points, 3D scene geometry must be available in some form. To allow also for illumination changes of the scene or for augmentation, in addition surface reflectance properties must be estimated, Figure 17.11. If the scene is diffusely reflecting, the reconstruction of one consistent texture map suffices. Different approaches exist to estimate a consistent diffuse texture atlas from multi-view imagery given 3D geometry, illumination, and camera parameters [Wang et al. 01, Lempitsky and Ivanov 07, Gal et al. 10]. 3D geometry and consistent texture may also be estimated simultaneously [Matsuyama et al. 04, Starck and Hilton 07b, Liu et al. 10b, Schwartz et al. 11b, Autodesk 12, Agisoft 12, Hypr3D Development Team 12, Nguyen et al. 12].

In constrast to Lambertian objects, comprehensive reconstruction of scenes containing specular, glossy, semi-transparent, or mirroring surfaces is considerably more difficult [Ihrke et al. 10]. For non-Lambertian surfaces, the ability to recover normal directions accurately varies greatly with both actual surface BRDF and illumination pattern and may even be ill-posed [DŹmura 91]. If 3D scene geometry is known, inverse rendering allows recovering illumination and/or BRDF, represented in spherical harmonics basis functions [Ramamoorthi and Hanrahan 01a]. Alternatively, parameterized reflection models may be fitted to match captured multi-view scene

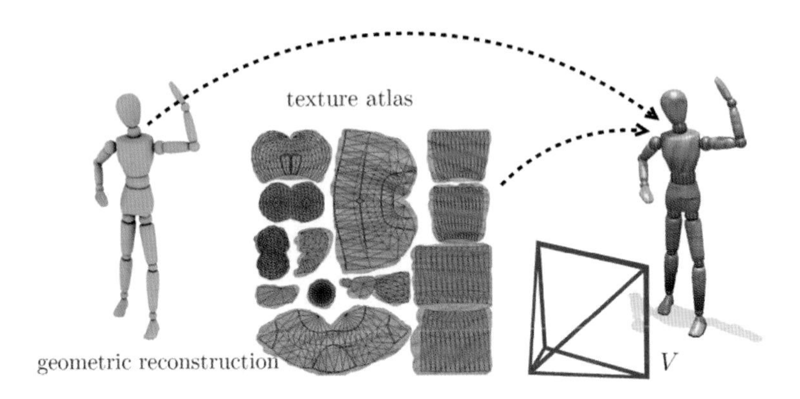

Figure 17.11: Comprehensive reconstruction: if both surface geometry and reflectance properties of the scene can be modeled from the input imagery, the traditional 3D rendering pipeline can be utilized to render the scene from any viewpoint and under arbitrary illumination.

appearance, either for known [Theobalt et al. 07] or unknown [Li et al. 13] scene illumination. If scene appearance from only a single viewpoint is to be varied for different illumination conditions, high-speed recording under time-multiplexed illumination allows relighting the scene [Wenger et al. 05].

Image-Based Object and Scene Manipulation

Image- and video-based rendering approaches concentrate on view interpolation. In recent times, image-based methods have been proposed also for modeling and rendering appearance variations. A set of parameters is used to span a domain to search and warp the images of an object. For articulated objects, for example, such a parameter set can be defined by a skeletal pose representation [Xu et al. 11, Hilsmann et al. 13] (Chapter 11), or by silhouette shape features [Hauswiesner et al. 13]. For facial expressions, facial feature locations in the image can be used [Zhang et al. 06]. During synthesis, a temporal coherent matching strategy is used to identify the *nearest* database image(s) to the given configuration of pose or facial expression in descriptor space. The retrieved images are then used to synthesize an image for this pose configuration. The best matching database image is mapped onto an animated 3D model of a human. As the retrieved image might not show exactly the same pose as required, fine-scale warping of the rendered image is necessary. Other approaches synthesize articulated human body poses by a convex combination of example poses [Casas et al. 14].

The interpolation domain for skeletal poses or facial expressions is much more complex and of higher dimensionality than the interpolation domain necessary for view synthesis. Possible expressions/poses that can be synthesized from the database are restricted to the convex hull of examples, and sampling the space such that every possible expression/pose can be synthesized becomes intractable. This limitation has been addressed by splitting up the object, i.e., the face or human body, into subregions that are assumed to be more or less independent. Each of these regions then has its own descriptor space and is synthesized from different example images. To produce the final image, the subregion images are seamlessly blended.

Compositing, Augmentation, and Consolidation with Traditional Computer Graphics

The paramount goal of image- and video-based rendering techniques is to capture real environments and synthesize novel views. However, if further interaction with or editing of the scene is required several additional problems occur, many of them still unsolved. One challenge is realistic compositing of image- and video-based rendering results with those from traditional 3D rendering approaches. It is as apparent that there are many applications where real content is to be complemented with synthetic assets, or the acquired 3D data is (partly) used in otherwise synthetic scenes. For example, few blockbuster movies nowadays are not augmented with 3D renderings in the form of special effects (Chapter 20). Also live TV sports broadcasts and other application areas rely on convincingly augmenting real-world footage with synthetic 3D-rendered content (Chapters 21, 23).

Arguably the simplest solution to combining image- or video-based rendering with traditional 3D rendering is to 3D-reconstruct the scene from the recorded video footage (Part II), augment the scene in 3D world space, and 3D-render it again. However, image-based 3D reconstruction methods have numerous limitations. A reconstruction method may be applicable only to single objects and may require a controlled environment for capture, or recorded scene appearance is not factorized into lighting and reflectance preventing photo-realistic augmentation with other 3D models, or the scene is optically too complex for faithful reconstruction. These restrictions give rise to several interesting research challenges toward realistic augmentation of real-world scenes with synthetic 3D graphics. Among the different aspects that need to be considered are correctly matching the illumination of the virtual object to that of the real-world scene as well as color bleeding and shadows cast by the virtual object onto the real-world scene, and vice versa. Fortunately, the task of augmenting a complex, dynamic real-world scene with virtual objects can, in essence, be reduced to consistently augmenting each video frame separately.

Illumination Reconstruction The appearance of a synthetic object is determined by its 3D shape, surface reflectance characteristics, and illumination. If the goal is to augment some virtual object into an image, its 3D shape and reflectance properties are known. Only the lighting conditions of the scene in the image are unknown and must be reconstructed. One simplification that is often made is that only the far-field illumination of the scene is reconstructed, i.e., the scene is assumed to be illuminated by a hemisphere infinitely far away so the illuminating light distribution can be represented as an environment map. Scene illumination can be acquired directly by placing a so-called light probe, often a mirroring sphere, into the real-world scene and taking a photo of it [Debevec 98]. Alternatively, a fish-eye lens can be used [Sato et al. 99]. If the recorded scene is not accessible anymore, illumination may be interactively estimated using some (coarse) scene geometry proxy [Karsch et al. 11]. In many cases, illumination estimation is tightly coupled with the reconstruction of overall scene appearance in a joint optimization approach [Kholgade et al. 14, Rogge et al. 14, Hara et al. 08, Haber et al. 09]. Regularly, surface reflectance of the real-world scene is assumed to be Lambertian [Ramamoorthi and Hanrahan 01b]. Illumination and reflectance reconstruction from photos and videos remains to be an active research area in computer graphics and computer vision.

Realistic Rendering In order to achieve consistent overall appearance when compositing virtual objects into real-world footage, not only direct scene illumination must be known but also inter-object light transport between all objects in the scene has to be computed. Inter-object light transport gives rise not only to cast shadows but also to such visually important yet subtle effects as indirect illumination, color bleeding, and caustics. In essence, taking these effects into account amounts to computing the rendering equation for the entire augmented 3D scene (Chapter 6).

Although stunningly realistic images can be rendered interactively (e.g., in video games), the light transport in these scenes is often approximated or based on simplifying assumptions [Ritschel et al. 12]. Rendering photorealistic images offline (i.e., without tight computational budgets but with significantly higher quality demands) is nowadays almost exclusively done using (Markov Chain) Monte Carlo methods. These methods share the concept of stochastically constructing paths that connect the sensor of a virtual camera to a virtual light source and computing the energy reaching the sensor's pixels. This process can be done in many different ways: sampling only from the sensor or the light sources (path tracing [Kajiya 86] or light tracing [Arvo 86]), sampling from both sides with deterministic connections (bidirectional path tracing [Lafortune and Willems 93, Veach and Guibas 94]), mutating paths with Metropolis light transport [Veach and Guibas 97], or density estimation of path vertices (photon mapping [Jensen 96]). Going beyond pure light transport, additional realism can be achieved by simulating the effects of actual cameras such as depth-of-field [Lee et al. 10b, Kán 12], lens flares [Hullin et al. 11], or accurate simulation of lens models [Hanika and Dachsbacher 14].

Global illumination has seen tremendous progress in the last decades [Pharr and Humphreys 10, Křivánek et al. 13, Dachsbacher et al. 13]. Nevertheless, not all techniques are equally well suited for all scene settings which can require specifically tailored solutions [Veach and Guibas 97, Jakob and Marschner 12, Kaplanyan et al. 14, Hachisuka et al. 08, Kaplanyan and Dachsbacher 13a, Kaplanyan and Dachsbacher 13b, Georgiev et al. 12, Hachisuka et al. 12]. The demands on the rendering methods increase with the richness of detail, accuracy, and the spectrum of materials.

Compositing The most successful algorithm to cope with the insertion of virtual objects into a given scene is presumably the differential rendering technique first proposed in [Fournier et al. 93] and made popular by [Debevec 98]. The idea behind differential rendering is to compute the light interaction between the scene and the virtual object, i.e., the near-field illumination (for example, shadows cast on the ground), by making use of a coarse (hand-made) representation of the scene surrounding the virtual object. The scene is rendered once with and once without the virtual object to be augmented, and the difference is applied to the original input image. Given an input image I_{bg}, the (potentially manually created) 3D scene geometry proxy is global illumination-rendered from the same viewpoint as I_{bg}, once without the virtual object and once with the object inserted, resulting in images I_{noobj} and I_{obj}, respectively. Additionally, an object matte α is computed to mark pixels depicting the virtual object as 1 and all remaining pixels as 0. The final composite is then computed via

$$I_{\text{final}} = \alpha \cdot I_{\text{obj}} + (\mathbf{1} - \alpha) \cdot (I_{\text{bg}} + (I_{\text{obj}} - I_{\text{noobj}})) \, .$$

In this form, it becomes clear that for each object the pixel value is simply copied from the rendered image I_{obj}. For the remaining pixels, if the virtual object does not affect its surrounding, I_{obj} and I_{noobj} are equal and the result is equal to I_{bg}. If I_{obj} is darker than I_{noobj} light is subtracted from the input photograph, introducing shadowed regions. On the other hand, if I_{obj} is brighter than I_{noobj} intensity is added signifying, for example, caustics. Several improvements of this technique have been proposed, e.g., for moving objects [Drettakis et al. 97], using final gathering [Loscos et al. 99], making use of differential photon mapping for refractions [Grosch 05] or taking near-field illumination into account [Grosch et al. 07].

17.5 Summary

Image- and video-based rendering can be categorized into purely image-based approaches, which directly synthesize new images by re-sampling and interpolating the captured data, and geometry-assisted approaches that exploit reconstructed depth information or higher-level models of the scene to guide the image synthesis process. Beyond pure view interpolation of static scenes, approaches for dynamic scenes and objects allow synthesizing new images in a space-time continuum. The categorization into purely image-based and geometry-assisted approaches is not to be understood as a fixed classification but rather aims at giving an overview on existing methods. Many approaches do not fit exactly into one single category but can be located somewhere in between. Combining image- and video-based rendering with traditional 3D rendering is an active field of research. The augmentation of real world-acquired scenes with virtually created content frequently requires specifically tailored methods and solutions.

18 Stereo 3D and Viewing Experience

Kai Ruhl

18.1 Introduction

In the last decade, technological advances in stereo 3D (S3D) have attracted renewed interest of both academia and industry, including the first commercially viable 3D content market for binocular video. Previous attempts at producing S3D content go back as far as the 19th century, using black and white photography providing one image per eye on static glasses. The 20th century saw first attempts toward S3D movies for a mass audience, but technological drawbacks had hindered its acceptance.

Today, S3D is well established particularly in the movie industry. Digital distribution and projection systems coupled with inexpensive polarization glasses that ameliorate user comfort issues have been crucial aspects. Equally important, content creators have gained more experience in producing high-quality S3D footage that enhances storytelling instead of relying on catchy effects. Toolmakers are able to effectively use the computational power of standard computer electronics to provide S3D content production and handling capabilities to a wider variety of movie production companies, while digital S3D camera systems provide footage that facilitates efficient post-production. In general, the entire production and delivery pipeline has matured to an unprecedented level, opening up many new research avenues (Figure 18.1).

S3D for gaming is another emerging mass market. However, the viewing experience has not yet reached sufficient acceptance levels as decreased

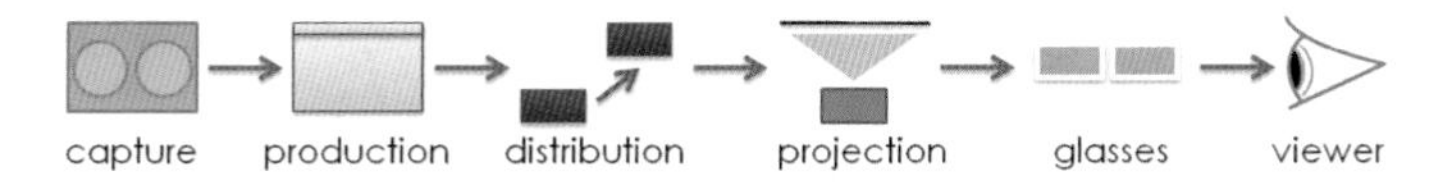

Figure 18.1: For 3D viewing, most of the traditional movie content production and presentation pipeline must be upgraded.

Table 18.1: Both monocular and binocular visual perception use 3D visual cues to construct a spatial model of a scene.

Monocular 3D Cues	Binocular 3D Cues
Size consistency	Convergence
Perspective consistency	Disparity
Motion parallax	
Focus	

willingness of users to wear glasses, a wider viewing angle distribution, and less isolated surroundings increase technological difficulties. This applies even more to 3D on mobile devices and to head-mounted displays, which only recently reached good levels of maturity.

This chapter gives an overview of contemporary S3D production and presentation, and goes into detail on 3D movies. Section 18.2 describes 3D cues that are used by the human visual system (HVS) to synthesize a 3D impression. Subsequently, current 3D viewing hardware and the way it produces these 3D cues are outlined in Section 18.3. Section 18.4 is devoted to the currently predominant stereoscopic S3D displays and their inherent challenges and limitations. Section 18.5 gives an overview of 3D video representation possibilities, both synthetic and from real-world data. Section 18.6 focuses on S3D post-production and depth estimation/correction from real-world data. Finally (Section 18.7) describes delivery of 3D content.

18.2 Stereo Perception and the Human Visual System

The human visual system (HVS) uses both monocular and binocular cues to create a natural 3D impression. Since the limited inter-ocular baseline makes stereo perception viable only for short to medium distances, monocular cues are at least as important as binocular ones and have been used by film and other media producers since the beginning of photography. Binocular cues are unavailable for a minority of the population (<10%) due to disorders known as phoria and tropia [Read and Bohr 14].

Monocular cues of a static view start with perspective and size consistency and the relative ordering of shapes. Adding observer or object movement, motion parallax is a strong cue, especially for objects at longer distances. With the eyes being finite-sized aperture instead of pinhole cameras, accommodation of the eye for sharpness is the another cue: If one eye focuses on an object of attention, only objects on that focal plane remain

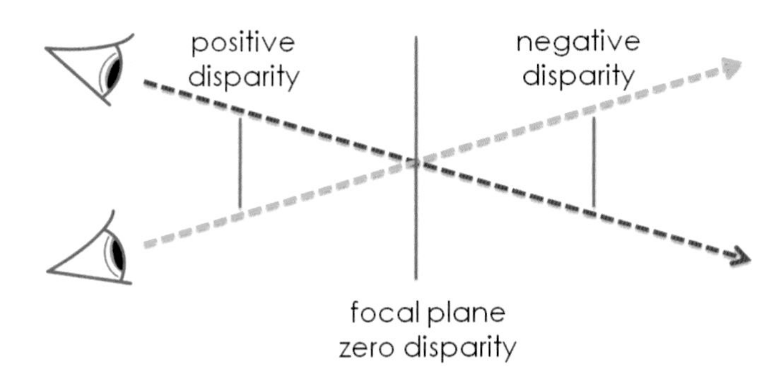

Figure 18.2: Positive, zero, and negative disparity form the stereoscopic cue of binocular vision that is based on object distance.

sharp, while objects in front or behind that plane are blurred. 2D film makers and photographers have long used this effect (called "depth-of-field") to guide the viewers' attention and produce an impression of depth, to the point that viewers have come to expect some amount of depth-of-field from professional footage.

Binocular cues use triangulation from both eyes to combine visual information and synthesize 3D structure from object relations, using only the fovea centralis, a small area on the retina with the highest receptor density and color sensitivity. Convergence is the cue that evaluates the angle between the eyes necessary to project the point of interest onto the fovea of each eye, and is the dominant cue for shorter distances. Retinal disparity is the second binocular cue that analyzes the differences between images projected into the left and right eye. Objects on the focal plane will produce the same image in both eyes; objects in front of the plane have positive disparity, and objects behind the plane have negative disparity, as shown in Figure 18.2. However, retinal disparity only works in a relatively limited depth range around the focal plane, known as Panum's area [Fender and Julesz 67]. Outside of that limited range of disparity fusion, the HVS is receiving double images (diplopia).

In a real-world setting, monocular and binocular cues are always consistent, providing a concerted viewing experience since there are no contradictions between cues extracted from the scene. The brain's cognitive processes weight and combine all this information in a meaningful way. For example, the influence of convergence is low for far-away objects, when the eye rays are almost parallel; while relative shifts from motion parallax would be weighted higher in this case.

Only a tiny portion of a natural scene is actually perceived clearly (i.e., on the fovea and in focus) at each time instance. As viewers look around any given scene, they combine different views of the same scene to assemble a mental model, all the while checking against their experience of the real world. Display systems which are not true 3D (unlike, e.g., holographic displays) have difficulties providing this experience.

Given a person's propensity to look around, content providers for display systems that cannot provide all cues in a consistent manner must steer viewer attention to keep it within focus and convergence (see also Section 18.4). An environment like a movie theater, where viewers are willing to remain seated in a dark room with mostly undivided attention toward a screen filling most of their field-of-view, is very conducive for content creators, while living-room (TV, gaming) or on-the-road scenarios (mobile devices) present additional challenges.

18.3 3D Displays

An ideal 3D display would replicate all monocular and binocular visual cues to provide the exact same viewing experience as if watching the original natural scene—essentially a holodeck whose extent covers line-of-sight. Given that molecular replication is infeasible with today's technology, other substitutes must be found.

Currently, holographic and volumetric displays come closest to providing "true" 3D, in the sense that the pixel or voxel location is physically where the natural scene would be [Yaraş et al. 10, Grossman and Balakrishnan 06]. However they are still in an early technological stage, featuring both relatively low resolution, low frame-rate and limited alpha definition (i.e., the ability to adjust solidity/translucency). Content creation for this type of display has also not yet been explored thoroughly, nor has data delivery where each frame must transport the information necessary for an entire volume.

A more easily exploitable and hence more developed technique is the use of two separate 2D displays, one for each eye. Head mounted displays (HMD), shutter glasses, polarized glasses, and autostereoscopic displays fall into this category.

HMDs are the most stable of these solutions in the respect that they eliminate crosstalk (i.e., the "bleeding" of content destined for one eye into the other) completely. However the displayed image must align to the (micro-) head movements of the wearer to avoid nausea caused by visual/inner ear balance inconsistencies, an effect prevalent in previous generation devices. As the displays are located in close proximity to the eyes, they must feature high resolution in a small panel. Recently, display panels have reached a stage where the resolution comes close to being practically applicable, and advances in real-time tracking have achieved a sufficient amount of visual/balance consistency, as demonstrated e.g., by the Oculus Rift.[1]

In scenarios where wearing a comparatively heavy HMD is impractical, shutter glasses coupled with a 2D display synchronized to them is a feasible solution that has had some commercial products, e.g., Nvidia 3D Vision,[2] mostly for small-audience applications since the glasses are still comparatively expensive. For mass audiences, passive systems such as polarized glasses are an inexpensive, practical possibility. The latest generation used in today's cinemas feature circular polarization, which allows viewers to tilt their heads to some degree without causing adverse effects. Compared to shutter glasses, today's polarized glasses still cause some crosstalk (typically less than 10% [Wang et al. 11b]).

Another emerging technology is autostereoscopic displays which provide multiple viewpoints (based on viewing angle) without requiring glasses [Urey et al. 11]. Moving around the center of a scene is possible to some degree, enabling motion parallax cues. On the other side, objects at infinity are not possible from all viewing angles since the scene needs to be centered somewhere. This requires major capture style adjustments from producers. Furthermore, the alignment of viewing angle slices to the viewer's eye is currently not entirely satisfactory and viewer position is crucial. Smaller autostereoscopic displays have been used in commercial mobile devices like phones (e.g., LG Optimus 3D, HTC Evo 3D) and gaming consoles (e.g., Nintendo 3DS), but again alignment issues make use cumbersome, i.e., the viewer has to hold the device in a certain way for best visual results.

Viewer acceptance of a certain display technology also depends on the setting. Within a movie theater, acceptance of glasses is higher than in an often more social scenario involving a TV in the living room. At the other end of the spectrum, acceptance of glasses is very low for mobile devices that are frequently operated in public.

18.4 3D Perception on 2D Displays

Currently, one of the predominant forms of 3D production and presentation is centered around S3D movies inside a theater, using polarized glasses. Creatives have understood how to design a compelling viewing experience while minimizing eye fatigue, focusing on expressing depth artistically to support the narrative and storytelling. This requires an understanding of the limitations of the 3D experience from a 2D screen plane.

[1] Oculus Rift – http://www.oculusvr.com/

[2] Nvidia 3D Vision – http://www.nvidia.com/object/3d-vision-main.html

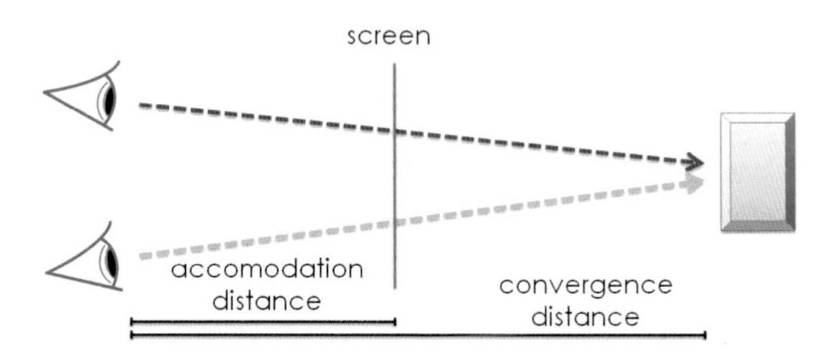

Figure 18.3: Accommodation distance is not equal to convergence distance in a S3D projection on a 2D screen, causing contradicting visual cues.

At the core, using a 2D display at a fixed distance to the viewer cannot produce a natural impression, but is instead an illusion for the human visual system that produces inconsistencies which need to be concealed, if the viewer is not to be overstressed. In particular, the accommodation (focal sharpness) distance is not equal to the convergence distance (Figure 18.3, "convergence-accomodation conflict").

Since the HVS will only tolerate a limited amount of inconsistency between cues, this limits the usable depth around the screen plane. First, the eyes' focus must always be on the screen in order to see sharp images. Now if a viewer converges on an object too far away from the screen, and the maximum sharpness is still on the screen plane, the monocular focus cue and the binocular convergence cue are inconsistent. The brain has to constantly fight against the contradicting input and may get stressed over time.

Content producers therefore carefully plan the use of the z-space, called "stereo budget" (or "depth budget"). Keeping the main action close to the screen at most times ensures the least stressful viewing experience for the audience. This can be characterized as a S3D comfort zone (Figure 18.4). Viewers will receive minimal stress within the green zone, and increasingly toward the red zone. In this manner, creatives today pay a lot of attention to the mindful and expressive depth design of each scene shot.

Additionally, the temporal aspects of depth must be considered. Scene transitions should limit aggressive jumps in depth between salient objects, as the forceful quick realignment of the eyes can be stressful. On the other hand, gentle guidance can stretch the comfort zone, as done, e.g., in a scene of the movie "Hugo"[3] where the main character slowly runs out of the screen while narrative climax, music, and depth-of-field encourage viewers

[3]Hugo movie – http://www.imdb.com/title/tt0970179/

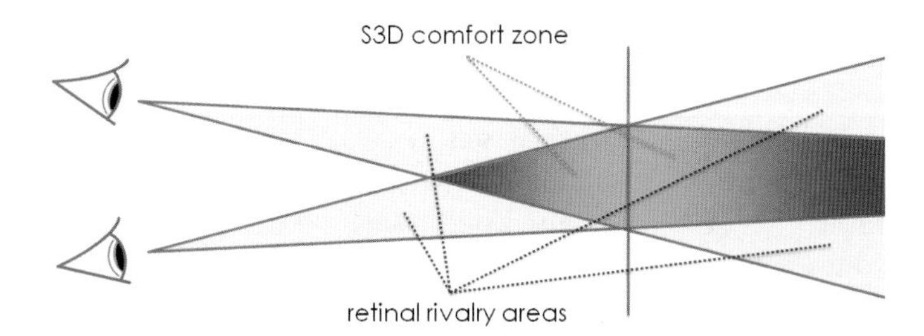

Figure 18.4: Viewer comfort is maximized by staying within the S3D comfort zone.

to fix all attention on the character. This leads to a final depth position that would be considered painful to the eyes when trying to converge on it without any setup.

Cuts should also be not too frequent because adjusting to a new scene requires more time when processing binocular cues [Berends and Erkelens 01]. Therefore, 3D movies typically feature more sweeping shots, contain fewer cuts in action scenes, and keep the depth at the same level during transitions. Applying this style is challenging for trailers, which typically have to include a multitude of cuts. Another factor is the adversarial situation of motion blur in the context of S3D, since depth is perceived more precisely at sharp object boundaries which are absent in the presence of motion blur, encouraging creatives to either increase frame rate or choose a different shooting angle or distance.

Finally, an entire class of post-production issues comes up with stereoscopic footage. The images must be aligned such that vertical disparity is eliminated, since natural scenes never produce it and it is thus perceived as stressful. Even slight inter-camera rotation or lens distortion can introduce this negative effect. Minor differences between camera sensors and lenses remain even after white balancing (more so if a mirror rig is used), and must be fixed via color grading.

18.5 3D Video Representation

Like all media, S3D defines a pipeline ranging from capture to display. When compared to standard 2D video, the additional data implies that 3D is more complex and thus bandwidth-demanding through all stages of processing (Figure 18.1). Even in the simplest case, an additional frame for the second eye has to be produced, transmitted, and displayed. While binocular S3D is currently the only commercial manifestation, other variants appear

Table 18.2: 3D scene representations can be classified as image-, depth-, or geometry-based.

Image-based	Depth-based	Model-based
Lightfield	S3D one + depth	Video textures
Lumigraph	Oriented billboards	3D mesh models
S3D two-frames	Layered depth videos	

either as part of post-production or as prototypes. The variant closest to binocular S3D uses one image plus depth map and renders the scene at the user's display. This confers a fine-tuning advantage to an individual viewer, e.g., interocular baseline adjustment or depth range remapping. Also, this representation is more compressible, an important property when streaming. On the other hand, computational cost for the display system increases, and the required inpainting at object borders and less-than-perfect depth maps may produce errors that the producers cannot remove in advance.

Taking a broader view, many other 3D representations exist which can be classified as a continuum between the two bounds of image-based vs. model-based forms (Table 18.2 and Figure 17.1).

One end consists of purely image-based approaches like dynamic light fields [Levoy and Hanrahan 96, Buehler et al. 01] or multi-view video, which do not use geometry at all and typically require many cameras for dynamic scenes. The other end is represented by approaches that use the full 3D information of a scene, e.g., by reconstructing 3D models from multi-view footage or by creating them manually by modeling and animation. In between are representations that borrow from both sides. Classical two-view S3D is sparsely image-based and borders on depth-based systems, with one-view plus depth[4] on the other side of that border [Fehn 03]. Layered depth videos are still fronto-parallel [Zitnick et al. 04], while oriented billboards [Kilner et al. 06] can already be seen as very coarse 3D video textures [Schödl et al. 00].

Different 3D representations imply different advantages and disadvantages to content producers, as shown in Table 18.3. Once chosen, the type of data is decisive for the design of the entire processing pipeline: It constrains acquisition systems and post-production algorithms, and determines rendering algorithms and view synthesis range.

As a general guideline, reconstruction costs (both computational and in terms of manual post-production refinement) are lowest for image-based systems and highest for model-based ones; conversely, the data storage and transmission demands are highest for images and lowest for models.

[4]known as depth image based rendering (DIBR)

Table 18.3: Different 3D representations imply different reconstruction efforts and amounts of data.

	Image-based	Depth-based	Model-based
Capture	Many cameras Dense sampling	Few cameras	No cameras for manual modeling or many cameras for reconstruction
Reconstruction	No effort	Depth estimation error-prone	Manual modeling labor-intensive Full 3D reconstruction error-prone, high effort for perfect results
Model insertion	Difficult	Simple if depth estimate accurate	Trivial
Coding	High bitrate, depends on number of views	Medium bitrate	Comparably low bitrate, very efficient compression possible
Rendering	Light field rendering or view interpolation, closest to footage	Depth-supported view interpolation, typically close to footage	Classical computer graphics, close to footage when rendering with high resources
Field of view	Narrow, view interpolation only	Narrow, limited extrapolation possible	Full freedom of viewing angle

Content producers must consider both filming and post-production effort, and it seems that two-view S3D movies are currently the most cost-efficient.

18.6 S3D Video Production from Real-World Data

In the 1950s, still in the analog film world, high-quality S3D production was arduous. Analog film from two cameras with limited synchronization caused errors through lens and sensor abberations; analog frames could only barely be color-calibrated by chemical means; and inserting models in post-production was only possible with physical layering, making the inclusion of temporally coherent 3D models very demanding.

Today, stereoscopic capture still suffers from lens/sensor abberations, caused, e.g., by the beam-splitting mirror in two-camera stereo rigs; lens distortion, sensor miscalibration, and inter-camera positioning errors are still present, though to a lesser degree. However, correcting these errors is much easier with digital footage. Typically in the first phase of post-production, artists remove errors with the help of commercial tools, including lens undistortion, color balancing, and vertical disparity removal, before performing depth estimation wherever needed. Because an estimated depth map is often not perfect, manual corrections are necessary and usually represent considerable effort. Alternatively, in 2D-to-3D conversion cases, a depth map may be produced manually from rotoscoping and depth map painting, possibly aided by semi-automatic approaches [Wang et al. 11c].

Depth itself is a dimension that can be edited to enhance the viewing experience as well as the narrative. Artists can use a stereoscopic analyzer like the one contained within TheFoundry Ocula[5] to visualize the depth distribution over a captured shot, and provide corrections if depth parameters get out of the S3D comfort zone (Figure 18.4); in cases where too much disparity would induce viewer discomfort, the cameras can be moved closer through disparity scaling. Capture-time stereoscopic analyzers for use with cameras also exist [Zilly et al. 10].

Another basic correction known as shift convergence adjusts the screen plane, initially determined by inter-camera positioning and their convergence. In post-production, the scene can be shifted forward or backward relative to the screen by shifting the left and right images toward or away from each other [Stelmach et al. 03]. However at the border of the images this leads to disocclusions, which can be circumvented by using bigger sensors capturing more pixels than actually needed on screen, or by cropping or inpainting.

Non-linear depth remapping is another tool-supported technique used to enhance storytelling [Lang et al. 10], e.g., for deep landscapes or to express the relative importance of actors, objects, and motions. For example, in one scene of the movie "Gravity",[6] the main character reaches out of the screen for a crucial handbook which is just floating away. Her arm has been depth-remapped to more than twice the length it could plausibly have, in order to emphasize the criticality of the moment.

A typical production can also feature the insertion of 3D models, be it actors, objects, or environmental effects like fog or fire. The amount

[5]TheFoundry Ocula – http://www.thefoundry.co.uk/products/ocula/
[6]Gravity movie – http://www.imdb.com/title/tt1454468/

of computer-generated graphics with respect to real-world footage is very flexible: Typical cases involve green-screen filmed actors being combined with completely modeled surroundings and effects, or real-world footage being enhanced by virtual objects.

All these depth-based frame editing steps have in common that the depth maps must be of high quality, otherwise depth abberations, damaged textures, and object boundaries must be repaired frame-by-frame in image space in another post-processing step. A typical approach is to fix depth map errors by manual depth map painting, where segmentation from rotoscoping can be re-used as well. This production step is labor-intensive and can generate substantial costs.

Alternatively, the depth can also be estimated and corrected interactively [Ruhl et al. 13], as described in the following. In this approach, initial depth estimation is based on fast cost volume filtering [Rhemann et al. 11], a real-time capable method to estimate depth from a stereo image pair. Given left and right views $I_l(\mathbf{x})$ and $I_r(\mathbf{x})$ with pixel coordinates $\mathbf{x} = (x, y) \in \Omega$ and RGB color values in the range $[0, 1]$, the method aims to attain for each $\mathbf{x}$ an optimal depth $Z_l(\mathbf{x}) \in [d_{\text{min}}, d_{\text{max}}]$, discretized to labels $d \in D = \{d_{\text{min}}, .., d_{\text{max}}\}$ from a set D of depth values.

Toward this purpose, a 3-dimensional cost volume $C_l(\mathbf{x}, d)$ for one (e.g., the left) view I_l is constructed. The first two dimensions of C_l are the image size, and the third dimension constitutes the number of depth labels. Each entry within the cost volume is initially a truncated sum of absolute differences (SAD) between the views, using a projection $\pi(\mathbf{x}, d)$ from left to right view based on epipolar geometry from standard calibration [Snavely et al. 06] instead of using disparity.

$$\begin{aligned} C_l(\mathbf{x}, d) = \quad & (1-\alpha)\cdot \quad \min(\tau_1, ||I_l(\mathbf{x})) - I_r(\pi(\mathbf{x}, d))||) \\ + \quad & \alpha\cdot \quad \min(\tau_2, \|\nabla I_l(\mathbf{x})) - \nabla I_r(\pi(\mathbf{x}, d))\|) \end{aligned} \tag{18.1}$$

A value of $\alpha = 0.11$ is used to favor the color term over the gradient term and $\tau_1 = 0.03$ and $\tau_2 = 0.008$ to favor only very exact matches. With the data term set, the next step is to perform a weighted filtering on C_l to arrive at a smoothed cost volume C_l':

$$C_l'(\mathbf{x}, d) = \sum_{\mathbf{x}' \in N_r(\mathbf{x})} W_{\mathbf{x},\mathbf{x}'}(I_l(\mathbf{x}')) \cdot C_l(\mathbf{x}', d) \tag{18.2}$$

The filter weights $W_{\mathbf{x},\mathbf{x}'}$ depend on the guidance image I_l [He et al. 10] similar in spirit to the anisotropic smoothness found in many variational approaches, and are computed on pairs of pixels $(\mathbf{x}, \mathbf{x}')$ in a neighborhood N_r within a filter radius r. Cost filtering is performed on each depth layer, but not between depth layers since there is no guide in the depth direction available.

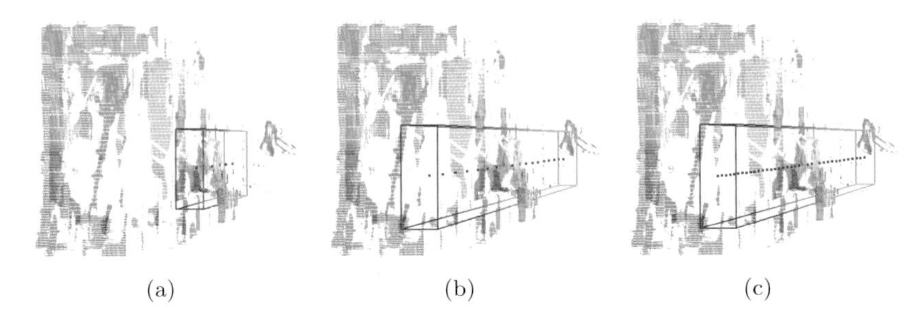

Figure 18.5: Cost block within the initial cost volume, visualized as green-blue bounding box; z-layers are shown as red dots in the center. (a) Cost block with a default size in z-direction; (b) the user has increased the z-extent of the cost block; (c) the user has increased the number of z-labels within the cost block.

Runtime is independent of filter radius r (9–24 are sensible radii, depending on image size) when using weighted box filters based on summed area tables, instead of evaluating the weights naively. The OpenCL implementation of [Ruhl et al. 13] uses a tile-based sliding-window variant which works in $O(n)$ on the GPU [Hosni et al. 11a]. Finally, the depth map Z_l is chosen by seeking the depth label with minimal cost per pixel:

$$Z_l(\mathbf{x}) = \arg\min_d C_l'(\mathbf{x}, d) \tag{18.3}$$

Since the results of any stereo algorithm are not perfect, errors cannot be avoided completely particularly on challenging natural scenes which tend to produce artifacts, among them:

Occluded regions. Objects that are occluded differently in the two views can lose significant overlap, preventing unambiguous matching. In a typical stereo configuration, this happens prominently for any object's left and right edges, which are each only visible in one camera. The closer the object is to the camera, the more pronounced the effect becomes. Automated algorithms cannot avoid this error since the information is simply not available; symmetric estimation is at least able to identify occluded regions, but inpainting is still largely heuristic. A human user, on the other hand, is able to provide plausible depth information for those non-visible parts by intuition about the object's shape.

Ill-textured regions. The majority of stereo algorithms for natural scenes (as opposed to controlled lab settings) rely on the color constancy assumption, which may be violated by lighting or camera sensor differences, noise, specular reflectance, translucent objects, caustics, etc. This impedes the matching of objects between both views. Largely uniform or repeating regions in conjunction with different occlusion boundaries in the two views (e.g., columned halls, gratings) are also not solvable with the available information. Again, a human user can assess which objects belong together, and thus distinguish between true and false matches.

The question now is how to integrate human scene understanding in a way that minimizes interaction times. Currently, the most common way is to use image editing tools to select a region via rotoscoping or segmentation, and then use stamp, cloning, and other tools to assign better depth labels.[7] The approach by [Ruhl et al. 13] also starts with a segmentation/mask in 2D image space, but instead of cloning without validation of the resulting depth, a possible range in z-direction is assigned, forming a 3-dimensional "cost block" $K_l \subseteq C_l$ (Figure 18.5). In the first two dimensions, the cost block is a bounding box around the masked or selected pixels and restricts $\mathbf{x}$ to come from $\Omega' \subseteq \Omega$. In the third dimension, the cost block is centered around the median depth of the selection $med(Z_l(\mathbf{x}'))$ with $\mathbf{x}' \in \Omega'$ to expunge outliers (other strategies are also possible) and has some extent that restricts d to come from $D' \subseteq D$. The initial extent in z-direction can either be a fixed parameter or some percentile of $Z_l(\mathbf{x}')$.

In a 3D view of the scene (Figure 18.5), both the current depth estimate and the cost block K_l are visualized. An artist can now shift the cost block along the z-axis until the estimation "snaps" the depth to the most plausible position. With each editing step, $Z_l(\mathbf{x}')$ is locally re-evaluated for all pixels in the mask, providing visual feedback in real-time. The z-extent of the cost block can be widened if objects in the selected area do not fit into it, or narrowed to eliminate superfluous estimates. As a third option, the depth label subset D' can be subdivided to include more depth labels, even to the point where $|D'| > |D|$. This increases the accuracy of z-values but takes longer to compute when the cost block is large.

Using cost blocks does not solve the problem of ill-defined regions in a mathematical sense. Instead, it merely reduces the effect of incorrect cost computation: In a narrowed set of labels d', the cost block $K_l(\mathbf{x}', d')$ merely evaluates to a more plausible depth $Z_l(\mathbf{x}')$, since a search window $I_l(\mathbf{x}')$ has a much lower probability of being matched to a randomly low-cost window $I_r(\pi(\mathbf{x}', d'))$. In the worst case when no support information can be found within filter radius r, the final depth Z_l will be essentially random, but still within the bounds of D'. If implausible, an artist can still narrow down the z-extent of K_l to a thin slice.

In essence, the user interaction cuts away large superfluous blocks from the cost volume, rather than refining the stereo matching itself. Due to

[7] http://www.fxguide.com/featured/art-of-stereo-conversion-2d-to-3d-2012/

Table 18.4: S3D coding standards will facilitate wider adoption of S3D content.

Format/Standard	Properties
Frame compatible coding	Works on existing 2D infrastructure 50% loss of resolution Not a standard
Blu-Ray 3D	Stereo high MVC profile Uses both temporal/inter-view redundancy Backwards compatible to H.264/AVC Standard since 2009
HEVC	Uses both temporal/inter-view redundancy Successor to H.264/MPEG-4 AVC Upcoming standard

the snapping behavior, artists can save considerable effort during depth map correction, and still switch back to depth map painting in case of any remaining failure cases. In the end, a high-quality depth map is available for all further stages of the S3D production and presentation pipeline.

18.7 S3D Delivery

For any type of data delivery, efficient coding and transmission is crucial, most noticeably when streaming individually (and less so when pre-loaded as in a movie theater setting). In order to achieve wide adoption, coding standards are necessary for S3D [Smolic et al. 09], complementing standards such as MPEG or VCEG (H.26x) for 2D movies. A number of coding format standards already exist for some of the 3D representations introduced in Section 18.5, where "representation" determines the format of the data and "coding" refers to the efficient compression of this data. Table 18.4 outlines a selection of current and future formats and standards.

One of the simplest coding formats regarding S3D is frame-compatible coding, whose principle is to embed a S3D stream side-by-side into a 2D stream [Vetro et al. 11]. As input, it takes a two-view stereo video and downsamples each frame by half (either vertically or horizontally). It then combines them back into a single-stream video of the same size as a single-view one, which can then be transmitted and decoded on existing broadcast infrastructure. Because of its simplicity, this approach was used in the first S3D broadcast channels, since only high-level syntax at the receiver is needed to separate the two streams and upsample them again for rendering. However, the loss in resolution is as noticeable as in interleaved 2D video, and is being considered as one of the reasons for the limited success of S3D on home television so far. TV production cost is another factor.

More recent S3D or multi-view video (MVV) approaches encode the full pixel resolution, and exploit spatial (inter-view) as well as temporal (inter-frame) redundancy for efficient compression. Classic 2D video coding makes use of temporal redundancies, i.e., not all pixels will change in neighboring frames, and many movements are regular and can thus be predicted for some number of frames. In S3D video, subsequent frames are not only related temporally but also spatially: Barring occlusion and reflectance effects, objects at the screen plane (same depth) show the same appearance, and objects in front or behind are horizontally shifted [Vetro et al. 11].

Where 2D video uses motion estimation and compensation, S3D allows for additional inter-view disparity estimation and compensation, as e.g., proposed for the 3D-Blu-Ray specification, where the second view typically adds 25-50% bitrate to the first one. Another example is the upcoming HEVC standard, which also features temporal/inter-view prediction [Müller et al. 13].

18.8 Summary

S3D video has come a long way, with a mature production/presentation pipeline and commercially successful products available today. A professional S3D movie ameliorates any adverse side-effects on the human visual system to induce a pleasent viewing experience. While an entire continuum of 3D scene representations exists, efficient hardware, tools, and processes from capture to encoding are most prevalent for two-view S3D videos enhanced by manual modeling. Other approaches are more experimental and do not have cost-effective target hardware at the current time, but are sometimes used for special effects. The perfect theoretical S3D experience featuring full view and focus flexibility is yet to be achieved.

19
Visual Quality Assessment

Holly Rushmeier

19.1 Introduction

The ultimate success of real-world visual computing applications depends on how it is perceived by its users. Integrating an understanding of visual perception into modeling systems and evaluating visual quality of rendered output have always been an essential part of computer graphics research. An example of this from the 1970s is Phong's seminal presentation of a then new reflectance model that used observation of real object highlights as motivation, and a real photograph of a sphere as validation [Phong 75]. Systematic user studies to evaluate and refine methods began to appear in the 1980s, such as Atherton and Caporeal's study of the discretization and shading of curved surfaces [Atherton and Caporael 85].

Visual quality assessment methods build on models from the extensive literature on human visual perception that have been published over more than a century. Perceptual experiments that addressed engineering applications involving human–computer interaction have resulted in useful compendiums of individual models that computer graphics researchers and practitioners can draw on [Boff et al. 88]. For many applications, however, existing human visual models are not yet adequate to address the perceptual questions of computer graphics. There is still a need in many cases to run human subject experiments [Ferwerda 08].

Methods and techniques from the study of human vision are applied both to the development of individual components of visual computing and the development of full systems. Visual assessment determines the accuracy that is needed for geometric reconstruction (Chapter 10) and motion reconstruction (Chapter 11). Visual measures are needed to determine the accuracy of illumination simulations that are used to render the reconstructed geometry and motion (Chapter 6). Alternative assessment techniques are needed for image-based rendering systems (Chapter 17). Image-based rendering systems use natural images to leverage nature's solution to part of the modeling and simulation of a new scene, but may introduce artifacts in the warping and re-sampling processes required to generate new images.

Visual quality assessment for real-world visual computing is problematic because artifacts can arise as the result of individual choices in building a system. To tackle these complex system challenges, traditional models and experiments are gradually being replaced by new tools, new concepts of what aspect of perception is to be measured, the use of crowd sourcing, and the use of new instrumentation.

In this chapter resources are given for perceptual modeling and experimentation that computer graphics researchers can draw on. Examples are given of how visual quality assessment techniques have been applied in illumination simulation, geometric modeling, motion simulation and image-based rendering. Finally, innovative approaches developed in computer graphics for quality assessment are discussed.

19.2 Models from Human Perception

The existing literature in human visual perception is used in two ways in computer graphics. First, computational models of individual aspects of perception are adapted from the biological vision literature to inform algorithms. Second, experimental techniques for studying perception in human subjects are applied to problems that uniquely arise in computer graphics. Human subject experiments are used both to explore new aspects of perception for which models are needed in graphics systems, and to evaluate final rendered results from complex systems.

Computational models used in the design of graphics systems have been drawn from basic models of human vision that were originally developed for printing, photography, and television. Most prominently, graphics has used models of color perception. Radiometry and color have already been discussed in the context of camera systems in Chapter 1. Real-world systems scatter light at wavelengths in the visible spectrum of electromagnetic radiation that ranges from 400 nm to 700 nm. The accuracy needed in spectral values is dictated by human perception of color. Human sensitivity to spectral variation is limited by the response of the eye's cones (at normal daylight conditions) and rods (under low light conditions). Formulae for the conversion of spectral distributions to standard color spaces was developed by researchers in the color and illumination fields in the early twentieth century. Different spectra but with the same coordinates in a standard space (such as the CIE XYZ space) appear identical to humans. In addition to this fundamental insight, there are numerous other models of color appearance that have been and could further be exploited such as color constancy and the change in hue with luminance. Fairchild describes these various effects as well as comprehensive models to characterize color appearance [Fairchild 13].

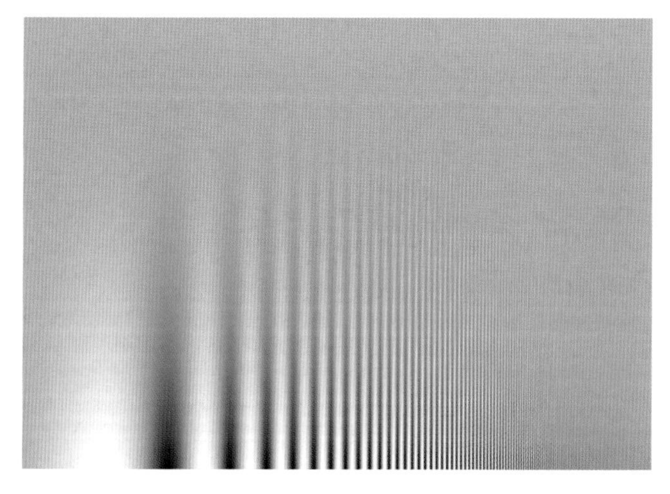

Figure 19.1: The contrast sensitivity function for human vision is illustrated by a spatial variation in intensity, with spatial frequency increasing on the horizontal axis, and contrast decreasing on the vertical axis. The illustration demonstrates that the visibility of a spatial variation is reduced at both very low and very high spatial frequencies.

In addition to color, another fundamental model that applies across many applications is the human contrast sensitivity function (CSF), illustrated in Figure 19.1. Essentially, human sensitivity to contrast is a function of the spatial frequency of the variation, with sensitivity peaking at about 5 cycles per visual degree. This set a lower limit graphics as to how spatially accurate renderings have to be. It is also important for understanding how artifacts that are invisible at one viewing distance become visible at another. For example, for a synthetic texture low spatial variations in intensity are not visible when an image is viewed closeup, but become distracting regular patterns when the image is viewed from a distance where the spatial variations increase in frequency per visual degree. Related to the CSF is the spatial masking effect. The presence of a pattern at a particular spatial frequency can mask other patterns at the same frequency. This is used in various ways to hide artifacts that could be computationally expensive to avoid [Ferwerda et al. 97].

The assessment of overall quality is a long-standing issue in photography and video, and numerous metrics have been developed over the years in the optics, photography, and video communities that combine basic perceptual models. A particularly influential metric in the computer graphics community has been daly's visible difference predictor [Daly 93]. Daly's model combines models of human sensitivity to luminance, contrast sensitivity, and masking to predict the perceivable differences in two images. This model has been used to drive iterative rendering methods [Myszkowski 98] and has served as a model for constructing image-level metrics from low-level vision models.

19.3 Experimental Techniques from Human Perception

Well-established techniques have been developed for running experiments to draw valid conclusions with respect to individual human perception phenomena. An overview of experimental methods and analysis for computer graphics have been summarized by Ferwerda [Ferwerda 08]. Different techniques are needed depending on whether threshold (just noticeable effects) or super-threshold (scaling effects) problems are being studied.

The fundamental techniques still in use today were developed in the nineteenth century. Gustav Fechner (1801–1887) described three ways to approach threshold effects. One, the method of adjustment, asks users to adjust a test display until it matches a standard. A second, the method of limits, displays pairs—a standard and a test. The test display is adjusted to approach the standard, and the user's comparison of the two, in the form of "is the test more or lest *foo* than the standard" (where *foo* is the effect being tested) is recorded. In the third, the method of constant stimuli, random test displays are shown along with the standard and the user's responses are recorded. Each approach has its inherent limitations and requires some different analysis. Super-threshold (scaling) effects can also be measured by various techniques including indirect rating, pair comparison, ranking and category scaling, as well as direct scaling. These methods range from asking directly for a rating of an impression (on, say, a scale of 0 to 100), to making multiple pairwise comparisons and using Louis Thurstone's (1887–1955) "law of comparative judgment" to derive a scale.

In general, before setting up a study, an investigator should either consult an experimental psychologist, or should carefully read the descriptions of many previous studies. The *ACM Transactions on Applied Perception* is an excellent source for descriptions of previous work, since it is particularly oriented to obtaining results that can be applied to computing. An investigator needs to pay attention to the controlled conditions and materials used, conducting the experiment with a protocol and ordering that prevents bias, and appropriate statistical analysis of the results. Controlled conditions require that calibrated display equipment is used. The experimental protocol typically includes randomized ordering of tests for the various subjects to avoid bias.

In the area of statistical analysis of human subject studies, a frequently misunderstood point is the number of subjects required by a study. As a rule of thumb, the study of very low-level effects, such as contrast perception for a sinosoidal display, require a small number of subjects. Higher level effects, such as color preferences, require larger numbers of subjects. The rationale for these numbers is that significant low-level effects are the same across the human population, and if they are found consistently for a small number of observers, they can be expected to be found for the general population. For higher level effects there is no magic number that is "enough" or typical. The adequacy of the number of subjects is indicated by the statistical significance computed for the results.

There is synergy between computer graphics and research in human perception. Computer graphics has made possible controlled conditions for investigating various types of perceptual phenomena. One result of this synergy is that a computational tool box has been developed for psychophysical experiments [Brainard 97]. This MATLAB toolbox[1] and lower level code is a long-term stable resource that offers many practical features for the mechanics of setting up an experiment.

Even simple experiments that ask observers to look at pictures and hit a button to respond are human subject experiments. Different procedures for being permitted to do such experiments apply in different countries. In the United States, it is typical that an experimental protocol that documents the measures to ensure subjects' safety and privacy must be submitted to an institutional review board (IRB) before the experiment can be conducted.

19.4 Evaluation in Lighting and Material Modeling

Perceptual methods have been used to extensively evaluate the accuracy of global illumination techniques. An early example is the Cornell box experiment conducted to evaluate radiosity solutions and color calculations [Meyer et al. 86]. The set up for this experiment is shown in Figure 19.2. Rather than establishing a scale or threshold, this work was concerned with whether an observer could reliably discern a simulation from a natural image of a simple scene. The experimental design was the classic "two alternative forced choice." Presented with two images, the observer was forced to make a selection of which was the simulated image. Obvious cues were eliminated by having the observer view both the 3D scene and the image on a display through view cameras. Observers choosing the simulation over the original approximately fifty percent of the time, the same as guessing, indicated the success of the simulation.

[1]ttp://psyctoolbox.org/

(a)

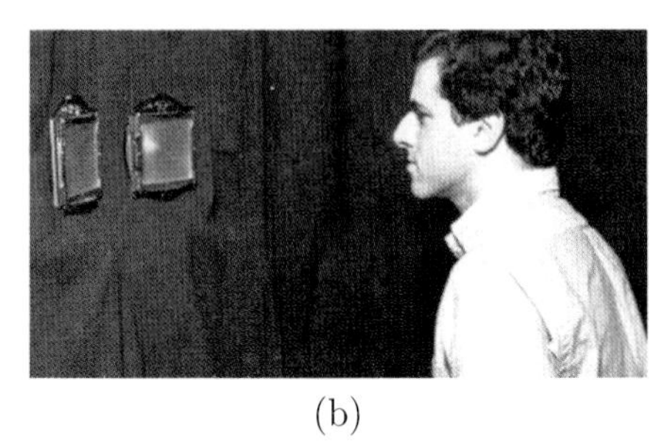

(b)

Figure 19.2: The Cornell box experiment was conducted to evaluate global illumination solution techniques. In (a) a real scene and its visual simulation are shown. Human subjects saw the real and simulated scenes through view cameras, as shown in (b). The use of the view cameras made it possible to obscure the obvious cues that indicated which was the real scene. Images from [Meyer et al. 86].

A more extensive dataset for evaluation was provided by Myszkowski et al. in "the Atrium study" [Drago and Myszkowski 01]. Both the Cornell box and Atrium used side-by-side comparisons and were focused on evaluating the ability of particular global illumination techniques in reproducing a natural image. An alternative approach is task-oriented evaluation of real and synthetic scenes introduced by McNamara et al. [McNamara et al. 00]. In a task-oriented evaluation, a user is asked to do the same task in physical and simulated environments. The goal is to find whether the user makes the same choices, whether correct or incorrect, in both the physical and simulated environments.

Local illumination models describe how individual materials scatter light. Perceptual studies to inform local models include a perceptual space for gloss [Pellacini et al. 00] and more recent studies of perceptual material evaluation [Fleming 14]. Pellacini et al. were able to establish a low dimensional gloss space by means of a scaling experiment where observers used sliders to indicate the magnitude of gloss difference between two renderings of a sphere on a checkerboard [Pellacini et al. 00]. Defining and parameterizing a perceptual space for more general scattering properties has proved more difficult because of the high dimensionality of the space and the infeasibility of asking observers to do the required very high number of comparisons. Furthermore, Fleming's more recent work suggests that humans don't recognize materials by estimating the parameters of scattering models. He suggests that human perception works with "statistical

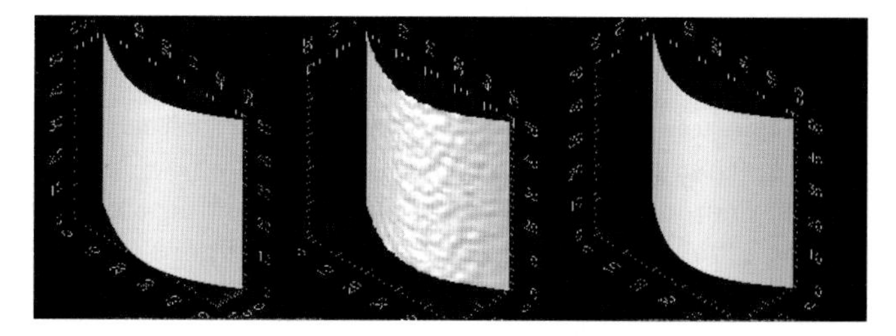

Figure 19.3: Positional accuracy in geometry is not necessarily equivalent to visual accuracy. On the left is a reference surface. The center surface is a vertex-by-vertex more accurate approximation of the reference surface than the surface on the right. However, the distribution of positional errors on the center surface introduce objectionable visual artifacts that are not visible on the right.

appearance models." Understanding what image appearance features are important in judging different classes of materials, and relating modeling parameters to the formation of these features will require much more experimental investigation.

Even if narrowing from the general problem of material appearance to the specific instance of rendering faces, there is still much more experimental work to be done. Fan et al. asked observers to classify each face in a collection of natural and computer generated images whether they were photographs of actual people, or simulations [Fan et al. 12]. In analyzing the results, Fan et al. found variability in the influence of fundamental rendering components such as color, small-scale geometric detail, and gloss. The study demonstrates that there are subtle interacting effects in modeling reflectance, geometry, and illumination that affect the perception of faces that are not yet well understood.

19.5 Geometric Modeling

Following Atherton and Caporeal's model study of geometric representation [Atherton and Caporael 85], studies have been performed to understand the requirements for modeling shape for realistic rendering. As shown in Figure 19.3, higher local positional accuracy doesn't necessarily produce visually more desirable results. Representations that introduce or remove salient, i.e., attention drawing, features are both objectionable.

Real-time rendering typically uses triangle meshes because of their compatibility with the hardware rendering pipeline. The 1990s and early 2000s saw a large volume of work on different techniques for mesh simplification to increase rendering efficiency, and this motivated the need for measuring mesh quality. Innovative work by Watson et al. used "naming time" to study simplification methods [Watson et al. 00]. Instead of comparing side-by-side renderings, model quality was judged by how long it took observers to name the object being represented. "Naming time" is a useful measure in that it does not require the user to compare directly to some "ideal" rendering of a model. In another alternative to side-by-side comparisons, Howlett et al. used gaze tracking to understand the saliency of mesh features, and the effect of simplification of more salient features [Howlett et al. 05]. Subsequently Kim et al. used eye tracking to evaluate various computational models of mesh salience [Kim et al. 10].

Finally, material properties and lighting both affect shape perception [Ferwerda et al. 04], meaning that errors in materials and geometric representations are related.

19.6 Motion

Perception has been used to understand acceptable limits in motion simulation and its presentation. It is well known that humans are sensitive to motion, and that simple animations showing bright dots moving can convey ideas such as a human walking [Hodgins et al. 98]. Since fully realized animation is expensive to compute, previewing a motion sequence in simplified form is desirable. However, in an early paper on human animation and perception Hodgins et al. presented human subject experiments that showed that people are able to detect more subtle variations in motion with full polygonal models [Hodgins et al. 98]. Doing side-by-side comparisons is even more difficult for animations than for static images. Experiments need to be designed so that they don't require too many judgments at a time and don't require judgments that are difficult for the experimenter to convey. In Hodgins et al.'s work the observers were asked to compare the motions of figures rendered with the same geometric models and to assess when they were able to detect subtle differences in motion. In later work in the same area, Hodgins, working with colleagues in neuroscience, found that there is a relationship between the perceived realism or naturalness of character motion and a character's anthropomorphism [Chaminade et al. 07].

The perception of individual character motion is different from the perception of motion in crowds. From practical industrial experience, realistic large crowd animation does not require every character to be distinct. Instancing is a heavily used technique to reduce computational expense. McDonnell et al. studied the visibility of instanced characters, or clones, in a crowd [McDonnell et al. 08]. The technique used was to present observers with sets of characters that were either appearance clones (i.e., same shape and clothing) or motion clones (walking with the same gait). The time for observers to spot the clones was recorded. It was found that motion clones were much harder to detect than appearance clones.

While humans are very sensitive to nuances in human and character motion, they can also detect errors in the physical motion of inanimate objects. O'Sullivan and Dingliana studied human sensitivity to the accuracy of collisions of simple objects [O'Sullivan and Dingliana 01]. In this work, observers were asked to make judgments about collisions between spheres. For example, they were shown collisions where varying gaps were left between the spheres rather than showing the spheres actually touching during the collisions. The observers were asked whether the spheres had touched during the collision. The result that the observers reported the spheres touching in cases where there was actually a gap suggested a level of detail algorithm for detecting collisions that allowed for error that would not be detected visually [O'Sullivan and Dingliana 01].

19.7 Image-Based Systems

Image-based systems rely on natural images for many component elements to be correct (Chapter 17). However, the source input images frequently need to be warped and resampled to generate new images. Techniques are needed to see if the operations on the images introduce visual artifacts.

An example of the evaluation of an image-based rendering system is the study by Vangorp et al. of perspective distortions in a virtual tour along city streets [Vangorp et al. 13]. The study outlined a model for how virtual tours are created. Multiple images of building facades are captured as a vehicle moves down a street. The images form a panorama which is then displayed for arbitrary views along the route by projecting the images onto planar proxies for each of the buildings. The problem is that if images aren't captured closely enough, the projected images will look obviously incorrect, with corners of buildings that should have faces meeting at 90 degrees making the distortions most obvious. The experimental procedure used had observers indicate their estimate of the angle of corners presented to them by adjusting the relative orientations of two boards on a hinge, rather than asking the observers to give a number. This allowed the observers to report their judgments more directly. The result of the study was a recommendation for how closely images needed to be captured given a particular depth variation for buildings along the street.

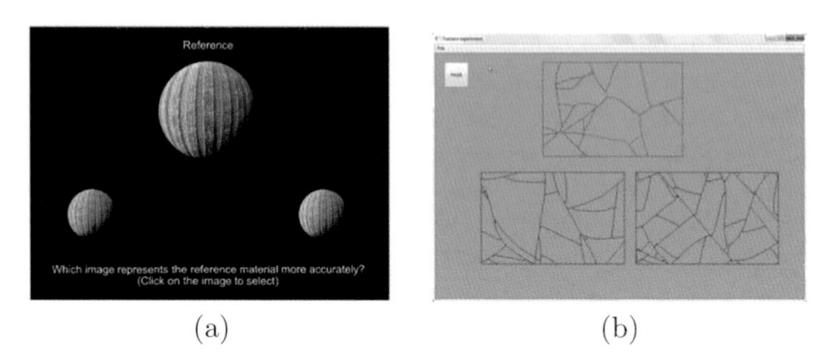

Figure 19.4: Examples of tests for visual equivalence. In (a) filtering techniques for bidirectional textures are tested for the visual equivalence of filtered and downsized images of an object covered by a complex material (image from [Jarabo et al. 14]). In (b) the visual equivalence of crack patterns are tested relative to a reference standard (image from [Glondu et al. 12]).

19.8 Beyond Classic Models and Experimental Techniques

Approaches for modeling and evaluation have in many ways moved past traditional perceptual models and psychophysical experiments. New ideas for visual assessment that have emerged in computer graphics include no-reference metrics, visual equivalence, crowd sourcing, and novel measurement techniques.

In general, the ideal image or animation is not available or is too expensive to compute. An assessment method that doesn't rely on a comparison is desirable. The idea of "no reference" metrics addresses this issue. For assessing the quality of a rendering given reliable geometric, material, and motion representations, a no-reference metric was developed by Herzog et al. [Herzog et al. 12]. The metric is built using machine learning on a database of good and bad renderings. The quality of the rendering is learned as a function of the data that a rendering algorithm has easily available—depth information and diffuse textures.

Taking the full problem of determining the quality of input representation and accuracy needed of the rendering technique, an important issue is that accuracy does not mean pixel-by-pixel equivalence to the "perfect image." It is a rare, and perhaps never occurring case, that images are needed that are pixel-perfect. Instead, what is needed is a picture that represents the same thing, made of the same stuff, in the same environment. Ramanarayanan referred to this type of similarity as "visual equivalence" [Ramanarayanan et al. 07]. For example, from work by Jarabo et al. [Jarabo et al. 14], an object covered by material rendered at two different spatial scales is shown in Figure 19.4(a). The method for filtering the material texture is being tested. It is not expected that the images are pixel-identical because they are at different scales. It is necessary, however, that a successful filtering technique produces an image that appears to be the same object made of the same material—that is, that they are visually equivalent. Another example, from Glondu et al. is shown in Figure 19.4(b) [Glondu et al. 12]. In this example, it is desired to create a material that has been cracked the same as an exemplar. An experiment was conducted to ask observers whether the cracks on the left or right looked like they were from the same process on the same material—that is, whether they were visually equivalent. In this second case, statistics were obtained to characterize when patterns were visually equivalent, and this was used to drive a simulation to create virtual materials that appeared to be equivalent to an example from a natural photograph.

A "weak link" in most perceptual experiments is that observers have to give an active response—verbalizing a response or hitting a button. Recent work has exploited alternative, passive techniques for measuring response. Lindeman et al. used electroencephalograph (EEG) [Lindemann et al. 11] to measure an observer's response to artifacts in image-based rendering. Advances in hardware that have made EEG economically accessible, and in signal processing techniques to deal with the noise in the EEG signal have made this a viable technique. Mustafa et al. demonstrated that certain types of video artifacts can produce consistent EEG responses [Mustafa et al. 12].

Another problematic issue in conducting experiments is recruiting an adequate number of diverse observers. For some cases crowd sourcing is a viable approach [Heer and Bostock 10]. Heer and Bostock found that for some basic perceptual experiments crowd sourcing produced the same results as controlled experiments. They hypothesized that the larger number of subjects that are possible in a crowd-sourced experiment offsets the lack of control in experimental conditions. In a sense, with a large group the diverse conditions are "averaged out" by the large number.

19.9 Summary

There is a long history of using perceptual insights and human subject experiments to create realistic images and evaluate them. Past work has built a lot on performing experiments in laboratory-controlled conditions, with an emphasis on traditional approaches such as two-alternatives-forced-choice. Recent work has moved beyond this to systems that predict perceptual impact by validated metrics that encompass a wide range of effects, the use of crowd sourcing, and by using systems to gather user responses that do not require active, conscious, user input. The problem, though, of producing and assessing realistic imagery remains challenging because of the non-linear, convoluted interaction of many perceptual effects simultaneously.

Part V

Applications

20
Facial Capture and Animation in Visual Effects

Darren Cosker, Peter Eisert, and Volker Helzle

20.1 Introduction

In recent years, there has been increasing interest in facial animation research from both academia and the entertainment industry. Visual effects and video game companies both want to deliver new audience experiences—whether that is a hyper-realistic human character [Duncan 09] or a fantasy creature driven by a human performer [Duncan 10]. Having more efficient ways of delivering high-quality animation, as well as increasing the visual realism of performances, has motivated a surge of innovative academic and industrial developments.

Central to many of these developments are key technical advances in computer vision and graphics. Of particular note are advances in multi-view stereo reconstruction, facial tracking and motion capture, dense non-rigid registration of meshes, measurement of skin rendering attributes (e.g., BRDFs for skin and skin subsurface scattering models), and sensing technology.

This chapter builds on concepts already described earlier in this book—such as 3D capture, rigging, and non-rigid registration—and takes a more practical look at how they might typically be applied in visual effects. First, methods and applications for facial static capture and rendering are considered, before dynamic capture is addressed. Finally, a case study is examined called *The Gathering* involving the creation of an animated face from animation to final composite.

20.2 Static Facial Realism and Capture

In today's world, static facial realism has reached a level where humans cannot distinguish 3D facial models from real photographs anymore.

Technology such as the Light-Stage [Ma et al. 07] allow the capture of highly detailed facial surface information. Coupled with sub-surface reflectance data [Donner et al. 08] facial models now display photo-realistic likenesses to real faces. Such technology is now widely used in modern motion pictures, e.g., *Spider Man 2* [Fordham 04] being one recent high-profile use of the Light-Stage. In these circumstances, the high-detail static scans are composited onto either a stunt-actor's body, or a digital double. This is where the actor is to be placed in situations that may not be practical or safe (such as explosions). However, it is still the case that the actor's expression is typically static in these situations, and close examination of such shots reveal a dead-like facial quality. An early high-profile example of facial replacement was in the *Matrix* sequels [Borshukov et al. 05], where passive facial scanning was used to obtain 3D faces with high detail facial texture. An important aspect of the use of facial scans for movies and video games is that faces must be renderable in a wide range of environments so that the face can be convincingly composited into the overall scene. Therefore, the UV map (texture data) is typically diffuse albedo. Skin detail is enhanced through high-resolution normal maps [Ma et al. 07] or geometry [Beeler et al. 10], and rendering is enhanced through sophisticated BRDFs/BSSRDFs modeling material properties [Donner and Jensen 06, Jensen et al. 01]. Acquisition of such reflection properties has advanced widely over recent years, resulting in highly detailed rendering [Donner et al. 08].

An important aspect to the realism of synthetic humans is the realistic rendering of hair, which has made significant progress in the last years. Single hair fibers have been modeled [Marschner et al. 03] as semi-transparent elliptical cylinders. By defining surface reflection as well as scattering inside the hair, the complex lighting characteristics of real hair with its view dependency, highlights, and color changes can be accurately reproduced with moderate rendering complexity [Ren et al. 10]. The possibilities to model several fibers up to a complete hairstyle range from NURBS surfaces via thin shell volumes to strain-based modeling by parameterized clusters, fluid flow, or vector and motion fields [Ward et al. 07]. In order to simulate the complex lighting interaction between strands of hair, Lokovich et al. [Lokovic and Veach 00] propose a deep shadow map which relates visibility to depth for each pixel, yielding realistic but computationally expensive self-shadowing. Approximation algorithms for making the simulation of multiple scattering among hair fibers tractable have been proposed using methods like photon mapping [Moon and Marschner 06b] or spherical harmonics [Moon and Marschner 06a].

Whereas laser scanning technology was initially the most accurate way to derive static facial detail, passive scanning technology using consumer hardware is now popular [Beeler et al. 10, Blumenthal-Barby and Eisert 14]. Multiple consumer-level SLR cameras are used to acquire high-detail images which provide strong features for stereo matching algorithms (Chapter 8), and can result in captures with skin pore (mesoscopic-level) facial details [Beeler et al. 10]. One aspect to consider when using such data is practicality, as the meshes can contain millions of vertices. This is a different approach to those methods currently considered in, for example, movies where low polygonal meshes are used along with high-detail normal maps to display facial meso-structure. There is therefore still a great deal of work to be done on using such technology practically for video games and modern VFX.

Many state-of-the-art stereo and multi-view approaches are local in the sense that they reconstruct the 3D location, and sometimes orientation, of isolated image patches [Furukawa and Ponce 10]. While this strategy is beneficial for parallelization, it requires a post-processing stage to generate a mesh. The reconstruction yields a point cloud with outliers which has to be filtered and meshed with appropriate algorithms (Chapter 10), such as Poisson meshing [Kazhdan et al. 06]. Smoothness priors are often only considered at the meshing stage. Local reconstruction is difficult to combine with efficient interactive tools. As each patch is unaware of its neighbors, the correction of a single mismatched patch by the user will not affect its neighbors, although they are likely to be erroneous as well. Therefore, [Blumenthal-Barby and Eisert 14] follow a similar approach as [Beeler et al. 10] but uses mesh-based deformable image alignment for the reconstruction of high-detail face geometry (including hair) from two or more SLR cameras (Figure 20.1). Instead of iteratively matching small image patches along the epipolar line, an entire view is warped to target views in an uncalibrated framework incorporating a mesh-based deformation model. The additional connectivity information enables the incorporation of surface-dependent smoothness priors and optional user guidance for robust and interactive geometry estimation [Schneider and Eisert 12].

Figure 20.1: Static reconstruction of the head including hair from two images [Blumenthal-Barby and Eisert 14].

Most digital face replacement in movies involves static face replacement, with the actor having little or no movement in facial expression. Although this might be satisfactory for a few frames, as soon as the face moves, or the shot continues for more than a few seconds, this illusion becomes hard to maintain. In the next section, the movement of faces is considered, especially with respect to maintaining an illusion of realism.

20.3 Dynamic Facial Capture and Animation

The holy grail of facial animation research is the portrayal of characters indistinguishable from real humans. This is extremely difficult since humans are experts in detecting the slightest flaws in faces. Even minor defects can break the illusion of realism. In the previous section, static faces are considered where realism has reached a point where it is impossible to distinguish computer graphics from real photographs. However, in order to display a synthetic human that is truly life-like, the movement of the face remains a major challenge.

Arguably, it is easier to convey dynamic realism in the play-back of actual recorded performances than to author a new animation. In order to highlight this, the acquisition of dynamic 3D facial sequences (termed here as 4D for brevity) is first considered.

There are now many commercial companies that market 4D facial capture systems, i.e., those that can obtain 3D mesh data at video recording rate (e.g., Dimensional Imaging,[1] 3DMD[2]). However, the focus here is primarily on academic research in this area. One of the first compelling uses of dynamic facial capture in movies was in the *Matrix* sequels [Borshukov et al. 05]. A passive stereo capture system was constructed where 3D mesh data can be acquired from a face at video rate along with high-resolution texture. The recorded sequences were then composited onto the actors in key action sequences. An extension of this system called *Universal Capture* was later used in several Electronic Arts (EA) promotions and video games (e.g., Tiger Woods Golf [Borshukov et al. 03]). Here, the system was made more robust by adding markers to the actor's face. This could be used to stabilize and track a canonical mesh (i.e., mesh with a known topology) through the captured sequence. Bickel et al. adopt a similar approach with the addition of extra facial paint to appropriately capture wrinkles on the face [Bickel et al. 07].

The use of markers has overcome previous issues related to tracking a face mesh using optical flow. Such methods are notorious to drift, caused

[1] http://www.di3d.com
[2] http://www.3dmd.com

primarily by fast facial changes, for example during speech. Early approaches to avoid the drift in markerless tracking over longer sequences are the incorporation of additional constraints from the silhouette [DeCarlo and Metaxas 00] or the use of an analysis-by-synthesis estimation as in [Eisert 03]. Both methods ensure that estimates are referred to a global reference and avoid error accumulation over time. This approach is also followed by Bradley et al., who propose a multi-view stereo capture system comprising 14 HD cameras mosaiced together [Bradley et al. 10]. This results in a highly detailed set of images upon which to apply optical flow for mesh tracking. Referencing the initial frames of the sequence results in improved mesh stabilization over time. Expanding further on this work, Beeler et al. introduced the concept of anchor frames for stabilizing 4D passive facial capture [Beeler et al. 11]. In this work, neutral frames in the sequences are searched for and then used to essentially reinitialize mesh tracking where possible. This also has the added benefit of offering robustness to certain facial self-occlusions (e.g., as caused by the lips). Although having a lower geometric resolution than previous, passive static capture work [Beeler et al. 10]—which includes approximated skin pore geometry—the extension to 4D including the impressive temporal mesh coherence is a high current benchmark in contemporary facial capture research and development.

One highly successful recent demonstration of the use of 4D capture in industry is from the video game LA Noire.[3] Hundreds of hours of actor footage were recorded in a controlled lighting environment. Key 3D character scenes were then composited with the volumetric facial performances resulting in highly detailed and realistic results. Another high profile use of 3D technology for industrial use was by Alexander et al. [Alexander et al. 10]. The Digital Emily Project was a collaboration between Image-metrics and USC using Light-Stage technology to capture high detail normal map and surface reflectance properties from an actor's face [Hawkins et al. 07]. A facial blendshape rig was constructed from captured 3D data and then matched to the performance of the actor using proprietary Image-metrics markerless facial capture technology. Blendshapes are facial poses of different expressions—from stereotypical (happy, sad) to extremely subtle (narrow eyes). The term *rig* is used to describe the complete facial model with all its control parameters. The degrees of freedom of the facial rig are a function of the number and complexity of the blendshapes, and new facial poses are created by combining blendshapes with different weights (Chapter 13). More recently, the Digital Ira Project [von der Pahlen et al. 14] demonstrated how high levels of static and dynamic realism could be animated and rendered in real-time. Thirty high-resolution facial scans were

[3]http://www.rockstargames.com/lanoire

captured using the new Light-Stage X system [Ghosh et al. 11], providing data for a facial rig. Video performance was then captured of an actor and used to animate the facial performance, combined with sophisticated real-time rendering of multiple effects [Jimenez et al. 12].

While the LA Noire production is a high-profile use of 4D capture, it is essentially playback of the recorded data. On the other hand, the Digital Emily and Ira projects demonstrate a degree of performance-driven animation, or retargeting. This type of animation is highly popular in academia and industry, where a performer animates a *puppet* via motion capture or speech (audio only or phonemes). In the case of the both the Digital Emily and Ira projects, the rigs are tracked and animated directly from the actor reference video footage. However, in other cases it is often necessary to retarget between two different rigs one created as a likeness to the actor's face (which is tracked to the input performance) and a second (often a creature or non-human character) animated from output controls of the first rig. In such cases, a rule-based or example-based mapping must be learned between the two rigs. This is a current active area of academic and industry research [Bhat et al. 13]. While it is not the intention of this chapter to give a detailed review of retargeting methods, the excellent course material in [Havaldar et al. 06] encompasses many of the ideas in this area still used today. The aim here is rather to make the distinction between direct playback of captured volumetric animation and the creation of realistic character animation given some reference (e.g., actor performance). However, one important point to make is that even given the best tracking or analysis of a human performer, *automatic* animation of a rig to a level satisfactory for visual effects is still an open problem. Typically, after automatically animating a face in this way, an artist is still required to spend considerable time matching and adding secondary rig movements to the reference performance. In video games, this process can be a hindrance—where hundreds of hours of generated performance may be required for delivery under a short time constraint. In this scenario, a lower level of quality than VFX may therefore be acceptable, as generating VFX quality for current video game productions would add an unrealistic burden on third-party facial animation production or in-house game studios.

The movie *The Curious Case of Benjamin Button* [Duncan 09] contains another successful example of human realistic performance-driven animation and retargeting. MOVA[4] performance capture technology was used to collect 3D scans of Brad Pitt's face and used for blendshape rig construction. Animation was then carried out with the aid of markerless performance mapping from reference footage of the actor. The movie *Avatar* [Duncan 10] also pushed forward the technology of facial

[4]http://www.mova.com

performance capture and retargeting. Although the characters were not human, the movie demonstrated that modern techniques involving motion capture and artistic input could be used for producing large volumes of high-quality performances. The production involved the use of head-mounted cameras, targeted at the actor's face for recording the movements of painted markers. These movements were transferred into a combination of blendshapes per-frame, and the resulting animation used to *block out* an initial animation as a first pass for artists—who later edited and enhanced the performance with the aid of additional video reference (akin to the method previously described earlier in this chapter).

While marker-based motion capture techniques are widely popular e.g., using commercial optical capture systems or painted markers (e.g., Vicon's CARA system[5]), markerless methods provide the potential to capture areas of the face where marker placement is too obtrusive. In addition, it raises the possibility of obtaining a dense capture field for the face, for example based on skin pores. Where the facial rig is based on blendshapes [Havaldar et al. 06], the aim is to optimize a set of weights that approximate the positions of the markers. In marker-less systems, such markers might be located using image-based deformable tracking techniques such as active appearance models [Cootes et al. 98]. Another alternative is to fit the blendshapes to 4D surface data [Weise et al. 09, Valgaerts et al. 12]. This latter method has also been shown to work with consumer 2.5D capture devices such as the Kinect [Weise et al. 11]. However, whether this technology alone can provide the fidelity required for VFX to move beyond optical or marker-based methods remains unclear. It may therefore be sensible in the future to consider a combined approach: markers, high quality RGB, and depth sensors.

In the examples so far, facial dynamics have been captured and replayed, often with considerable artistic manual intervention [Alexander et al. 10]. However, the concept of using such data to author entirely novel performances without reference footage remains a difficult challenge. The success of such methods still largely depends on artistic talent. Advances in interactive facial models, and new methods to create efficient rigs, are promising avenues for improvement. In the last part of this section, some recent advances in blendshape rig construction are briefly considered that could help animators use performance capture data more efficiently and provide better artist tools.

One challenge is how to create effective blendshapes. A standard approach in modern VFX is based on action units (AUs) from the facial action coding system (FACS) [Ekman and Friesen 78], e.g., in *Monsters House* [Havaldar et al. 06] and *Watchmen* [Fordham 09]. Having a FACS

[5]http://www.vicon.com/System/Cara

basis can potentially provide a mapping between different facial rigs. This can be especially useful if one blendshape model is based on actor facial scans and fitted to an actor, and then the weights are transferred onto a puppet model, perhaps of a creature. More recently, Li et al. considered creating blendshape rigs given only a few example expressions and a generic blendshape rig with a wider number of expressions [Li et al. 10]. Such systems can potentially reduce artist time when manually sculpting blendshapes for rigs, and also for reusing existing blendshape models when new rig creating is required. Facial rigs in movies can potentially become very large, with hundreds of blendshapes for *hero rigs*, i.e., rigs required to deliver close-up expressive performances [Fordham 03]. Any technique for increasing efficiency is therefore of high value to industry.

Facial animation bases, or blendshape bases, are also not restricted to artistically sculpted facial expressions or captured 3D scans. Principal component analysis (PCA) also offers a basis for animating faces. However, although this basis is orthogonal—meaning that each expression has a unique solution with respect to the basis—these are often not intuitive enough for artistic animation. In order to address this, Tena et al. [Tena et al. 11] recently proposed a region-based PCA modeling approach that allows more intuitive direct manipulation of local facial regions. Their method also highlights how solving for expression weights locally can provide better approximation of motion capture data. Ultimately however, what an artist will desire of the facial model is a set of controls that are both intuitive and also orthogonal such that altering one expression does not interfere too much with others. To counter this, blendshape rigs become highly complex, with additional shapes (corrective blend shapes) included to counter interference cases. In an ideal world such correctives would not be necessary given the extra work burden they impose, and future work is still required to address this core problem.

Given the discussion of static and dynamic facial capture, the next section considers a case study where facial models based on the likeness of real people were animated and composited onto real footage. This builds on many of the ideas expressed previously in this chapter, and also highlights many of the practical and real-world constraints such a project imposes on animators and technical directors.

20.4 Case Study: The Gathering

The Gathering is a final year short film of Filmakademie Baden-Wuerttemberg.[6] The protagonists in this short film were created digitally

[6]http://www.svendreesbach.com/the-gathering/

based on photo references. The process relied completely on artistic skills since no real reference for 3D scanning or reflectance measurement could be employed. The budget and time constraints of the project demanded to create all assets digitally.

Once the digital models were completed, their facial animation rigs were created by applying the *Adaptable Setup for Performance Driven Facial Animation*[7] [Helzle et al. 04]. This extension for Autodesk Maya[8] enables a rigging artist to apply a generalized library of muscle group movements conforming to the FACS [Ekman and Friesen 78] system to any humanoid geometry. The deformations are driven by a dense data model which includes the non-linear characteristic of facial actions. Compared to static modeling and interpolation of blendshapes, this approach allows for fast and flexible control over the individual facial deformations. The toolset allows complete control in how this data is applied and adjusts to the physiognomies of the geometry. The rigging artist has to manually apply facial landmarks that drive the deformation. The approach has its limitations mainly with respect to the amount and influence of the 69 deformation objects. One way to overcome this limitation was with the use of a limited number of corrective blendshapes, i.e., new blendshapes that trigger other key blendshape combinations to alleviate unwanted or unnatural movements.

Custom extensions to the toolset allow controlling the stickiness of the lips as they part when speaking. This effect is due to moisture on the lips, causing them to open from the inside to the outside as the lips part. Furthermore, the effect of the eyes bulging the upper or lower lids as the gaze changes was realized using a complex constellation of additional deformation rig objects. A fast animation rig allowed quick iterations when animating the sequences and kept the animation artists motivated. All facial animation was realized by rotoscoping the movements recorded from the real actors on set. The head movement could be extracted by rigid body tracking of the markers applied to the actors' heads as shown in the top left of Figure 20.2.

The animated models were rendered using Newteks Lightwave 3D[9] software package. The top right of Figure 20.2 shows the raw rendering which included additional information like motion vectors, reflection and diffuse values embedded inside Open-EXR files, which were provided for compositing. The final compositing was accomplished in The Foundry's Nuke[10] software, integrating the CGI elements into the plates (lower left of Figure 20.2)

[7]http://fat.research.animationsinstitut.de
[8]http://www.autodesk.com/maya
[9]http://www.lightwave3d.com/
[10]http://www.thefoundry.co.uk/nuke/

Figure 20.2: The Gathering: Case study on facial animation. ©Filmakademie Baden-Wuerttemberg, The Gathering, 2011.

before final coloring and additional effects like cigarette smoke were added (lower right of Figure 20.2).

The Gathering shows that it is possible to create convincing digital faces by relying mostly on artistic skills and powerful tools for facial animation. However, it also highlights that the complexities of creating a face and its movements digitally demand a wide set of skills.

20.5 Summary

Capturing real faces with modern technologies like Light-Stage provides 3D face models with high geometric detail and sophisticated material properties that enable face synthesis that is almost indistinguishable from a real picture. While static face models can be used for replacing an actor's face for a few frames, often dynamic face capturing is also desired, which is still a challenging task due to the sensitivity of a human observer for subtle inconsistencies in facial motion. Facial dynamics are usually modeled by a blendshape rig, either from scans, multiview images, or manual work of an artist, while animation data is often derived from marker-based or, more recently, markerless motion capture systems. As shown in the presented case study, convincing digital faces can be animated with such techniques. However, creating realistic facial animations and models still requires significant manual work by artists, leaving room for novel algorithms and toolsets to simplify and automate the process. In terms of research challenges, there is still a large scope for further work in this area. Creating rigs is a time-consuming process, and methods to automate this at production quality are highly desirable. Another core area for future work is in the retargeting of actor faces to new creatures. One challenge in achieving this aim, however, lies perhaps in the contrast between the academic and industrial worlds: academia often has the time to focus on algorithms that could solve this problem while lacking the complex rigs required to test their ideas. Conversely, industry has the expertise to produce such rigs but given practical movie constraints often does not have the time to focus on the algorithms. It is therefore unsurprising that the best advances in this area have been from academic and industrial collaboration.

21
Television and Live Broadcasting

Graham Thomas, Philippe Bekaert, and
Robert Dawes

21.1 Introduction

A popular current use of visual computing in live television broadcasting is in the production of graphics and effects for sports analysis. Applications can help the broadcaster illustrate, analyze, and explain sporting events by the generation of images and graphics that can be incorporated in the broadcast, providing visual support to the commentators and pundits. While currently limited to sport, in the future these techniques might spread to other areas of TV broadcasting, to other genres and for use in other aspects of production. Some broadcasters are already experimenting with free-viewpoint, panoramic, and omni-directional video to see how these technologies might change the traditional methods of live TV production.

This chapter looks at some of the challenges involved in developing visual computing tools for use in television coverage and at the techniques that some example systems have developed to tackle them.

21.2 Sports Graphics

Sports directors often refer to "telling a story" to the viewer and are keen to use the best tools available to bring out the key points in a clear, visually interesting, and succinct way. Graphics are a key tool for the TV program maker to help explain sporting events to viewers. The earliest systems, known as telestrators, allowed the pundit or "color commentator" to draw on top of a still image to illustrate some aspect of their analysis. More modern and advanced systems make use of image- or sensor-based systems to extract information about the scene and generate more detailed analysis or visually impressive graphics.

Examples of current state-of-the art graphical analysis systems which involve aspects of visual computing include tools for drawing offside lines and other contextual information into a soccer pitch for tracking the movement of balls in 3D. Other tools provide new views on the action, such as zooming into the image to provide a close-up or overlaying multiple races or heats (such as in downhill skiing) on top of one another so direct comparisons can be made between them. Virtual views of the action can be generated either to provide a new angle on the action that the broadcast camera didn't offer, or to provide a smooth way of transitioning between two different cameras without jumping from one to the other.

These graphics technologies have become so important and integral to their respective sports that some are now used as aids for the judges or referees. For example, the positional information generated for the America's Cup yacht race graphics is used by the event's umpires, and in soccer the Hawk-Eye system[1] is used by the English Premier League for deciding when a ball has crossed the goal line.

21.3 Challenges for Visual Computing in Sports Broadcasting

In some ways visual computing techniques are ideal for use in sports coverage. It is very hard to add extra physical equipment such as sensors into the sporting areas themselves. They might get in the way of the sport being played or affect the performances of the players and athletes. Instead, unobtrusive cameras can be placed around the area of action. Very often the cameras already in place for the television broadcast can be used.

However, there are many significant challenges to the use of visual computing in TV sports coverage that are not found in the laboratory. The environment is out of the control of the system designers; the rules and practices of the sports, such as the type and color of the clothes of the participants, and the size, shape, and fabric of the arenas are normally long established and not necessarily conducive to easy scene analysis. The event can take place over a large area and the locations of the action cannot always be easily predicted. The participants may wear baggy clothing or all wear the same clothing and will often occlude one another. Most sports are played outdoors without controlled lighting or weather conditions and as a result the appearance can change dramatically. Some sporting bodies may be prepared to make changes to their sport if it can be justified by improved television coverage. Perhaps most famously, the fluorescent yellow of tennis balls was chosen to ensure greater visibility on color television.

[1] http://www.hawkeyeinnovations.co.uk/

Sporting events can also be chaotic, with hundreds of people around performing their individual roles. The environment is dynamic, so while a piece of the arena might form a useful calibration point in the morning, it may no longer be there in the afternoon. Similarly, furniture or advertising hoardings may be moved around as the event progresses. Athletics stadiums, for example, will construct or dismantle equipment based on whatever event is happening at that time. This even includes seemingly fixed objects such as the distance markers on the javelin field which consist of pinned down white tape that will disappear when the discus starts from the other end of the field. Some events such as the long jump can even change location when the direction of wind changes.

Broadcasting has historically used interlaced cameras where a series of pairs of fields captured at different times make up each frame. One field contains all the odd lines in the image, the other contains all the even lines. This effectively doubles the temporal resolution of the video, but each field only has half the vertical resolution of the full frame. Combining the two fields into one frame can introduce artifacts because of the time difference between them. The use of interlace is declining but this is still how pictures are transmitted to the home, and if a video feed is made available from a TV camera or recording then it will generally be interlaced. As a result any analysis or graphics systems have to be able to cope with interlace fields rather than the progressive frames as might be expected in the lab.

The use of zoom lenses on broadcast cameras inevitably makes intrinsic calibration more difficult. Lens distortion, image center point, and chromatic aberration can all vary as a function of both zoom and focus settings, and in the absence of data from lens sensors it can be difficult to accurately determine their values. Coverage of sports events also involves a large range of zooms. It is not unusual for a camera covering athletics, for example, to move from a field of 3 or 4 degrees to 45 degrees in a single sequence.

Broadcast cameras have a control known as aperture correction or sometimes "detail," The aperture correction is used to "sharpen" an image and is one element that distinguishes the "TV-look" from "film-look." Effectively, the correction emphasizes high-frequency image components and is therefore a high-boost filter. A high level of aperture correction causes a significant color shift to pixels close to luminance edges. This can affect an area of about 2-3 pixels around contour edges and leads to incorrect segmentation results using color-based segmentation methods. The segmentation can be improved by compensating for the effects of the aperture correction [Grau and Easterbrook 08].

Camera calibration is difficult in uncontrolled scenes which typically have few features with accurately known 3D positions. Although lines on sports pitches are notionally in well-defined positions, pitches are rarely flat (being deliberately domed to improve drainage) and lines are often not painted in exactly the right positions. Techniques such as 3D laser scanning sometimes need to be used to build an accurate model of the environment which can then be used for camera calibration. When calibrating sparsely spaced cameras with a wide baseline, features in the scene can change their appearance significantly between camera viewpoints, making reliable feature matching between cameras difficult. Cameras can often get moved accidentally so calibration may need to be repeated.

Sports broadcasting has its own methods and workflows, into which any graphics system must try and fit. Live sport in particular requires real-time or very fast turnaround results, and the conventions of the coverage may limit the camera choices available. The system also has to be robust enough that the sports director can be confident that it won't break when potentially millions of people are watching.

Unless extra cameras can be installed at a venue, the event will only be available as a video feed from the main broadcast cameras. This means that the operators of any analysis or graphics system will have little or no control over the camera positions and movements. Communication to and from the camera will normally have to pass through the broadcaster's lines of communication with requests for camera movements being passed to the director and then to the camera operator. Camera operators will often cover more than just the game and may start recording the crowd or other interesting events that have no direct relevance to the sport.

A successful system should also offer good value-for-money to the broadcaster compared to other ways of enhancing sports coverage. Sports producers have many ways to improve their coverage and to add significant value. For example, they can add extra cameras around the ground—including special cameras such as infrared heat cameras for cricket analysis, flying cameras suspended from wires attached to the stands, or some of the cameras discussed in Section 21.7 and Chapter 3—microphones on the umpires to hear their decisions and orders or even, and technically rather less interesting, they can just hire a new pundit to provide the analysis.

21.4 Foreground Segmentation

For many visual computing systems it is important to get a high-quality segmentation of the participants from their environment.[2] For example graphics may want to be drawn on the pitch or arena such that they appear

[2]Segmentation is histrionically called keying in the broadcast industry and matting in the movie industry.

to be under the players. If the background is of a reasonably uniform color then some relatively simple image processing can be used. For example, with soccer the players are normally stood on green grass so a color-based segmentation algorithm (known as a chroma-keyer) can be employed such that graphics only appear in regions that are suitably grass-colored. The result is that nothing is drawn on the players themselves but may be drawn on either side of them, giving the impression the graphics are behind him or her.

The color of grass on pitches can vary significantly. This is due to uneven illumination and anisotropic effects in the grass caused by the process of lawn-mowing in alternating directions. After periods of bad weather or late in the season patches of mud can appear to further discolor the pitch. In some applications the fact that the chroma-keyer will not generate a key for discolored areas can actually be an advantage as this adds to the realism that the graphics are painted on the grass, as long as such areas are not large. However, these challenges for the chroma-keyer can give a segmentation that is too noisy to produce the high-quality key needed for applications that require scene reconstruction such as free viewpoint video. Difference keying can produce better results and is also able to segment pitch lines, logos, and other markings of the background. The background model or "background plate" can be created by either taking a picture of the scene without any foreground objects or, if this is not possible, the background plate can be generated by applying a temporal median filter over a sequence to remove moving foreground objects.

For some applications, such as those that draw graphics onto live video, the keying and graphics rendering processes most be performed in real-time.

21.5 Camera Calibration for Sports Events

Early telestrator systems allowed 2D annotations to be drawn on top of a video still. However, if the camera's position, pan, tilt, and zoom are known (together with off-line lens calibration to relate zoom ring position to focal length) then graphics can be overlaid on the image such that they appear in the same orientation as the scene. This is known as registration (Chapter 7). For example, off-side lines can be drawn on the pitch such that they appear to be painted on the soccer field. This camera calibration, combined with knowledge of the geometry of the scene allow measurements such as distance to be taken and displayed. Figures 21.1 and 21.2 illustrates this. American football down lines, for example, must be placed at specific distances along the pitch. Often, some assumptions will be made about the scene to help translate 2D knowledge into 3D. For example, when finding the position of a soccer player it is assumed that he is stood on the

Figure 21.1: Player positions and movements along with distance measurements are added to the scene as if drawn on the pitch.

Figure 21.2: Graphics added to the triple jump to aid viewers in judging the distance of a jump. The graphics are drawn on screen as the camera follows the athlete into the pit.

ground so his height off the ground plane will be zero. As a result the other two dimensions can be calculated using the player's 2D position in the image. Multiple calibrated cameras allow for more sophisticated 3D effects, discussed in Section 21.6.

21.5.1 Image-Based Calibration and Moving Cameras

Calibration is most easily obtained from a static camera of known position and orientation. However, it is often desirable to be able to move the camera. Indeed if a broadcast camera is being used it will be moving. As the camera moves, many modern graphics systems such as Piero[3] retain the position of the drawn graphics relative to the real world. This has the effect of making them look even more like they are part of the scene. This is sometimes referred to as tied-to-pitch graphics. Such systems can be applied to many sports, not just those with soccer style pitches, including swimming, running events, and long jump, adding lines to indicate world records or qualifying distances. They can also be used to add logos or sponsorship to the scene either on the pitch or court or to the side as virtual billboards.

To produce this effect the movement of the camera must be tracked so that the calibration information about camera pose can be updated. This was originally achieved using a mechanical sensor on the camera and mount which measured the changing pan, tilt, and zoom. However, there are now several systems, including Piero, that are entirely image-based. This has the benefit that the same system can be used on multiple camera feeds without the need for expensive equipment on each camera or the need to be located near the camera itself. Indeed, such image-based systems can work just as well on recorded material without the need to work live.

21.5.2 Line-Based Tracking

One method of tracking camera pose is to use the lines on the pitch. Sports such as rugby, soccer, and tennis have regular pitch markings with known (or easily measurable) geometry which can be used to calculate the camera's position and pan, tilt, and zoom. The lines on the pitch are in known positions in the real world. If corresponding lines are visible in the camera view the correspondence can be used to compute camera pose. Typically, calibration charts are used (Section 1.6). In the case of sports analysis, however, the use of lines is more practical. With sports coverage it is generally not practical to get access to the camera or to be able to place objects into its view. Instead it is more reliable to use what is known to be visible to the camera in the scene. This has the added benefit that calibration can be performed after the fact from recordings, even if there had been no intention to use the video at the time of filming. A minimum of four lines (which cannot all be parallel) must be visible to solve for the unknowns and fully compute the camera position and pose, fewer are needed if the pose is known. One implementation of this approach is described in [Thomas 07] and forms the basis of the pitch line tracking features within the Piero and

[3]http://www.redbeemedia.com/piero/piero

tOG-Sports[4] graphics systems. The tracking process is split into stages. First, the position of the camera is estimated, then the tracker is started with a view of the pitch from which it must calculate an initial pose. From then on the tracker runs from frame to frame at full video rate.

Most cameras covering events such as soccer generally remain in fixed positions during a match. Indeed, the positions can remain almost unchanged between different matches at the same ground, as the camera mounting points are often rigidly fixed to the stadium structure. It therefore makes sense to use this prior knowledge to compute an accurate camera position which is then used as a constraint during the subsequent tracking process.

Estimating the position of a camera from a set of features in a single image can be a poorly-constrained problem, particularly if focal length also needs to be estimated. Changes to the focal length have a very similar effect on the image to moving the camera along the direction of view. To improve accuracy, multiple images can be used to solve for a common camera position value. The pose for all images is computed simultaneously, and the position is constrained to a common value for all the images. The images should cover a wide range of different pan angles (e.g., in soccer it should cover both goal areas) so that each image's line of position/focal length uncertainty lies in a different direction. In this way the differing views provide complementary information about the pose and solving for all of them together allows for a significant reduction in the ambiguity.

This process can be repeated using images from all the cameras feeds onto which the production team want to add virtual graphics. For soccer, this is likely to include the camera on the center line, the cameras in line with the two 18 yard lines, and possibly the cameras behind the goals. The computed camera positions are then stored for future use.

The frame-to-frame tracking process uses the pose estimate from the previous image and searches a window of the image centered on each predicted line position for points likely to correspond to pitch lines. A straight line is fitted through each set of points, and an iterative minimization process is used to find the values of pan, tilt, and focal length that minimize the distance in the image between the ends of each observed line and the corresponding line in the model when projected into the camera image. The observed and projected lines are highlighted in Figure 21.3. In such a way the lines can be tracked from frame to frame in an efficient and robust manner.

The operators of the graphics software need to be able to start working on generating their analysis clips very quickly after a suitable incident has occurred. For example, if a goal is scored just before half time in a

[4]http://rtsw.co.uk/products/tog-sports/tog-sports-lite/

Figure 21.3: The pitch model is projected onto the scene highlighting the lines being tracked. The red lines represent the projected model while the yellow lines are the lines found in the image.

soccer match then a clip to aid analysis and discussion of the goal might be required within around 5 minutes. As such they cannot spend time calibrating the camera with image correspondences every time they want to make a clip. Similarly, if the system is tracking on live footage it needs to get the correct pose very quickly once it starts to receive the video—calibration would again take too long. The system must be able to initialize to a suitable camera pose as quickly as possible when presented with a new scene. Various methods can be used, including classifier based systems that recognize the scene [Dawes et al. 09] or by matching features using the descriptors described in Section 7.2. A very reliable method is to use the camera position saved earlier in the process and then perform an exhaustive search of all possible pan, tilt, and zoom values to find the one that best matches the lines visible in the scene. This, however, would be very slow so the process is sped up by performing the search in Hough space after first generating a Hough image based on lines in the video image [Thomas 07].

21.5.3 Feature-Based Tracking

For sports such as athletics, the camera image will generally show limited numbers of well-defined lines, and those that are visible may be insufficient to allow the camera pose to be computed. For example, lines on a running track are generally all parallel and thus give no indication of the current distance along the track, making pose computation impossible from the lines alone. For events such as long jump, the main area of interest (the sand pit) has no lines in it at all, (see Figure 21.2). Thus for these kinds of events to accurately estimate the camera pose for the insertion of virtual graphics an alternative approach is needed.

One approach is to view the problem as a specific example of SLAM (simultaneous location and mapping [Smith and Cheeseman 86]), in which the pose of the camera and the 3D location of tracked image features are estimated as the camera moves. The system is given an initial pose for the camera and then uses image features to track the pose from frame-to-frame (Chapter 7). There are some special considerations to take into account in the context of sports coverage. Specifically, significant changes in camera focal length can occur. But at the same time, camera position is constrained since it is generally mounted on a fixed point. This is in contrast to most implementations of SLAM which assume a fixed focal length camera but allow full camera movement. A significant degree of motion blur can occur as motion speeds of 40-50 pixels per frame are not uncommon when covering sports events with tightly-zoomed-in cameras. The approach described in [Dawes et al. 09] is designed to meet these requirements. It uses a combination of fixed reference features to prevent long-term drift (whose image texture is always that taken from the first frame in which the feature was seen), and temporary features to allow non-static scene elements (such as faces in the crowd) to play a useful part in the tracking process. The image features are assigned an arbitrary depth, as their depth cannot be determined from a fixed viewpoint. Although it would be possible to treat the whole image as a single texture and determine a homography that maps it to a stored panoramic reference image (since the fixed position of the camera makes the 3D nature of the scene irrelevant), the presence of a large number of moving features (the athletes) makes it advantageous to consider the image as a set of separate features so that outlying features caused by athlete movement can be discarded using a technique such as RANSAC (Section 7.2).

21.6 3D Analysis

As described, the combination of foreground object segmentation and camera calibration can allow the approximate 3D positions of objects to be inferred. This simple approach can also be used to create a crude 3D model of the scene. To place the segmented players into a 3D model of a stadium they can be rendered on flat planes, sometimes know as billboards (Section 17.3), at the estimated locations (Figure 21.4). This allows the generation of virtual views of the game from locations other than those at which real

Figure 21.4: Player billboards are placed into a 3D model of the stadium.

cameras are placed. If this process has been conducted on two different cameras, a sequence can be rendered where the virtual camera moves from one real camera to the next. The two sets of flat player textures are blended into one another as the camera transitions. Systems such as Viz Libero[5] make use of this effect.

The simple player-modeling approach works well in many situations, but the use of a single camera for creating the models restricts the range of virtual camera movement. The planar nature of the players becomes apparent when the viewing direction changes by more than about 15 degrees from that of the original camera. One solution is to use pre-generated 3D player models, manually selected and positioned to match the view from the camera, as is the case with the Piero "3D Players" feature.[6] It can take a skilled operator several minutes to model such a scene, which is acceptable for a post-match analysis program, but too slow for use in an instant replay. The player models also lack realism. Some tweaking can be made to the flat textures to give them 3D character, such as altering the orientation of some part of the texture, particularly the limbs. However, this has only a limited effect. Very often the transition between two camera views, as described above, will be performed quickly to ensure that any oddities in the flat player models aren't too apparent [Kilner et al. 09] (Chapter 11).

[5]http://www.vizrt.com/products/viz_libero/
[6]http://www.redbeemedia.com/piero/solutions/3d-players

A multi-camera modeling approach provides an alternative. As described in Chapters 2, 12, and 17, multiple cameras arranged around the scene can be used to generate a 3D model. This process can be conducted over a video sequence to add a temporal element to the scene representation. The camera can then be entirely virtual and view the model of the scene from any point. This free viewpoint video enables sport presenters to explore interesting incidents by moving to new viewpoints, like a virtual flight down to pitch level or overhead [Hilton et al. 10, Goorts et al. 13]. This is a powerful tool to visualize spatial relationships between players and their tactics. It also allows for the transition between real cameras without the need to move to an entirely virtual world or try and work round the flat texture problems previously described. However, the few broadcast cameras used might see too little detail of the action or from angles too different to provide the basis for good 3D shape reconstruction. Extra specialist-cameras may be needed in the stadium which adds to the cost and logistical difficulty. Even with multiple cameras, the size of a typical soccer field and the distance of the stadium from the pitch mean that the image quality is not ideal and any small calibration inaccuracies can have a large effect. This can lead to low resolution models.

The regular problem of occlusion becomes more acute while covering team sports where players can easily block a camera's view of other players. The players can also contort themselves into shapes and interact with one another in ways that makes their 3D reconstruction particularly difficult. The crowd watching the games from the stands can also make segmentation particularly difficult from certain views. As a dynamic multi-colored texture they can sometimes prove indistinguishable from the players and ball.

The freeD system, developed by Replay Technologies,[7] avoids some of these issues by concentrating on small areas of known action, such as a piece of gymnastic equipment, the tee off in golf, or the plate in baseball. For recording a baseball sequence around 12 HD cameras were placed around the ground focused and framed on the batter.

Visual effects such as free viewpoint video face the problem of turnaround time. To generate a good quality free viewpoint video may currently take several minutes. However, live sports coverage would ideally like such effects to be available for immediate replay. Even in the case of highlights programming, when there is more time to create sequences, the time a sequence takes to render becomes an issue. Unless the effect is particularly spectacular and insightful, a TV production would rather generate several less complicated sequences than spend that time creating a single complicated one. As processing power improves and better cameras come

[7] http://replay-technologies.com/

onto the market, such effects can be generated far more quickly. The freeD system claims to be able to render out sequences in around 30 seconds.

Free viewpoint video also opens up new possibilities for how viewers would interact with sports coverage. An event could be delivered to the viewer over the web as a free viewpoint video allowing them to view the action from whatever viewpoint they choose [Budd et al. 12].

Hawk-Eye was one of the first commercially available visual computing systems used in sports coverage. Developed for tracking balls in 3D using a multi-camera system, it was first commercially applied in the tracking of cricket balls to aid the analysis featured in TV coverage of international cricket. Since then it has been applied to tennis and, more recently, soccer, among others. The system is deployed with a number of dedicated high-frame-rate, synchronized cameras arranged around and framed on the area of interest, such as the tennis court, goal mouth, cricket wicket, etc. There is normally space around the court or pitch (often on the roofs of the stands) to install the cameras. As static cameras they are easier to calibrate, and because they are not used to generate broadcast footage short shutter times and higher frame rates can be used to create the optimal pictures for processing. The problems of interlaced video can be entirely avoided by using progressive scan. There is a lot of prior knowledge about each sport that the system can use, including the size and appearance of the ball, its motion (once it is hit, its motion can be predicted using the laws of physics), and the area over which it needs to be tracked. The system first identifies possible balls in each camera image by looking for elliptical regions in the expected size range. Candidates for balls are linked with tracks across multiple frames, and plausible tracks are then matched between multiple cameras to generate a trajectory in 3D. This records the path and position of the ball during play and can even be used to predict the likely movement of the ball had it not struck another object or has spin applied. Other multi-camera analysis systems for balls in sporting events include QuesTec's Umpire Information System[8] for baseball and the similar Zone Evaluation System.

21.7 Virtual Broadcast Cameras and Second Screen

Panoramic, omni-directional, and free viewpoint video promise to change the way events are being captured for broadcasting. The common work flow today is to have one camera operator per broadcast camera. The main duty of this operator is to make sure that an event is captured with appropriate framing from the vantage point of the camera position. In high-end live

[8]http://www.questec.com/q2001/prod_uis.htm

productions, the exposure and color settings of the cameras are controlled by shading operators in the outside broadcast van, so a camera operator really can solely focus on framing. This takes particular skills of attention and context awareness in order to predict to a certain extent what is about to happen and how to perform the actual framing.

Different cameras provide raw footage from different vantage points. In soccer, for instance, the camera positions are standardized by FIFA, and assigned numbers so that everyone can refer to a camera position in a non-ambiguous way. But also different types of events are usually captured from established sets of camera positions. Major rock concerts, for instance, are captured by a camera on either side of the stage, a camera middle front stage (shooting in "frogs" perspective), a camera (often on a crane) from within the audience, and sometimes more cameras from back-stage corners.

The raw footage from each of these camera vantage points is presented to a director on an array of video monitors in the outside broadcast van. The director makes live decisions on what stream on the mosaic is dispatched for live broadcasting.

In offline broadcasting, one or more camera operators autonomously choose vantage points, frame an event and control exposure themselves, with the same goal of producing the best possible source footage that can be combined later into the final program. The process of putting together a program after filming has finished is known as post-production. During the editing stage of post-production, footage may be cropped slightly in order to correct framing.

This current work flow is dictated by the fact that camera pixels are expensive, and used to be very expensive. However, pixels are getting cheaper nowadays. As a consequence, it becomes feasible to capture events with a high-resolution camera, at a wide shooting angle, so that framing can be performed by an operator in an outside broadcast van, or in post-production for off-line productions. During the 2014 Super Bowl, for instance, 4K broadcast cameras were used in this way to generate full HD footage. The advent of 8K cameras promises a fourfold zooming capability without loss of quality, offering considerable freedom in reframing events. This comes at its highest benefit in the context of replays, in order to "tell the story" of what has happened at an event, shortly after a particular incident. Post-factum framing no longer requires the prediction skills of a good camera operator today.

Such "panoramic" virtual camera technology is already feasible today at a commercial level by stitching together images of closely spaced, full-HD, cameras. In the context of pitch sports, the stitching can be performed to sub-pixel accuracy today with relatively simple algorithms since the scene being captured is at a large distance compared to the inter-camera spacing

Figure 21.5: 16-times full-HD panoramic camera system for sports broadcasting (left), camera installed in American football stadium (middle), operator reframing footage while programming an instant replay using a joystick controller (right).

Figure 21.6: Omni-directional video capture for broadcast TV (left), director controlling framing and lens effects using a joystick controller in a montage booth (middle), resulting video clip frame with stereographic "virtual lens" (right).

(tens of meters versus centimeters), and depth variations are relatively small (Chapter 3).

Another application scenario of omni-directional cameras is to facilitate video capture on location. A compact omni-directional camera is installed on site and captures whatever happened around the location where it was set up. Actual framing of events takes place in post-production, in a montage or edit booth, at the broadcaster's facility. Creative decisions concerning the field of view or lens effects in order to frame an event are no longer taken in a predictive manner by a camera operator on location, but based on posterior knowledge of what has happened. This work flow may eventually reduce costs, while allowing additional creative freedom. The work flow was tested, among others, in the context of a popular prime time program in "1000 zonnen" by Belgian national broadcaster VRT, with two 1- to 4-minute items produced this way weekly over a period of three months.

In addition to allowing cost reduction and additional creative freedom, panoramic and omni-directional video capture techniques can also be used to provide content for second screen applications that complement a traditional broadcast watched on the main television. A second screen is a viewer's personal device such as a mobile phone or tablet which they may use while watching a program on the shared primary television. A second screen spectator can follow the same event as shown on the primary TV screen but the way the event is shown could be fundamentally different. On the primary screen, a director's cut may be shown, while on the secondary screen, a spectator can be allowed the freedom to "look around" at the event him/herself. Experience from actually bringing this into practice in the context of a popular Belgian talk show, and for sports coverage, showed that the primary screen is by far the best way to show details (close-ups) and steer the spectators' attention (as it always and exclusively is on primary screen), while the surround video content on the second screen helps to convey the ambiance of an event location better than by any other means. In the context of talk shows or political discussions, usually just the person speaking is shown on TV, while surround video on a second screen allows viewers to observe the body language of the others, which may be very informative on occasions.

21.8 Summary

Visual computing systems face various challenges if used in live television broadcasting and in sports coverage. Some systems are already in use, and the approaches they take to tackle the challenges are encouraging. In the next few years, as technology improves there is the potential for visual computing to play an increasingly important role in TV broadcasting. Free viewpoint video systems and wide angle cameras may change how television is produced and even create new ways for the viewer to consume and interact with TV productions.

22 Web-Based Delivery of 3D Mesh Data

Max Limper, Johannes Behr, and Dieter W. Fellner

22.1 Introduction

Visual computing applications are often dealing with 3D mesh data, in order to represent objects that have been captured within the real world. Such data can stem from various 3D scanning methods, which usually output point clouds and then apply some mesh generation method, for example by using the marching cubes algorithm, Delaunay meshing, or the *3D Snakes* method (Chapters 9 and 10). Depending on the size of the data, as well as on the target application, the meshed result might then be processed further, for example through simplification or remeshing methods, before a final, meshed 3D model is ready for presentation.

A trend that can be observed over the last few years is that there is, on the one hand, a growing number of devices with powerful graphics hardware, and on the other hand almost no device any more that is not potentially connected to the Internet, or to a local intranet. As a natural consequence, 3D Web technology serves more and more as a powerful integration platform for visual computing applications: Suddenly, with low-level graphics APIs like WebGL, it has become possible to deliver 3D mesh data to a wide variety of devices, ranging from desktop PCs to smart phones, in a consistent, platform-independent way. Figure 22.1 shows a conceptual overview of this situation.

22.2 3D Meshes vs. Image-Based Representations

There are several possible ways for visually presenting captured real-world scenes or objects to a user. In general, when considering the data transmission format, a distinction can be made according to two basic categories. On the one hand, 3D data may be stored for presentation as polygonal

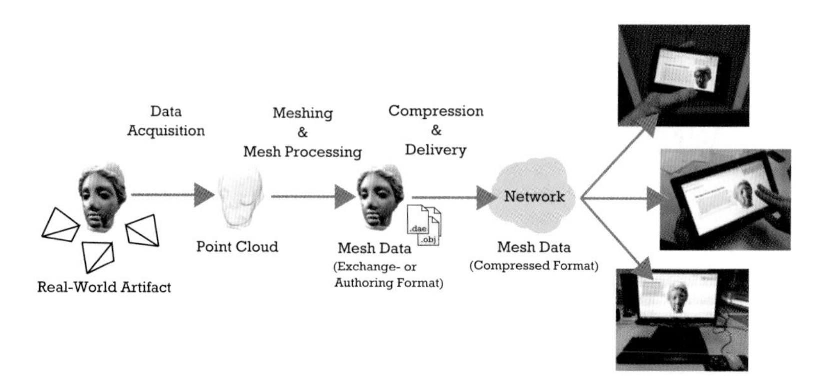

Figure 22.1: Pipeline for a typical visual computing application from the field of cultural heritage. Data is acquired through a 3D scanning method and then meshed. The meshed data may be further processed by simplification algorithms, for instance. It is then compressed for delivery over a network, transmitted, and finally presented on a client device.

mesh. In the most simple case, this mesh is simply a list of triangles, which represent the surface of an object. Appearance details (such as surface normals, displacement or colors) are usually represented at the vertex level, or in the form of texture images. The client application then generates images on-the-fly, using common 3D rendering techniques, as they are also used in computer games, for example. Within this section, such representations are referred to as *mesh-based representations.*

On the other hand, captured data may be directly represented via images. Instead of generating perspective views, along with matching lighting and appearance information, on the client side, all the client has to do is to display the correct image, or to interpolate between a set of given images. Methods that use such representations are usually referred to as image-based rendering (IBR) or video-based rendering (VBR). It is worth noting that there are also other ways of storing captured 3D data for presentation, such as point clouds or depth maps—an overview can be found in Chapter 17. In Figure 17.1, mesh-based 3D representations can be found at the very right edge. Both of the mentioned basic approaches for representing captured data for remote presentation, mesh-based representations as well as IBR/VBR solutions, have specific advantages and disadvantages.

Advantages of IBR and VBR methods. Rendering mesh-based representations in an efficient way requires full access to a dedicated 3D graphics API and hardware. In contrast, VBR or IBR solutions potentially have lower device requirements, and therefore might be a better fit for a maximum portability. If critical data (for instance, protected CAD product data) should be displayed over an unsecure network, security restrictions may also prohibit a delivery of mesh-based representations. With IBR or VBR solutions, deriving such critical data from the displayed images is a much harder (and error-prone) task, therefore these methods might be preferred in such cases. Another great advantage of VBR and IBR solutions is that they are basically independent from the complexity of the model and its appearance properties. Extremely high polygon counts, or the request for a very detailed view of complex material or illumination properties, might therefore also prohibit the use of mesh data, and lead to IBR or VBR solutions instead.

Advantages of mesh-based representations. Depending on the application, the user might not be satisfied with viewing a non-interactive scene, as is usually the case with IBR and VBR solutions. If the user wants, for example, to be able to change object materials, or to move objects or light sources around, IBR or VBR solutions cannot provide this degree of interactivity (Figure 22.4). In such cases, the necessary visual information has to be dynamically generated on the client side, using common 3D graphics techniques and mesh-based representations. If network bandwidth is a critical factor, using mesh-based 3D representations is also the method of choice, especially compared to approaches like light fields, which require the storage of huge amounts of data. As already noted in Chapter 17, this is due to the fact that an optimized, meshed 3D model of a scene, along with a texture atlas, makes very compact scene representations possible. One possible solution to achieve maximum quality along with a maximum degree of interactivity, on almost any client device, is the use of a dedicated server for remote rendering. However, the big disadvantage of this image- or video-based approach is its bad scalability: As soon as multiple clients connect, a dedicated rendering server or process must be maintained for each of them, which is not possible for many kinds of public, large-scale Web applications (such as online shops or exhibitions). Another problem in this context is the need for a connection with minimum latency, in order to be able to provide fluent user interaction. Therefore, mesh-based representations are usually preferred for interactive small and medium-size scenes, which should be accessable for a large number of clients.

Considering the specific advantages and disadvantages of IBR/IBR solutions and mesh-based representations, it strongly depends on the context of the application which representation is best-suited. Using service-oriented architectures (SOA) and RESTful APIs allows to deliver 3D assets along with context-specific application templates, as it is done by Instant3DHub [Jung et al. 12] or XML3DRepo [Doboš et al. 13], for example.

Client devices can then, for instance, automatically receive a specific representation, matching their CPU or GPU capabilities, the available bandwidth, and security-related contraints. This way, the decision between using IBR / VBR methods or using mesh-based representations can even be dynamically performed per client. A server might, for example, decide to share only images instead of real 3D information across unsecure networks, in order to prevent theft of proprietary 3D construction data. Providing an image stream for an interactive 3D experience, on the other hand, is only possible with server-side rendering, which requires a powerful server architecture to scale well, even for a small number of clients. Therefore, the approach of statically providing compact, mesh-based representations is usually the preferred way to deliver 3D assets on the Web, and already applied for a wide variety of use cases.

22.3 Application Scenarios

The field of application for high-performance 3D Web technology is growing every day, including many interesting scenarios from the field of real-world visual computing. One obvious use is the online presentation of scanned products for marketing purposes.

Another use case, which might be not that obvious, is shown in Figure 22.2(right). Here, research data from the CAESAR anthrophometric database has been made accessible as part of an interactive 3D Web application. Such fast, portable, and convenient access to the 3D mesh data allows quick looks on specific poses, without needing to explicitly download the corresponding files and then open them in a specialized viewing application.

One increasingly important class of real-world visual computing applications, using 3D Web technology, can be found in the domain of cultural heritage. Artifacts from this domain are mainly preserved in large archives, mostly belonging to museums all over the world. Usually, such archives contain too many artifacts to fit into a single exhibition, and they are often managed in traditional, paper-based archive infrastructures. The current shift toward digitized archive data also includes the storage and distribution of 3D scans for each artifact (Figure 22.1). This enables people around the world to remotely inspect collections in distant archives via the internet—in an ideal case, such inspection scenarios can be performed using regular Web browsers. An example from a recently initiated Web portal [Martínez et al. 14], using the open-source X3DOM framework for visualization, is depicted in Figure 22.2(left).

As a typical real-world visual computing application from this domain, Schwartz et al. have presented both, a capturing setup as well as a WebGL-

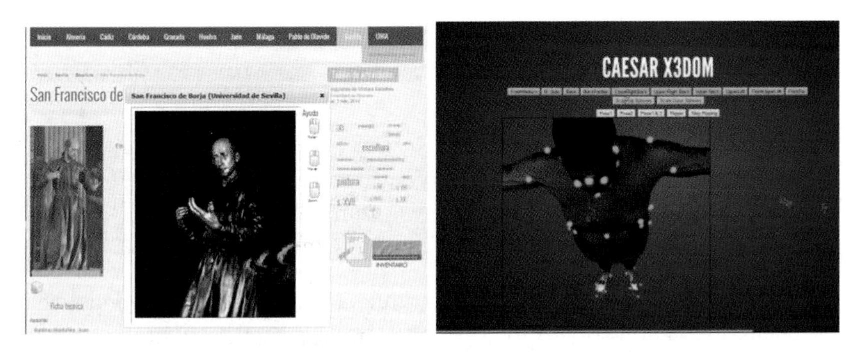

Figure 22.2: 3D Web technology, used for real-world visual computing applications. Left: A scanned cultural heritage object, which is part of an online exhibition. Right: Online inspection of the CAESAR antrophometric database.

based streaming framework, for acquisition and presentation of cultural heritage artifacts. Figure 22.3 gives an impression of their DOME II capturing setup, as well as of the 3D Web application that is used to present the results [Schwartz et al. 13, Schwartz et al. 11a]. Captured reflectance information is represented via bidirectional texture functions (BTF), which are approximated by texture images and progressively transmitted for presentation (Section 22.4). The user is able to interactively modify the lighting conditions inside the viewing application, which allows for a detailed inspection of not only geometry, but also material properties.

In a similar fashion, but using a less sophisticated material representation, the *Radiance Scaling* method enables the user to inspect surface details on an arbitrary 3D mesh [Vergne et al. 10] (Figure 22.4). During

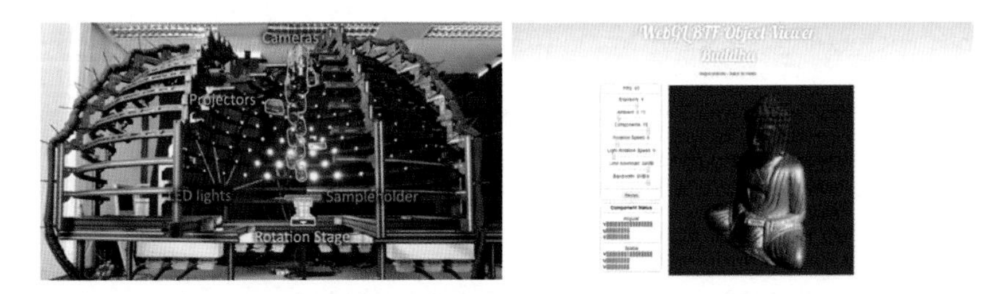

Figure 22.3: First and last stages of a pipeline for online presentation of cultural heritage, as presented by Schwartz et al.: DOME II setup for data acquisition (left), and Web-based BTF streaming and rendering framework (right).

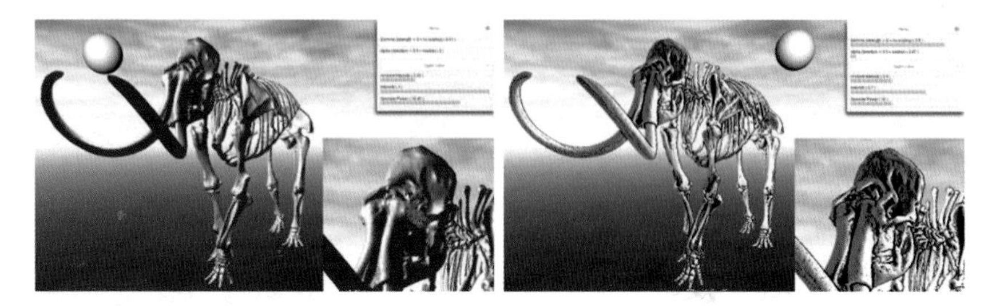

Figure 22.4: WebGL-based viewer, showing a wooly mammoth model that has been digitized and kindly provided by the Smithsonian Institution (see http://3d.si.edu/). The sphere acts as a handle for manipulating the position of the light source. Lighting parameters can be interactively modified to inspect different parts of the model, or to reveal surface details.

the shading stage of the rendering process, surface concavities and convexities are enhanced, depending on some parameters that can be interactively modified by the user. This technique only requires surface normals, and can therefore be used with any capturing setup, even if it only acquires geometry data.

22.4 Compression and Transmission

As soon as the acquired geometry has been reconstructed in the form of a polygonal mesh, additional mesh simplification or resampling steps might take place (Section 10.8). At this stage, mesh data is represented in the form of a topological and geometrical description, which can, for instance, be simply a text-based description in the form of a *Wavefront OBJ* (.obj) or *COLLADA* (.dae) file (Figure 22.1). This representation of the 3D mesh data (potentially enriched by texture data in the form of images) is then transferred to the Web-based delivery pipeline (Figure 22.1). To achieve a convenient user experience, it becomes necessary to apply compression and progressive transmission methods when delivering the data over a network. The conversion from the input format (e.g., from an .obj or .wrl file) to a compressed transmission format is either performed once (for static data) or as soon as needed (for dynamically changing data). The latter approach is, for example, taken by the transcoder component of the Instant3DHub architecture, as proposed by Jung et al. [Jung et al. 12]: As soon as a client requests a page that shows a specific mesh, the server decides whether the current compressed version of the requested mesh inside the server cache is up-to-date. If this is not the case, the outdated version inside the cache is directly replaced by compressing the input file again. Within this step, additional adaption to other parameters like network bandwidth, security constraints, or client device can also take place. A server could, for example, perform additional simplification steps to store representations at different level-of-detail (LOD) inside its cache. Depending on the client device that posts a request, the server might then pick a matching LOD for delivery.

Compressing Mesh Data

A crucial aspect of mobile and Web-based real-world visual computing applications dealing with 3D mesh data is efficient data compression and transmission. In contrast to common desktop 3D applications, the typical user scenario is radically different: users will browse to a specific URL where a 3D showcase should be available, or click a thumbnail in an online catalogue. As soon as the 3D Web application starts to load, users will typically expect *instant* results, as they are used to from other Web pages. This does not necessarily mean that everything should be directly loaded, but that a progressive refinement of the whole page must instantaneously take place.

Regarding compression methods, this means that a very important trade-off has to be made: On one hand, a good compression method could be necessary to deliver the mesh data in acceptable time over the network. On the other hand, a fast decompression is also crucial, since a complex decompression step could take more time than directly sending the uncompressed data would do [Hoppe 98, Pajarola and Rossignac 00, Limper et al. 13b]. This aspect becomes even more crucial for specific application setups, like mobile client devices (with less CPU or GPU power for decoding) in a fast company intranet (using a high bandwidth).

Therefore, only methods that allow a quick and straightforward decompression can be considered as candidate technology for Web-based and mobile 3D graphics applications. For an exhaustive overview over the wide field of research related to mesh compression in general, the survey of Peng et al. might serve as a good starting point [Peng et al. 05].

Lossy compression. The simplest method used to reduce the size of a mesh, and at the same time the most important one, is the use of quantized data. For the most popular quantization in Cartesian space, the idea is fairly simple: if the 3D mesh can be described with its vertex positions lying on a regular grid, without a visible loss of precision, then floating-point coordinates can be safely replaced by integer coordinates. The decoder will compute the original vertex positions from the normalized and quantized representation, using the original size of the mesh's bounding box that has been provided along with the data. Decoding can be performed in parallel on the GPU, inside a vertex shader. This approach has several advantages.

First, it leads to a significant reduction of memory that is needed to store the vertex buffers. Second, and even more important in a browser context, the time spent for CPU-based decoding steps becomes effectively zero, since clients simply need to pass the downloaded data chunks to the GPU, where the decoding is performed during rendering.

It strongly depends on the resolution of the model how many bits of precision are sufficient; typical values vary between 8 and 14 bits per vertex position component [Pajarola and Rossignac 00, Peng et al. 05, Lee et al. 10a].

A simple yet efficient approach to reduce the size of the compressed mesh is the subdivision into several sub-meshes. Such sub-meshes can be obtained from the original mesh with several partitioning methods, and the subdivision does not necessarily relate to manifold surface patches. A simple method is to use a kD-Tree to partition the mesh into regions that contain an approximately equal number of triangles. Another method for subdividing a mesh into multiple sub-meshes is to use a cache-optimization strategy to re-arrange the vertex data, and then cutting slices out of the resulting lists [Chun 12]. A third possibility is to use a hierarchical face clustering approach, where a cost function can be used to produce compact bounding boxes [Lee et al. 10a].

The use of several locally bounded, compact sub-meshes enables the encoder to achieve a very compact representation, since a lower quantization precision can be used to encode the vertex data within each sub-mesh, without having to sacrifice the overall quality of the full mesh after compression (the overhead of additional bounding box data is usually of neglectable size). However, subdividing a surface potentially introduces another problem, which occurs in the form of visible seams between the sub-meshes. To solve this problem, quantization is performed with respect to a common bounding box size, which is, for each spatial dimension, equal to the largest bounding box available among all sub-meshes [Sander and Mitchell 05, Lee et al. 10a].

Besides the vertex positions, other vertex attributes can also be stored in a quantized form for compression purposes. For RGBA colors, it is very common to use an 8 bit range per channel anyway. To store vertex normals in a quantized form, several approaches exist. Jung et al. use spherical coordinates (employing two angles on the sphere instead of 3D Cartesian coordinates) to store normal data for compact transmission, enabling decompression during rendering [Jung et al. 13]. Meyer et al. have provided a study on floating-point normal vectors and several quantized normal vector parameterizations, showing that an octahedron parameterization gives the best results, compared to 3D Cartesian or spherical parameterizations [Meyer et al. 10]. No matter which particular normal storage format is finally picked, it is always the main aim of an interactive 3D Web or mobile application to perform decompression on-the-fly during rendering (usually inside the vertex shader) in order to avoid slow CPU-based decoding.

Aiming at a browser-based inspection of human anatomy, Google conducted experiments that resulted in *WebGL-Loader*, a minimalistic library for compact 3D mesh transmission [Chun 12]. The most interesting aspect of this method is that the UTF-8 file format is a good alternative to binary formats, because it can be parsed very quickly within the browser, while also providing a simple variable-length encoding. The algorithm achieves a comparatively good compression and, combined with the native GZIP implementation of the browser, realizes fast decompression.

It is worth noting that limiting data precision to the minimum amount which is necessary does not only help to reduce transmission time. It has also been shown that it can improve geometric calculations and lighting computations, in terms of execution speed and energy consumption [Hao and Varshney 01, Pool et al. 11]. This aspect becomes especially important in the context of mobile graphics.

Loss-free compression. Besides lossy compression methods, there is a variety of loss-free compression algorithms that potentially fit well with a Web or mobile environment.

A very quick and efficient method is delta-coding and additional entropy-coding, like provided by GZIP compression [Chun 12]. Considering Web-based transmission, GZIP compression is available as a standard encoding method in HTTP, therefore applications can rely on fast, built-in browser implementations.

The open compressed triangle mesh (OpenCTM) format,[1] intended for fast exchange of compressed mesh data, uses LZMA compression, which is potentially superior to GZIP when applied to 3D binary mesh data. However, since there is no Web browser that supports LZMA natively, Web applications have to provide their own LZMA decompression implementation. This in turn might lead to the overall loading time (consisting of download time plus decode time) becoming much longer than for methods that rely on less sophisticated compression algorithms, but provide a faster decoding [Limper et al. 13b].

The Khronos group has recently considered several compression methods as candidate technology for the GL transmission format *glTF*. Among those, an important candidate technology to be included into the glTF specification is the TFAN approach, which has also been considered for ISO standardization within the MPEG-4 standard [Mamou et al. 09]. The idea is to provide good compression rates, support for all kinds of meshes (including non-manifold geometry), along with a low-complexity, fast

[1] http://openctm.sourceforge.net/

decoding method. The encoding algorithm first constructs a set of triangle fans from the input mesh data. The most frequently appearing triangle fan configurations are then used to efficiently encode the mesh connectivity, along with an arithmetic coder [Moffat et al. 95].

Progressive Mesh Data Compression

Since common Web applications demand an instantaneous user experience, the question arises how 3D mesh content can be progressively retrieved. In the ideal case, a compressed representation can be quickly delivered over the network, and at the same time be progressively decompressed without much overhead inside the client application.

A trivial yet efficient method to transmit a mesh in several stages is also provided with mesh partitioning approaches (Section 22.4). However, the term *progressive* is usually used for methods that progressively transmit the data of the full mesh. First approaches were made in this direction in the late 1990s, a long time before Web and mobile graphics APIs and the modern graphics pipeline were available. The pioneering work of Hoppe encodes mesh data in a compact and progressive structure, based on sequential edge collapse and vertex split operators [Hoppe 96]. Mesh data is initially provided in the form of a simplified, coarse mesh, which can be obtained using error-controlled edge collapse operations [Garland and Heckbert 97]. The original high-resolution mesh is then iteratively reconstructed by splitting vertices, which are read from the incoming stream of mesh data. In a later publication, Hoppe provided a short study on when additional compression of vertex data is worth the effort, depending on the decoding speed and bandwidth available [Hoppe 98]. In a similar manner, Pajarola and Rossignac stated that an efficient progressive mesh compression method has to balance between three contradictory constraints [Pajarola and Rossignac 00]. First, it needs to provide fine-grained progressive refinements. Second, high compression ratios should be achieved. Finally, the third important goal is fast decompression.

In the Web and mobile context, fast decompression has a very high priority. But even on desktop machines, sophisticated progressive mesh compression algorithms do not always pay off in terms of decode time compared to download time. Therefore, game developers have tried to port parts of Hoppe's method to the GPU [Svarovsky 99]. The focus of latter work is primarily shifted toward the optimization of rate-distortion performance [Peng et al. 05], which, however, mostly completely ignores the aspect of decode time. Lavoué et al. recently presented a modified progressive mesh compression algorithm which works relatively well on the Web [Lavoué et al. 13] (Figure 22.5). However, it is unable to deal with non-manifold geometry, and decoding steps still have to be performed on

Figure 22.5: Progressive loading, using a progressive mesh. The method produces very good approximations during refinement, while maintainig a good compression rate. On the downside, it is unable to deal with non-manifold geometry, and thereby introduces additional pre-processing steps, and decode time might also become a critical factor.

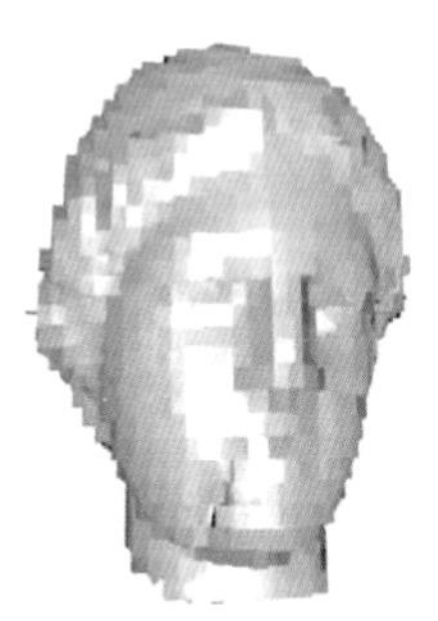

Figure 22.6: Progressive loading, using the POP buffer method. The idea is comparable to progressive PNG images, but transferred to 3D mesh data. It provides fast decoding without any CPU-based steps and is able to handle arbitrary triangle soups, but sacrifices compression performance and quality of the intermediate representations.

the CPU. An alternative, which has been specifically designed for the Web and mobile environment are POP Buffers resembling progressively loaded PNG images [Limper et al. 13a] (Figure 22.6). The idea of the algorithm is very straightforward: Since aggressive quantization of the geometry (for instance, with 5 or 6 bits) results in many triangles being collapsed to a line or even to a single point, it is possible to progressively sort out those triangles when using a coarser quantization resolution. The triangles can then be grouped according to the precision level where they first appear,

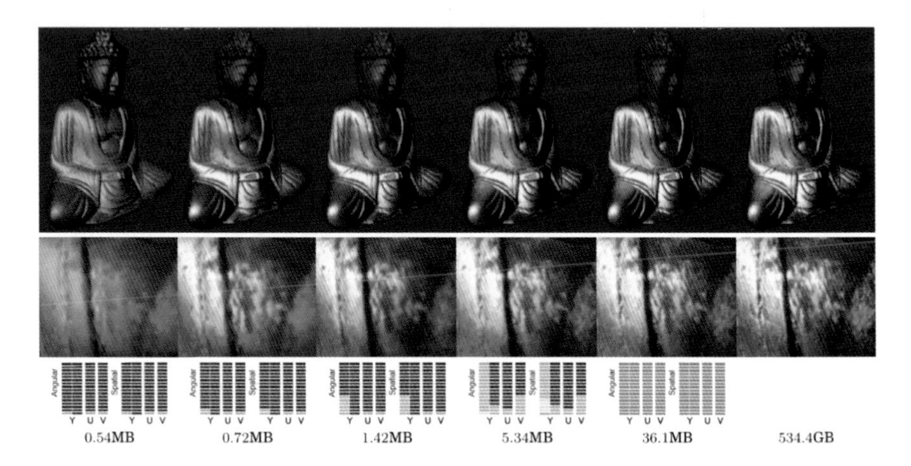

Figure 22.7: Web-based BTF streaming. The BTF data, which is used to shade the textured surface of a 3D mesh, is progressively transmitted as a compressed, decorrelated matrix factorization, with additional wavelet compression for the corresponding images.

which enables the creation of a nested, progressive structure. The method is able to handle arbitrary meshes, and it does not involve any CPU-based decoding steps. However, the quality of intermediate representations is not as good as for other progressive compression methods.

Schwartz et al. have presented a framework for progressively streaming bidirectional texture functions (BTFs), representing the appearance data of digitized artifacts [Schwartz et al. 11a]. The BTF data, which has been acquired from their DOME II capturing setup (Figure 22.3) is compressed using the decorrelated full matrix factorization (DFMF), as proposed by Müller [Müller 09]. Like other compression methods that are used in the Web and mobile context, this method provides a balanced trade-off between compression performance and decode time, and allows for real-time rendering on the client's GPU. The BTF is represented as a product of two functions, one depending on view- and light direction (angular function), and one depending on the spatial position of the sample (spatial function). By performing a singular value decomposition and streaming the resulting components of both functions, this information can be progressively transmitted to the client application, which produces high-quality results after just a few seconds (Figure 22.7). The corresponding images, which are used to store the component data, are compressed for faster transmission, using a wavelet compressor.

22.5 Summary

Remote presentation of real-world data, obtained through different scanning methods, has already become a common use case within many fields of application. Cultural heritage is only one example where the full potential of a visual computing pipeline has not been completely explored yet.

With the current technology trends, like cloud-based applications, the *Industry 4.0* initiative, and the Internet of things, there are also high chances that lightweight, mobile and Web-based 3D graphics applications will continue to gain further importance. Especially, the confluence of mobile vision and mobile graphics, performed on a single, Web-capable device—namely, the smart phone—bears high potential for many new, innovative, real-world visual computing applications. One important field of research in this context remains the search for a commonly accepted standard format for compression and progressive transmission of 3D mesh data.

23 Virtual Production

Volker Helzle, Oliver Grau, and Thomas Knop

23.1 Introduction

The concept of producing digital media content with software operated on affordable workstations has been established during the last two decades. It was mainly influenced by the gain of processing power and is applied in disciplines like visual effects (VFX), 3D character animation, visualization, simulation, or videos games, to name a few. The majority of work, however, is carried out in an offline process which separates the artist from other members of the creative team.

The concept of virtual production foresees to combine key aspects of media production in a real-time, or close to real-time, environment where creative decisions can be taken in direct consultation with other members of the team. Virtual production utilizes existing technology and concepts already established in television (chroma-keying, camera tracking, virtual studio), industrial visualization and design processes (virtual reality and augmented reality) or progress in capturing human motion, for example, in medical or sport applications.

The contemporary situation provides a unique opportunity to cause a major shift in how media is being produced based on the multitude of existing and upcoming hard- and software technologies. Those allow onset content interaction, visualization, and modification, with intuitive methods of controlling creative parameters. This can be achieved by combining modules like accelerated software algorithms for image synthesis, performance capture, sophisticated tracking of cameras and input devices, gesture recognition, virtual and augmented reality display devices, and capturing of additional information such as depth, geometry, or omni-directional video. In this chapter, different aspects of virtual production beginning with the already established virtual studio in television will be looked at. For instance, set extensions and filming on virtual locations offers a unique opportunity compared to the complexities involved in shooting at real locations. Finally, a case study production in which game engine technology and a motion capture system were used for pre-visualization and final render of an advertisement spot will be discussed.

23.2 Virtual Studios

Virtual studio systems were pioneered in broadcast in the 1990s. They were designed to be used in real-time in live broadcast productions. Early systems were using 2D digital video effects hardware systems (DVEs) combined with available studio chroma-keying equipment and mechanical camera sensors to capture pan, tilt, and zoom variations. With the advent of graphic hardware, like the Silicon Graphics Reality Engine™in the mid 1990s, it was possible to render 3D graphics in real-time at video frame rates. An early virtual studio system was developed in the EU-project Mona Lisa [Blonde et al. 96] and led to one of the first commercially available virtual studio system on the market.[1]

A specific problem of virtual studio systems is the tracking of camera movements. Mechanical sensors are only capable of capturing the movement of a TV camera on a pedestal, as pan and tilt[2] and in addition the zoom and focus settings of the camera lens. Early solutions to capture the 3D camera movement (6 DOF pose) were already based on visual tracking of special markers. Examples of this technology are the ORAD pattern-based system or the Vinten-Radamec Free-D system, developed by BBC [Thomas et al. 97]. Recent progress in computer vision in combination with inertial and gyro-sensors led to the development of real-time capable 6 DOF camera tracking without the requirement of special markers in the studio. Moreover, these systems work even outside a studio and are based on natural image features [Chandaria et al. 07] and are now available as commercial products, e.g., from Ncam[3] or SolidTrack.[4]

The take-up of virtual studio systems was initially slow. The hope that virtual techniques would save cost in set design was contrary to very expensive equipment and additional complexity in the preparation and production. With significantly reduced hardware cost virtual studio techniques are today used in many TV productions, in particular in news and factual shows.

Since the techniques developed for virtual studios are real-time capable by design, they were also used in early virtual production approaches in movies. In particular, virtual production systems were used in on-set pre-visualization in a number of movie productions involving real actors

[1]The ELSET™system developed by VAP.
[2]There is usually no roll on a studio camera pedestal.
[3]http://www.ncam-tech.com/
[4]http://www.solid-track.com/

interacting with otherwise invisible virtual actors by giving a real-time preview of the final composited scene [Rosenthal et al. 01]. The ORIGAMI project developed an actor-feedback system using view-dependent projection [Grau et al. 04]. Projector-based feedback was used in TV productions, like the BBC "BAMZOOKi" program for children. Many recent virtual production implementations make use of advances in display technology and are using bright projectors and displays directly in the studio to insert graphical or video content without the need for chroma-keying. This avoids problems of chroma-keying, like color back-spill on presenters and actors and also makes it easier for the presenters on set as they have visual feedback of the inserted content.

Further evolution of virtual studio techniques happened to some extent in techniques developed for use outside the studio space, in particular for sports visualization (Chapter 21).

23.3 Virtual Production for Cinema and TV

Virtual production employs a number of tools throughout the whole pipeline to make a movie or even smaller productions like commercials and drama series. These tools include software and hardware to create virtual assets[5] and/or capture reality with 3D scanning technology. Further, it includes software tools to manage the workflow and to visualize the virtual and virtualized components throughout the production process [Grau 05]. The workflow in movie and (high-end) TV production is divided into phases: The preparation and planning phase is based on the story board and script and implements or acquires all assets (virtual and real). In the on-set phase real assets and actors are captured (filmed) and in the post-production phase final quality renderings are produced and combined with the real camera footage in a process called compositing. The post-production phase is a very expensive step; in some cinema productions the combined cost for visual effects and post-production match the other production cost. It involves highly specialized and skilled jobs, like modeling, animation, virtual lighting, compositing, color grading, and linear editing of the final video/movie. Because of significantly lower budgets in TV, post-production is kept to a minimum and still the tendency in TV is to produce "in real-time."

It became more and more essential to the production team members to experience the sometimes fully or partially virtual scenes and assets. This has led to a drastic change in making moving pictures and resulted in

[5]In media production all items that are created in a creative process, including images, text, 3D content, and video clips are called assets.

higher efficiency, higher quality, and greater variety. It also challenges each team member to adapt to new technologies and methods such as set extensions, character animation, environmental controls, and pre-visualization [Nitsche 08].[6] Cinematography increasingly makes use of virtual data: Its elements and assets are created in the virtual realm all the way from the very first script until the very last frame on the screen.

Currently, there are barely any available systems to connect the different stages of a virtual production, and if so, they are proprietary and not accessible to other companies and parties. The vast variety in technical development (new cameras, new codecs, programming languages, interfaces, sensors, data transfers, etc.) renders the integration of different systems a very complicated task. The different approaches in producing moving pictures challenge at the same time vendors of technology to adapt their products for these needs, and demand from users new and more knowledge.

Using virtual production provides a direct and interactive visual feedback for everyone on the set, making decisions easier and more reliable. For instance, seeing the actual surrounding gives the director the chance to adjust camera framing and moves while shooting. Further, it gives actors visual feedback of virtual objects. Although actors are trained to perform without visual reference, it is hard to impossible to achieve a natural appearing interaction between real and virtual actors without on-set visualization, leading, e.g., to not matching eye-lines.

Besides these creative benefits, virtual production makes the film crew more independent of physical facts. For example, it is not necessary to wait until the right lighting situation outdoors arises, leaving the crew only a small time slot to shoot. The background as well as the fine-tuning of the lighting can be done later in a post-production step. This gives the film team much more time to retry a specific shot or to adjust it if needed. Furthermore, weather conditions or spatial limitations are almost no issue. It might be very costly to block a whole street or bring the whole team to the desired location in the Arctic. Using virtual production, the location in question can be filmed separately or created by artist on the PC, while the shooting of the scene takes place in the backyard or a film studio.

Apart from the direct interaction on set and the independence from time and space, virtual production can be used to decrease the time and cost that is needed for post-production. The actual amount of time that can be saved depends on the type and complexity of a specific production. It is possible to record the virtual background as well as the real foreground and compose both in real-time. This recorded material can be handed to the final cut, without being touched by any post-production. With TV and

[6] Also called previz.

movie productions getting visually more and more complex, this is rather an exception today.

Still, virtual production makes the post-production process easier in many ways. Recording a composed version of the shot beside the foreground gives the video editor, directors, producers, and 2D artists a solid and reliable impression of what the actual scene should look like. Using this reference can decrease the time that would be needed for an iterative process between director and artist until all scene elements are in their proper places. Misunderstandings are less likely to happen as the reference shot visually explains the scene.

Virtual production is a tool for the interaction between everyone involved in the production process. It is ideally designed for all different departments necessary to produce a film, TV show, commercial, live show, etc., and to connect those departments in an early stage and all through the whole process until the finished product.

Virtual production enables a very close relationship between clients and service providers, with the aim of ensuring that any questions and issues can be dealt with at a very early stage of production. The exchange of the (virtual) production data enables an efficient workflow between the departments. This can only be achieved with software and hardware systems that allow the exchange between the components involved in the production. Tasks can be split up after the idea or vision has been discussed and it is possible to imagine the result with the aid of pre-visualization.

Set Extensions and Virtual Backdrops in Commercials and Film

Filming at virtual locations can dramatically reduce costs compared to shooting on a real location. Critical production scenarios which allow filming only at short frames, like dawn or dusk, can be used continuously for an arbitrary duration in a virtual environment. Logistics of blocking streets for a production involving a dialogue in a driving car are no longer required if the scene can be convincingly produced in a virtual production scenario.

Moreover, this technique also allows situating film sets in locations that are inaccessible, or completely based on fictitious content. This applies for studio productions as well as for productions at outside locations where the real foreground is merged with virtual environment at medium to far distance. An example is shown in Figure 23.1 showcasing the real location which was shot at a former airfield in East Germany. The lower image displays the final scenery in which the location was extended to be situated in Guantanamo, Cuba.

For extending the scenery, digital assets and matte painting were created and combined with the filmed material. To fit the virtual elements into the filmed scene, it was necessary to match the set, the camera, and the

Figure 23.1: Set extension before and after. ©Filmakademie Baden-Wuerttemberg, 5 Jahre Leben, 2013.

lighting conditions. Additionally, the virtual camera needed to be as close as possible to the real one, including movement, lens effects, color, and even imperfections and artifacts caused by the real camera. The first step was to perform an exact match move to get information about the camera's movement and the lens geometry. Lighting and the surface definition were typically realized during, or directly after, the 3D asset creation process using digital content creation applications like Autodesk Maya or Maxon Cinema 3D. All 2D extensions, matte paintings, and cardboard styled assets were painted in tools like Adobe Photoshop or similar. These elements could also be animated or mixed with filmed footage. In the final step the generated assets were combined, visually matched and color grated in a compositing tool like the Foundry's Nuke or Adobe After Effects. For the given example the effective working time was about 20 man-days: 12 for the asset generation, 5 for the match move, 3 days compositing, and 1 day for revision requests.

A complete technical solution that enables virtual set production scenarios with realistic backdrops has been developed by Stargate Studios.[7] To meet the growing demand for low-cost alternatives to location shooting, Stargate Studios has developed the "next generation" stock footage library which specializes in virtual environments and customizable stock "locations." Branded by Stargate Studios in 2002 as the VIRTUAL BACKLOT (VB), the VB Library allows actors to literally walk into pre-shot footage—as if they were actually on location—without ever having to leave town.

The Virtual Backlot uses various techniques and methods to provide a range of options for creative production staff. The major components on producing in studio or outdoors are 3D rendering, camera tracking, and color keying (green and blue screen), all in real-time. Essential to the whole workflow is pre-production where, for example, the locations defined in the script are pre-recorded only with a small VFX team, instead of moving the whole production crew of up to 100 people to a designated destination. The recordings are called plates or background plates because they are used to replace the green screen.

Already a proven success in broadcast television, the VB "immersive environments" Library is beyond anything available in today's stock footage market. The environments vary from panoramic photographed and video sequences up to full 3D sceneries. Hundreds of hours of immersive footage have been acquired from multiple locations and successfully used by more than 165 feature, television, and commercial productions like the PanAm Demo[8] shown in Figure 23.2. As in 2014, 12 major television series are using VB on a regular basis. The cost model is so compelling that television networks are actually adopting creative content to match what is available in the VB library.

The most challenging and therefore also expensive piece of technology is the real-time component because its development is both time- and knowledge-consuming, and the market is still small. To register the movements of a film camera in real-time, only a few different technologies are available that allow the use in- and outdoors and are simple enough to be set up in as short as 1-3 hours on a stage or outdoors.

One of the most recent systems used in VB is the Lightcraft Previzion[9] system using Intersense technology. Here fiducial markers with circular barcodes are recorded by a small camera. The signal is processed in real-time to identify where each of the up to 300 circular barcodes is placed on the set and allow the real-time 3D render engine to define the current position and render a matching video stream according to the practical camera signal with only a short delay of 7 frames (@25 fps in 1920x1080).

[7] http://www.stargatestudios.net/
[8] http://vimeo.com/25483317
[9] http://www.lightcrafttech.com

Figure 23.2: PanAm Demo using Stargate VB, top to bottom: camera green screen material, on set previz (pre-composite), final composite after post-production. ©Stargate Studios.

A weaknesses of the Intersense tracking system is that its tracking camera has only a preset resolution, operational frame rate and light sensitivity. This means the fiducial markers have to be lit properly, have to be 100% fix (no movement is allowed). In some cases the limitation of only 300 fiducials allows for only a limited space to be tracked. The mounting of both fiducials and the Intersense camera on the production camera can be complicated.

Another extra procedure added to the production is the requirement to calibrate the lenses of the production camera before green screen shooting. This requires extra effort, and in the case of the Intersense system, a complex rig using pattern recognition on a large board of roughly the size of a mid-size desktop. To measure a prime lens (for example 35 mm Ultra Prime) takes from 30 to 60 minutes. To measure a zoom lens takes about 6-7 hours per lens. It is common to use, on average, film shooting 10 to 20 different lenses. Every single individual lens has to be calibrated because none is alike.

23.4 Real-Time Rendering with Game Engines

Game engines include reliable tools for real-time visualization addressing some of the needs of virtual production scenarios [Nitsche 08]. One solution currently in development is the Cinebox™system from the game developer studio Crytek™. The Cinebox™system is based on the Cryengine™[10] and provides extended functionality to meet filmmaking requirements. The software was used in the 3-minute branded-entertainment virtual production graduation short-film Dark Matter[11] [Y. Sahin and Backhaus 14].

The Dark Matter production combined digital video footage in 3k (ARRIRAW 16:9, 2880 x 1620 pixels) with real-time renderings from the game engine for live, on-set visualization, and the final render for the post-production. The game engine was the main and only tool for lighting and rendering of the short film. The workflow allowed having an editor on-set working with the produced media and preparing new versions of the short-film by the end of each production day, which could then be reviewed by the entire team.

One challenging issue in the workflow was to synchronize the game engine output, the live camera imagery, and the motion-capture data in a frame accurate way. This was accomplished by manually delaying the data streams using a custom software solution. Nevertheless, the game engine signal was not delivered at a fixed delay, which in minor cases resulted in asynchronous image output. A possible solution for this problem would be the integration of the live camera image directly in the game engine, thus reducing the complexity of syncing two sources to just one. Another aspect was camera tracking accuracy. Given that the motion capture systems precision was in the sub-millimeter range, the position definition of the real

[10]http://www.crytek.com/cryengine
[11]http://youtu.be/XlDL8hHVdDw

Figure 23.3: The Dark Matter production combined real camera footage with computer-generated imagery from game engine to provide interactive high-quality previsualization directly on a film set. ©Filmakademie Baden-Wuerttemberg, Dark Matter, 2014.

cameras film-back and the lenses nodal point was not sufficient. This led to a mismatching overlay of the real-footage and the CG elements. In post-production, the accuracy was still sufficient for most of the takes, and only 4 of the total of 33 sequences involving dynamic camera movement needed additional match move handling.

The alignment of the virtual with the real scene elements turned out to be another important issue. Lens and camera parameters were encoded as metadata [ALE 14] located in non-picture regions of the individual image frames, known as vertical ancillary data (VANC), to match the virtual lenses with the real ones. Real-time, high-quality chroma-keying and extraction of the actors, and practical assets have been other major obstacles to overcome in rendering the final images directly on set. In traditional VFX productions this is established as a classical post-production process, whereas in this production a dedicated hardware chroma-keying device was used that combined the game engine output with the keyed camera signal.

The Dark Matter short-film showcased an exemplary production pipeline for generating interactive, high-quality previsualization directly on the film set. Lighting parameters of the real set could be adjusted in the virtual scenery on the fly. The director of photography was able to preview the composed video signal directly in the camera. The visual consistency between on-set and final image was extremely close due to the continuous use of the same game engine technology throughout the entire production process. Exchanging and integrating assets in the post process was also possible in real-time. This resulted in a much more efficient post-production phase requiring only half of the time compared to a traditional workflow. Figure 23.3 shows the live camera, the real-time output of the game engine combined with the camera feed, and the final image.

23.5 Summary

Virtual production processes promise a new, more collaborative and interactive form of filmmaking. Advances in visual computing, in particular in sensing real-world parameters of cameras, assets and actors on one hand and progress in rendering and visualization on the other side show currently a major impact on the production process. With more visual computing tools to come it is expected that the process will be using more and more virtual techniques.

The state-of-the-art today is in most cases a rather complex custom development of all required technological aspects. Further development toward open standards will help to overcome some of these obstacles and allow for a more seamless exchange of data between individual hard- and software solutions. There are only a few turn-key solutions which can be picked up effortlessly. Game engines provide flexible, real-time tools, and their rendering quality and capabilities are constantly improving. Many film productions, however, demand physically correct image synthesis which in most cases relies on ray tracing. The future will show if these two can be combined in a meaningful way to support virtual production processes.

Preparing for virtual productions will also require a lot of rethinking of how assets and creative decisions will be made. The new creative freedom to make visual decisions directly on-set might also mix up established production pipelines (where most decisions are taken in post-production). Virtual production tools are also highly capable of democratizing the filmmaking process to individuals, independent filmmakers, tight budget, and academic productions.

Bibliography

[Adabala et al. 03] Neeharika Adabala, Nadia Magnenat-Thalmann, and Guangzheng Fei. "Visualization of Woven Cloth." In *Proc. of the 14th Eurographics Workshop on Rendering*, pp. 178–185. Eurographics Association, 2003.

[Adam et al. 09] Michael Adam, Christoph Jung, Stefan Roth, and Guido Brunnett. "Real-Time Stereo-Image Stitching Using GPU-Based Belief Propagation." In *Proc. Vision, Modeling and Visualization (VMV)*. Braunschweig, Germany, 2009.

[Addison et al. 95] Alonzo C. Addison, Douglas MacLeod, Gerald Margolis, Beit Hashoah, Michael Naimark, and Hans-Peter Schwarz. "Museums without Walls (panel session): New Media for New Museums." In *Proceedings of the 22nd Annual Conference on Computer Graphics and Interactive Techniques, SIGGRAPH '95*, pp. 480–481. New York, NY, USA: ACM, 1995.

[Adobe Systems Inc. 13] Adobe Systems Inc. "After Effects." 2013. Http://www.adobe.com/products/aftereffects.html.

[Agarwal and Triggs 06] Ankur Agarwal and Bill Triggs. "Recovering 3D Human Pose from Monocular Images." *IEEE Transactions on PAMI* 28:1 (2006), 44–58.

[Agarwal et al. 10] Sameer Agarwal, Noah Snavely, Steven M. Seitz, and Richard Szeliski. "Bundle Adjustment in the Large." In *Computer Vision–ECCV 2010*, pp. 29–42. Springer, 2010.

[Agarwal et al. 11] Sameer Agarwal, Yasutaka Furukawa, Noah Snavely, Ian Simon, Brian Curless, Steven M. Seitz, and Richard Szeliski. "Building Rome in a Day." *Comm. ACM.* 2011.

[Aggarwal and Ahuja 04] Manoj Aggarwal and Narendra Ahuja. "Split Aperture Imaging for High Dynamic Range." *International Journal of Computer Vision* 58:1 (2004), 7–17.

[Agisoft 12] Agisoft. "PhotoScan." http://www.agisoft.ru/, 2012.

[Agrawal et al. 06] Amit K. Agrawal, Ramesh Raskar, and Rama Chellappa. "What Is the Range of Surface Reconstructions from a Gradient Field?" In *ECCV*, pp. 578–591, 2006.

[Ahmed et al. 01] Amr Ahmed, Farzin Mokhtarian, and Adrian Hilton. "Parametric Motion Blending through Wavelet Analysis." In *EUROGRAPHICS. Proceedings of Short Presentations*, pp. 347–353, 2001.

[Alahi et al. 12] Alexandre Alahi, Raphael Ortiz, and Pierre Vandergheynst. "Freak: Fast Retina Keypoint." In *Computer Vision and Pattern Recognition (CVPR), 2012 IEEE Conference on*, pp. 510–517. IEEE, 2012.

[ALE 14] "Metadata in the ALEXA—Digital Workflow Solutions." Technical report, Arnold & Richter Cine Technik GmbH & Co. Betriebs KG, 2014. Available online (http://www.arri.com/camera/alexa/downloads).

[Alexa and Adamson 09] Marc Alexa and Anders Adamson. "Interpolatory Point Set Surfaces—Convexity and Hermite Data." *ACM Transactions on Graphics* 28:2 (2009), 20:1–20:10.

[Alexa et al. 01] Marc Alexa, Johannes Behr, Daniel Cohen-Or, Shachar Fleishman, David Levin, and Claudio T. Silva. "Point Set Surfaces." In *Proceedings of the IEEE Conference on Visualization '01*, pp. 21–28, 2001.

[Alexa et al. 04] M. Alexa, S. Rusinkiewicz, and Anders Adamson. "On Normals and Projection Operators for Surfaces Defined by Point Sets." In *Proceedings of the Eurographics Symposium on Point-Based Graphics 2004*, pp. 149–155, 2004.

[Alexander et al. 10] Oleg Alexander, Mike Rogers, William Lambeth, Jen-Yuan Chiang, Wan-Chun Ma, Chuan-Chang Wang, and Paul Debevec. "The Digital Emily Project: Achieving a Photorealistic Digital Actor." *IEEE Computer Graphics and Applications*.

[Allen et al. 02] B. Allen, B. Curless, and Z. Popović. "Articulated Body Deformation from Range Scan Data." *ACM Transactions on Graphics (Proceedings of SIGGRAPH)* 21:3 (2002), 612–619.

[Allen et al. 03] B. Allen, B. Curless, and Z. Popović. "The Space of Human Body Shapes: Reconstruction and Parameterization from Range Scans." *ACM Transactions on Graphics* 22:3 (2003), 587–594. Proceedings of SIGGRAPH.

[Allen et al. 06] B. Allen, B. Curless, Z. Popović, and A. Hertzmann. "Learning a Correlated Model of Identity and Pose-Dependent Body Shape Variation for Real-Time Synthesis." In *Symposium on Computer Animation*, pp. 147–156, 2006.

[Alvarez et al. 00] Luis Alvarez, Joachim Weickert, and Javier Sánchez. "Reliable Estimation of Dense Optical Flow Fields with Large Displacements." *International Journal of Computer Vision* 39:1 (2000), 41–56.

[Alvarez et al. 07] Luis Alvarez, Rachid Deriche, Théo Papadopoulo, and Javier Sánchez. "Symmetrical Dense Optical Flow Estimation with Occlusions Detection." *International Journal of Computer Vision* 75:3 (2007), 371–385.

[Amenta et al. 01] Nina Amenta, Sunghee Choi, and Krishna Kolluri. "The Power Crust." In *Proceedings of the 6th ACM Symposium on Solid Modeling*, pp. 249–260, 2001.

[Anandan 89] P. Anandan. "A Computational Framework and an Algorithm for the Measurement of Visual Motion." *International Journal of Computer Vision* 2:3 (1989), 283–310.

[Anderson et al. 11] Robert Anderson, Björn Stenger, and Roberto Cipolla. "Color Photometric Stereo for Multicolored Surfaces." In *ICCV*, 2011.

[Anguelov et al. 04a] D. Anguelov, D. Koller, H. C. Pang, P. Srinivasan, and S. Thrun. "Recovering Articulated Object Models from 3D Range Data." In *Proceedings of AUAI*, pp. 18–26, 2004.

[Anguelov et al. 04b] Dragomir Anguelov, Praveen Srinivasan, Hoi-Cheung Pang, Daphne Koller, Sebastian Thrun, and James Davis. "The Correlated Correspondence Algorithm for Unsupervised Registration of Nonrigid Surfaces." In *NIPS*, 2004.

[Anguelov et al. 05] Dragomir Anguelov, Praveen Srinivasan, Daphne Koller, Sebastian Thrun, Jim Rodgers, and James Davis. "SCAPE: Shape Completion and Animation of People." *ACM Transactions on Graphics (Proc. SIGGRAPH)* 24:3 (2005), 408–416.

[Anisimov et al. 77] Sergei I. Anisimov, V. A. Benderski, and G. Farkas. "Nonlinear Photoelectric Emission from Metals Induced by a Laser Radiation." *Soviet Physics Uspekhi* 20:6 (1977), 467. Available online (http://stacks.iop.org/0038-5670/20/i=6/a=R01).

[Arikan and Forsyth 02] O. Arikan and D.A. Forsyth. "Synthesizing Constrained Motions from Examples." In *Proc. ACM SIGGRAPH*, 2002.

[Arvo 86] James Arvo. "Backward Ray Tracing." In *Developments in Ray Tracing, ACM SIGGRAPH Course Notes*, pp. 259–263, 1986.

[Ashikmin et al. 00] Michael Ashikmin, Simon Premože, and Peter Shirley. "A Microfacet-Based BRDF Generator." In *Proc. of the 27th Annual Conference on Computer Graphics and Interactive Techniques*, pp. 65–74. ACM Press/Addison-Wesley Publishing Co., 2000.

[Atherton and Caporael 85] Peter R. Atherton and Linnda R. Caporael. "A Subjective Judgment Study of Polygon Based Curved Surface Imagery." *SIGCHI Bull.* 16:4 (1985), 27–34. Available online (http://doi.acm.org/10.1145/1165385.317462).

[Aujay et al. 07] G. Aujay, F. Hétroy, F. Lazarus, and C. Depraz. "Harmonic Skeleton for Realistic Character Animation." In *Proceedings of the ACM Symposium on Computer Animation (SCA)*, pp. 151–160, 2007.

[Autodesk 12] Autodesk. "123D Catch." http://www.123dapp.com/catch, 2012.

[Azouz et al. 06] Z. Ben Azouz, C. Shu, and A. Mantel. "Automatic Locating of Anthropometric Landmarks on 3D Human Models." In *Symposium on 3D Data Processing, Visualization, and Transmission*, pp. 750–757, 2006.

[Baak et al. 13] Andreas Baak, Meinard Müller, Gaurav Bharaj, Hans-Peter Seidel, and Christian Theobalt. "A Data-Driven Approach for Real-Time Full Body Pose Reconstruction from a Depth Camera." In *Consumer Depth Cameras for Computer Vision*, pp. 71–98. Springer, 2013.

[Badler et al. 87] N. I. Badler, K. H. Manoochehri, and Baraff D. "Multi-Dimensional Input Techniques and Articulated Figure Positioning by Multiple Constraints." In *Proc. of SI3D*, pp. 151–169, 1987.

[Baker et al. 11] Simon Baker, Daniel Scharstein, J. P. Lewis, Stefan Roth, Michael J. Black, and Richard Szeliski. "A Database and Evaluation Methodology for Optical Flow." *Int. J. Comput. Vision* 92:1 (2011), 1–31.

[Balan and Black 08] A. Balan and M. Black. "The Naked Truth: Estimating Body Shape under Clothing." In *European Conference on Computer Vision*, pp. 15–29, 2008.

[Balan et al. 07] A. Balan, L. Sigal, M. Black, J. Davis, and H. Haussecker. "Detailed Human Shape and Pose from Images." In *Conference on Computer Vision and Pattern Recognition*, pp. 1–8, 2007.

[Ballan et al. 10] Luca Ballan, Gabriel J. Brostow, Jens Puwein, and Marc Pollefeys. "Unstructured Video-Based Rendering: Interactive Exploration of Casually Captured Videos." *ACM Trans. on Graphics (Proc. SIGGRAPH)* 29:3 (2010), 87:1–87:11.

[Ballan et al. 12] Luca Ballan, Aparna Taneja, Jürgen Gall, Luc Van Gool, and Marc Pollefeys. "Motion Capture of Hands in Action Using Discriminative Salient Points." In *Computer Vision–ECCV 2012*, pp. 640–653. Springer, 2012.

[Baraff and Witkin 98] David Baraff and Andrew Witkin. "Large Steps in Cloth Simulation." In *Proc. 25th Annual Conf. on Computer Graphics and Interactive Techniques, SIGGRAPH '98*, pp. 43–54. New York, NY, USA: ACM, 1998.

[Baran and Popović 07] Ilya Baran and Jovan Popović. "Automatic Rigging and Animation of 3D Characters." *ACM Transactions on Graphics (Proc. SIGGRAPH)* 26:3 (2007), 72.

[Baran et al. 09] I. Baran, D Vlasic, E. Grinspun, and J. Popovic. "Semantic Deformation Transfer." *ACM Transactions on Graphics (Proc. ACM SIGGRAPH 2009)* 28:3.

[Barron et al. 94] John L. Barron, David J. Fleet, and Steven S. Beauchemin. "Performance of Optical Flow Techniques." *Int. J. Comput. Vision* 12:1 (1994), 43–77.

[Bartczak and Koch 09] Bogumil Bartczak and Reinhard Koch. "Dense Depth Maps from Low Resolution Time-of-Flight Depth and High Resolution Color Views." In *ISVC (2)*, pp. 228–239, 2009.

[Baumgart 74] Bruce Guenther Baumgart. "Geometric modeling for computer vision." Ph.D. thesis, Stanford, CA, USA, 1974. AAI7506806.

[Bay et al. 06] Herbert Bay, Tinne Tuytelaars, and Luc Van Gool. "Surf: Speeded Up Robust Features." In *Computer Vision–ECCV 2006*, pp. 404–417. Springer, 2006.

[Beardsley et al. 97] Paul A. Beardsley, Andrew Zisserman, and David W. Murray. "Sequential Updating of Projective and Affine Structure from Motion." *International Journal of Computer Vision* 23:3 (1997), 235–259.

[Beder and Steffen 06] Christian Beder and Richard Steffen. "Determining an Initial Image Pair for Fixing the Scale of a 3D Reconstruction from an Image Sequence." In *Pattern Recognition*, *4174*, edited by K. Franke, K.-R. Müller, B. Nickolay, and R. Schäfer, pp. 657–666. Springer, 2006.

[Beder et al. 07] Christian Beder, Bogumil Bartczak, and Reinhard Koch. "A Combined Approach for Estimating Patchlets from PMD Depth Images and Stereo Intensity Images." In *DAGM-Symposium*, pp. 11–20, 2007.

[Beeler et al. 10] Thabo Beeler, Bernd Bickel, Paul Beardsley, Bob Sumner, and Markus Gross. "High-Quality Single-Shot Capture of Facial Geometry." *ACM Transactions on Graphics (TOG)* 29:4 (2010), 40.

[Beeler et al. 11] Thabo Beeler, Fabian Hahn, Derek Bradley, Bernd Bickel, Paul Beardsley, Craig Gotsman, Robert W. Sumner, and Markus Gross. "High-Quality Passive Facial Performance Capture Using Anchor Frames." *ACM Transactions on Graphics (Proc. SIGGRAPH)* 30 (2011), 75:1–75:10.

[Beier and Neely 92] Thaddeus Beier and Shawn Neely. "Feature-Based Image Metamorphosis." *Computer Graphics (Proc. of SIGGRAPH'93)* 26:2 (1992), 35–42.

[Belbachir et al. 12] Ahmed Nabil Belbachir, Manfred Mayerhofer, Daniel Matolin, and Joseph Colineau. "360SCAN: High-Speed Rotating Line Sensor for Real-Time 360 Panoramic Vision." In *Proc. 6th Int. Conf on Distributed Smart Cameras (ICDSC)*, pp. 1–6. Hongkong, 2012.

[Bentley 75] Jon Louis Bentley. "Multidimensional Binary Search Trees Used for Associative Searching." *Communication of the ACM* 18:9 (1975), 509–517.

[Berends and Erkelens 01] Ellen M Berends and Casper J Erkelens. "Adaptation to Disparity but Not to Perceived Depth." *Vision Research* 41:7 (2001), 883–892.

[Berent and Dragotti 06] J. Berent and P.L. Dragotti. "Segmentation of Epipolar-Plane Image Volumes with Occlusion and Disocclusion Competition." In *IEEE 8th Workshop on Multimedia Signal Processing*, pp. 182–185, 2006.

[Berger et al. 11] Kai Berger, Kai Ruhl, Christian Brümmer, Yannic Schröder, Alexander Scholz, and Marcus Magnor. "Markerless Motion Capture Using Multiple Color-Depth Sensors." In *Vision, Modeling, and Visualization (VMV)*, pp. 317–324, 2011.

[Berger et al. 14] M. Berger, A. Tagliasacchi, L. M. Seversky, P. Alliez, J. A. Levine, A. Sharf, and C. T. Silva. "State of the Art in Surface Reconstruction from Point Clouds." In *EUROGRAPHICS State-of-the-Art Report*, 2014.

[Besl and McKay 92] P. Besl and N. McKay. "A Method for Registration of 3-D Shapes." *IEEE Trans. Pattern Analysis and Machine Intelligence (PAMI)* 14:2 (1992), 239–258.

[Bhat et al. 03] Kiran. S. Bhat, Christopher. D. Twigg, Jessica. K. Hodgins, Pradeep. K. Khosla, Zoran. Popović, and Steven. M. Seitz. "Estimating Cloth Simulation Parameters from Video." In *Proc. ACM SIGGRAPH/Eurographics Symp. on Computer Animation*, pp. 37–51. Eurographics Association, 2003.

[Bhat et al. 13] Kiran Bhat, Rony Goldenthal, Yuting Ye, Ronald Mallet, and Michael Koperwas. "High Fidelity Facial Animation Capture and Retargeting with Contours." *ACM SIGGRAPH/Eurographics Symposium on Computer Animation*, pp. 7–14.

[Bickel et al. 07] Bernd Bickel, Mario Botsch, Roland Angst, Wojciech Matusik, Miguel Otaduy, Hanspeter Pfister, and Markus Gross. "Multi-Scale Capture of Facial Geometry and Motion." *ACM Transactions on Graphics* 26:3 (2007).

[Birchfield and Tomasi 98] Stan Birchfield and Carlo Tomasi. "A Pixel Dissimilarity Measure That Is Insensitive to Image Sampling." *IEEE Trans. Pattern Anal. Mach. Intell.* 20:4 (1998), 401–406.

[Bishop and Favaro 12] T. Bishop and P. Favaro. "The Light Field Camera: Extended Depth of Field, Aliasing, and Superresolution." *IEEE Transactions on Pattern Analysis and Machine Intelligence* 34:5 (2012), 972–986.

[Blache et al. 14] Ludovic Blache, Céline Loscos, Olivier Nocent, and Laurent Lucas. "3D Volume Matching for Mesh Animation of Moving Actors." In *Eurographics Workshop on 3D Object Retrieval (3DOR)*, pp. 69–76. Eurographics, 2014.

[Black and Anandan 96] Michael J. Black and P. Anandan. "The Robust Estimation of Multiple Motions: Parametric and Piecewise-smooth Flow Fields." *Comput. Vis. Image Underst.* 63:1 (1996), 75–104.

[Bleyer and Breiteneder 13] Michael Bleyer and Christian Breiteneder. "Stereo Matching—State-of-the-Art and Research Challenges." In *Advanced Topics in Computer Vision*, pp. 143–179, 2013.

[Bleyer and Gelautz 07] Michael Bleyer and Margrit Gelautz. "Graph-cut-based Stereo Matching Using Image Segmentation with Symmetrical Treatment of Occlusions." *Image Commun.* 22:2 (2007), 127–143.

[Bleyer and Gelautz 08] Michael Bleyer and Margrit Gelautz. "Simple but Effective Tree Structures for Dynamic Programming-Based Stereo Matching." In *Conference on Computer Vision Theory and Applications*, pp. 415–422, 2008.

[Bleyer et al. 11a] Michael Bleyer, Christoph Rhemann, and Carsten Rother. "PatchMatch Stereo—Stereo Matching with Slanted Support Windows." In *British Machine Vision Conference*, pp. 14.1–14.11, 2011.

[Bleyer et al. 11b] Michael Bleyer, Carsten Rother, Pushmeet Kohli, Daniel Scharstein, and Sudipta Sinha. "Object Stereo—Joint Stereo Matching an Object Segmentation." In *Computer Vision and Pattern Recognition*, 2011.

[Bleyer et al. 12] Michael Bleyer, Christoph Rhemann, and Carsten Rother. "Extracting 3D Scene Consistent Object Proposals and Depth from Stereo Images." In *European Conference on Computer Vision*, 2012.

[Blinn and Newell 76] James F. Blinn and Martin E. Newell. "Texture and Reflection in Computer Generated Images." *Commun. ACM* 19:10 (1976), 542–547.

[Blonde et al. 96] Laurent Blonde, Wolfgang Niem, Yakup Paker, Matthias Buck, Ricardo Galli, Wolfgang Schmidt, and Graham Thomas. "A Virtual Studio for Live Broadcasting: The Mona Lisa Project." *IEEE Multimedia* 3:2 (1996), 18–29.

[Blumenthal-Barby and Eisert 14] David Blumenthal-Barby and Peter Eisert. "High-Resolution Depth for Binocular Image-Based Modelling." *Computers & Graphics* 39 (2014), 89–100.

[Boff et al. 88] K.R. Boff and J.E. Lincoln. *Engineering Data Compendium–Human Perception and Performance, Vol. I, Sensory Process and Perception.* New York: John Wiley & Sons, 1988.

[Bolles et al. 87] R.C. Bolles, H.H. Baker, and D.H. Marimont. "Epipolar-Plane Image Analysis: An Approach to Determining Structure from Motion." *International Journal of Computer Vision* 1:1 (1987), 7–55.

[Bonnard et al. 13] J. Bonnard, G. Valette, C. Loscos, and J.-M. Nourrit. "3D HDR Acquisition and Restitution." In *3D Video: From Capture to Diffusion.* John Wiley & Sons, 2013.

[Borshukov et al. 03] George Borshukov, Jefferson Montgomery, and John Hable. *Playable Universal Capture, GPU Gems 3.* Addison-Wesley, 2003.

[Borshukov et al. 05] George Borshukov, Dan Piponi, Oystein Larsen, John Peter Lewis, and Christina Tempelaar-Lietz. "Universal Capture—Image-Based Facial Animation for "The Matrix Reloaded."" In *SIGGRAPH Course Notes*, 2005.

[Botsch and Sorkine 08] M. Botsch and O. Sorkine. "On Linear Variational Surface Deformation Methods." *IEEE Transactions on Visualization and Computer Graphics* 4:1 (2008), 213–230.

[Botsch et al. 10] Mario Botsch, Leif Kobbelt, Mark Pauly, Pierre Alliez, and Bruno Levy. *Polygon Mesh Processing.* A. K. Peters/CRC Press, 2010.

[Bougnoux 98] Sylvain Bougnoux. "From Projective to Euclidean Space under Any Practical Situation, a Criticism of Self-Calibration." In *Computer Vision, 1998. Sixth International Conference on*, pp. 790–796. IEEE, 1998.

[Bouguet 08] Jean-Yves Bouguet. "Camera Calibration Toolbox for MATLAB®.", 2008. Available online (http://www.vision.caltech.edu/bouguetj/calib_doc/.).

[Bouman et al. 13] Katherine L. Bouman, Bei Xiao, Peter Battaglia, and William T. Freeman. "Estimating the Material Properties of Fabric from Video." In *Proc. Int. Conf. on Computer Vision*, pp. 1984–1991, 2013.

[Boykov et al. 01] Yuri Boykov, Olga Veksler, and Ramin Zabih. "Fast Approximate Energy Minimization via Graph Cuts." *IEEE Trans. Pattern Anal. Mach. Intell.* 23:11 (2001), 1222–1239.

[Bradley et al. 08] Derek Bradley, Tiberiu Popa, Alla Sheffer, Wolfgang Heidrich, and Tamy Boubekeur. "Markerless Garment Capture." *ACM Trans. Graph.* 27:3 (2008), 99:1–99:9.

[Bradley et al. 10] Derek Bradley, Wolfgang Heidrich, Tiberiu Popa, and Alla Sheffer. "High Resolution Passive Facial Performance Capture." *ACM Transactions on Graphics (Proc. SIGGRAPH)* 29:4 (2010), 41:1–41:10.

[Bradski 00] Gary Bradski. "The OpenCV Library." *Doctor Dobbs Journal* 25:11 (2000), 120–126.

[Brainard 97] David H Brainard. "The Psychophysics Toolbox." *Spatial Vision* 10:4 (1997), 433–436.

[Breen et al. 94] David E. Breen, Donald H. House, and Michael J. Wozny. "Predicting the Drape of Woven Cloth Using Interacting Particles." In *Proc. of the 21st Annual Conference on Computer Graphics and Interactive Techniques, SIGGRAPH '94*, pp. 365–372. New York, NY, USA: ACM, 1994.

[Bregler et al. 04] C. Bregler, J. Malik, and K. Pullen. "Twist Based Acquisition and Tracking of Animal and Human Kinematics." *IJCV* 56:3 (2004), 179–194.

[Bridson et al. 03] Robert Bridson, Sebastian Marino, and Ronald Fedkiw. "Simulation of Clothing with Folds and Wrinkles." In *Proc. of the 2003 ACM SIGGRAPH/Eurographics Symposium on Computer Animation, SCA '03*, pp. 28–36. Eurographics Association, 2003.

[Bronstein et al. 07] Alex M. Bronstein, Michael M. Bronstein, and Ron Kimmel. "Calculus of Non-Rigid Surfaces for Geometry and Texture Manipulation." *IEEE Transactions on VCG* 13(5) (2007), 902–913.

[Brown and Rusinkiewicz 07] Benedict J. Brown and Szymon Rusinkiewicz. "Global Non-rigid Alignment of 3-D Scans." *ACM Transactions on Graphics* 26:3.

[Brown 66] D. C. Brown. "Decentering Distortion of Lenses." *Photogrammetric Engineering* 32:3 (1966), 444–462.

[Brox and Malik 11] Thomas Brox and Jitendra Malik. "Large Displacement Optical Flow: Descriptor Matching in Variational Motion Estimation." *IEEE Trans. Pattern Anal. Mach. Intell.* 33:3 (2011), 500–513.

[Brox et al. 04] Thomas Brox, Andrés Bruhn, Nils Papenberg, and Joachim Weickert. "High Accuracy Optical Flow Estimation based on a Theory for Warping." In *European Conference on Computer Vision (ECCV)*, Lecture Notes in Computer Science, 3024, pp. 25–36. Springer, 2004.

[Buades et al. 05] Antoni Buades, Bartomeu Coll, and J-M. Morel. "A Non-Local Algorithm for Image Denoising." In *IEEE CVPR*, pp. 60–65, 2005.

[Budd et al. 12] Chris Budd, Oliver Grau, and Peter Schübel. "Web Delivery of Free-Viewpoint Video of Sport Events."

[Budd et al. 13] Chris Budd, Peng Huang, Martin Klaudiny, and Adrian Hilton. "Global Non-rigid Alignment of Surface Sequences." *International Journal of Computer Vision* 102:1-3 (2013), 256–270.

[Buehler et al. 01] Chris Buehler, Michael Bosse, Leonard McMillan, Steven Gortler, and Michael Cohen. "Unstructured Lumigraph Rendering." In *ACM SIGGRAPH*, pp. 425–432, 2001.

[Butler et al. 12] Alex Butler, Shahram Izadi, Otmar Hilliges, David Molyneaux, Steve Hodges, and David Kim. "Shake'n'Sense: Reducing Structured Light Interference when Multiple Depth Cameras Overlap." In *Proceedings of Human Factors in Computing Systems (ACM CHI).* New York, NY, USA: ACM, 2012.

[Cagniart et al. 10] Cedric Cagniart, Edmond Boyer, and Slobodan Ilic. "Probabilistic Deformable Surface Tracking from Multiple Videos." In *ECCV*, pp. 326–339, 2010.

[Calonder et al. 10] Michael Calonder, Vincent Lepetit, Christoph Strecha, and Pascal Fua. "Brief: Binary Robust Independent Elementary Features." In *Computer Vision–ECCV 2010*, pp. 778–792. Springer, 2010.

[Carranza et al. 03] J. Carranza, C. Theobalt, M. Magnor, and H.-.P. Seidel. "Free-Viewpoint Video of Human Actors." *ACM Transactions on Graphics (Proc. ACM SIGGRAPH)* 22:3 (2003), 569–577.

[Casas et al. 12a] Dan Casas, Margara Tejera, Jean-Yves Guillemaut, and Adrian Hilton. "4D Parametric Motion Graphs for Interactive Animation." In *Proceedings of the ACM SIGGRAPH Symposium on Interactive 3D Graphics and Games*, pp. 103–110. ACM, 2012.

[Casas et al. 12b] Dan Casas, Margara Tejera, Jean-Yves Guillemaut, and Adrian Hilton. "Parametric Animation of Performance-Captured Mesh Sequences." *Journal of Visualization and Computer Animation* 23:2 (2012), 101–111.

[Casas et al. 13] Dan Casas, Margara Tejera, Jean-Yves Guillemaut, and Adrian Hilton. "Interactive Animation of 4D Performance Capture." *IEEE Transactions on Visualization and Computer Graphics (TVCG)* 19:5 (2013), 762–773.

[Casas et al. 14] Dan Casas, Marco Volino, John Collomosse, and Adrian Hilton. "4D Video Textures for Interactive Character Appearance." *Computer Graphics Forum (Proc. EUROGRAPHICS 2014)* 33:2.

[Chai et al. 00] Jin-Xiang Chai, Xin Tong, Shing-Chow Chan, and Heung-Yeung Shum. "Plenoptic Sampling." In *Proceedings of the 27th Annual Conference on Computer Graphics and Interactive Techniques*, pp. 307–318. ACM Press/Addison-Wesley Publishing Co., 2000.

[Chaminade et al. 07] Thierry Chaminade, Jessica Hodgins, and Mitsuo Kawato. "Anthropomorphism Influences Perception of Computer-Animated Characters Actions." *Social Cognitive and Affective Neuroscience* 2:3 (2007), 206–216.

[Chan et al. 08] Derek Chan, Hylke Buisman, Christian Theobalt, and Sebastian Thrun. "A Noise-Aware Filter for Real-Time Depth Upsampling." In *ECCV Workshop on Multi-Camera and Multi-Modal Sensor Fusion Algorithms and Applications*, edited by Andrea Cavallaro and Hamid Aghajan, pp. 1–12. Marseille, France, 2008.

[Chandaria et al. 07] Jigna Chandaria, Graham A. Thomas, and Didier Stricker. *J. Real-Time Image Processing* 2:2-3 (2007), 69–79.

[Chatterjee et al. 11] Priyam Chatterjee, Neel Joshi, Sing Bing Kang, and Yasuyuki Matsushita. "Noise Suppression in Low-Light Images through Joint Denoising and Demosaicing." In *IEEE CVPR*, pp. 321–328, 2011.

[Chaurasia et al. 11] Gaurav Chaurasia, Olga Sorkine, and George Drettakis. "Silhouette-Aware Warping for Image-Based Rendering." *Computer Graphics Forum* 30:4 (2011), 1223–1232.

[Chazal and Lieuter 05] F. Chazal and A. Lieuter. "The λ Medial Axis." *Graphical Models* 67:4 (2005), 304–331.

[Chen and Williams 93] Shenchang Eric Chen and Lance Williams. "View Interpolation for Image Synthesis." In *Proc. of ACM SIGGRAPH'93*, pp. 279–288. New York: ACM Press/ACM SIGGRAPH, 1993.

[Chen et al. 03] Chia-Yen Chen, Reinhard Klette, and Chi-Fa Chen. "Shape from Photometric Stereo and Contours." In *CAIP*, pp. 377–384, 2003.

[Chen et al. 13a] J. Chen, D. Bautembach, and S. Izadi. "Scalable Real-time Volumetric Surface Reconstruction." *ACM Trans. Graph. (Proc. SIGGRAPH)* 32:4 (2013), 113:1–113:16.

[Chen et al. 13b] Y. Chen, Z. Liu, and Z. Zhang. "Tensor-Based Human Body Modeling." In *Conference on Computer Vision and Pattern Recognition*, 2013.

[Chen 95] Shenchang Eric Chen. "QuickTime VR: An Image-Based Approach to Virtual Environment Navigation." In *Conference on Computer Graphics and Interactive Techniques (SIGGRAPH)*, pp. 29–38, 1995.

[Choi and Ko 02] Kwang-Jin Choi and Hyeong-Seok Ko. "Stable but Responsive Cloth." *ACM Trans. Graph.* 21:3 (2002), 604–611.

[Choi and Ko 05] Kwang-Jin Choi and Hyeong-Seok Ko. "Research Problems in Clothing Simulation." *Comput. Aided Des.* 37:6 (2005), 585–592.

[Chum et al. 07] Ondrej Chum, James Philbin, Josef Sivic, Michael Isard, and Andrew Zisserman. "Total Recall: Automatic Query Expansion with a Generative Feature Model for Object Retrieval." In *Computer Vision, 2007. ICCV 2007. IEEE 11th International Conference on*, pp. 1–8. IEEE, 2007.

[Chun 12] Won Chun. "WebGL Models: End-to-End." In *OpenGL Insights*, edited by Patrick Cozzi and Christophe Riccio, pp. 431–454. CRC Press, 2012.

[Collins 96] Robert T. Collins. "A Space-Sweep Approach to True Multi-Image Matching." In *Computer Vision and Pattern Recognition*, pp. 358–363. IEEE, 1996.

[Cootes et al. 98] Tim Cootes, Gareth J. Edwards, and Christopher J.Taylor. "Active Appearance Models." In *Proc. European Conference on Computer Vision ECCV98*, pp. 484–498, 1998.

[Cordier and Magnenat-Thalmann 05] Frederic Cordier and Nadia Magnenat-Thalmann. "A Data-Driven Approach for Real-Time Clothes Simulation." *Comput. Graph. Forum* 24 (2005), 173–183.

[Criminisi et al. 03] A. Criminisi, P. Perez, and K. Toyama. "Object Removal by Exemplar-Based Inpainting." In *Computer Vision and Pattern Recognition, 2003. Proceedings. 2003 IEEE Computer Society Conference on*, 2, 2, pp. 721–728. IEEE, 2003.

[Criminisi et al. 05] A. Criminisi, S.B. Kang, R. Swaminathan, R. Szeliski, and P. Anandan. "Extracting Layers and Analyzing their Specular Properties using Epipolar-Plane-Image Analysis." *Computer Vision and Image Understanding* 97:1 (2005), 51–85.

[Cryer et al. 95] James Edwin Cryer, Ping-Sing Tsai, and Mubarak Shah. "Integration of Shape from Shading and Stereo." *Pattern Recognition Society* 28:7 (1995), 1033–1043.

[Cui et al. 10] Yan Cui, Sebastian Schuon, Derek Chan, Sebastian Thrun, and Christian Theobalt. "3D Shape Scanning with a Time-of-Flight Camera." In *Proc. IEEE Conf. Computer Vision and Pattern Recognition (CVPR)*, pp. 1173–1180. IEEE, 2010.

[Curless and Levoy 96] B. Curless and M. Levoy. "A Volumetric Method for Building Complex Models from Range Images." In *Proc. Comp. Graph. & Interact. Techn.*, pp. 303–312, 1996.

[Dabov et al. 07] K. Dabov, A. Foi, V. Katkovnik, and K. Egiazarian. "Image Denoising by Sparse 3-D Transform-Domain Collaborative Filtering." *IEEE Transactions on Image Processing* 16:8 (2007), 2080–2095.

[Dachsbacher et al. 13] Carsten Dachsbacher, Jaroslav Křivánek, Milos Hasan, Adam Arbree, Bruce Walter, and Jan Novák. "Scalable Realistic Rendering with Many-Light Methods." *Computer Graphics Forum* DOI: 10.1111/cgf.12256.

[Daly 93] Scott Daly. *Digital Images and Human Vision.* edited by Andrew B. Watson, The Visible Differences Predictor: An Algorithm for the Assessment of Image Fidelity, pp. 179–206. Cambridge, MA, USA: MIT Press, 1993.

[Danette Allen et al. 01] B. Danette Allen, Gary Bishop, and Greg Welch. "Tracking: Beyond 15 Minutes of Thought." Technical Report Course 11, SIGGRAPH 2001, 2001.

[Dansereau et al. 13] D.G. Dansereau, O. Pizarro, and S.B. Williams. "Calibration and Rectification for Lenselet-Based Plenoptic Cameras." In *Proc. International Conference on Computer Vision and Pattern Recognition*, 2013.

[Davis et al. 12] Abe Davis, Marc Levoy, and Frdo Durand. "Unstructured Light Fields." *Comput. Graph. Forum* 31:2 (2012), 305–314.

[Dawes et al. 09] Robert Dawes, Jigna Chandaria, and Graham Thomas. "Image-Based Camera Tracking for Athletics." In *Broadband Multimedia Systems and Broadcasting, 2009. BMSB'09. IEEE International Symposium on*, pp. 1–6. IEEE, 2009.

[de Aguiar and Ukita 14] E. de Aguiar and N. Ukita. "Representing Mesh-Based Character Animations." *Computers and Graphics* 38:0 (2014), 10–17.

[de Aguiar et al. 07] Edilson de Aguiar, Christian Theobalt, Carsten Stoll, and Hans-Peter Seidel. "Marker-Less Deformable Mesh Tracking for Human Shape and Motion Capture." In *IEEE CVPR*, 2007.

[de Aguiar et al. 08a] E. de Aguiar, C. Theobalt, S. Thrun, and H.-P. Seidel. "Automatic Conversion of Mesh Animations into Skeleton-Based Animations." *Computer Graphics Forum* 27:2.

[de Aguiar et al. 08b] E.de de Aguiar, C. Stoll, C. Theobalt, N. Ahmed, H.-P. Seidel, and S. Thrun. "Performance Capture from Sparse Multi-View Video." *ACM Transactions on Graphics (Proc. SIGGRAPH 2008)* 27:3 (2008), 1–10.

[de Aguiar et al. 10] E. de Aguiar, L. Sigal, A. Treuille, and J. K. Hodgins. "Stable Spaces for Real-Time Clothing." *ACM Transactions on Graphics (Proc. SIGGRAPH)* 9:4 (2010), Art.106.

[Debevec and Malik 97] Paul E. Debevec and Jitendra Malik. "Recovering High Dynamic Range Radiance Maps from Photographs." In *Proceedings of ACM Siggraph '97 (Computer Graphics)*, pp. 369–378, 1997.

[Debevec et al. 96] Paul E. Debevec, Camillo J. Taylor, and Jitendra Malik. "Modeling and Rendering Architecture from Photographs: A Hybrid Geometry- and Image-Based Approach." In *Proceedings of the 23rd Annual Conference on Computer Graphics and Interactive Techniques, SIGGRAPH '96*, pp. 11–20. New York, NY, USA: ACM, 1996.

[Debevec et al. 98] Paul Debevec, Yizhou Yu, and George Borshukov. "Efficient View-Dependent Image-Based Rendering with Projective Texture-Mapping." In *Rendering Techniques 98, Eurographics*, edited by George Drettakis and Nelson Max, pp. 105–116. Springer Vienna, 1998.

[Debevec et al. 00] Paul Debevec, Tim Hawkins, Chris Tchou, Haarm-Pieter Duiker, Westley Sarokin, and Mark Sagar. "Acquiring the Reflectance Field of a Human Face." In *SIGGRAPH '00*, pp. 145–156, 2000.

[Debevec 98] Paul Debevec. "Rendering Synthetic Objects into Real Scenes: Bridging Traditional and Image-Based Graphics with Global Illumination and High Dynamic Range Photography." In *Proceedings of the 25th Annual Conference on Computer Graphics and Interactive Techniques, SIGGRAPH '98*, pp. 189–198. New York, NY, USA: ACM, 1998.

[DeCarlo and Metaxas 00] Douglas DeCarlo and Dimitris Metaxas. "Optical Flow Constraints on Deformable Models with Applications to Face Tracking." *International Journal of Computer Vision* 38:2 (2000), 99–127.

[Decaudin et al. 06] Philippe Decaudin, Dan Julius, Jamie Wither, Laurence Boissieux, Alla Sheffer, and Marie-Paule Cani. "Virtual Garments: A Fully Geometric Approach for Clothing Design." *Computer Graphics Forum* 25:3.

[Dellaert et al. 00] Frank Dellaert, Steven M. Seitz, Charles E. Thorpe, and Sebastian Thrun. "Structure from Motion without Correspondence." In *Computer Vision and Pattern Recognition, 2000. Proceedings. IEEE Conference on*, 2, 2, pp. 557–564. IEEE, 2000.

[DeMenthon and Davis 95] D. DeMenthon and L.S. Davis. "Model-Based Object Pose in 25 Lines of Code." *International Journal of Computer Vision* :15 (1995), 123–141.

[Dempster et al. 77] A. P. Dempster, N. M. Laird, and D. B. Rubin. "Maximum Likelihood from Incomplete Data via the EM Algorithm." *Journal of the Royal Statistical Society, Series B* 39:1 (1977), 1–38.

[Deng et al. 05] Yi Deng, Qiong Yang, Xueyin Lin, and Xiaoou Tang. "A Symmetric Patch-Based Correspondence Model for Occlusion Handling." In *International Conference on Computer Vision*, pp. 1316–1322, 2005.

[Deutscher et al. 00] J. Deutscher, A. Blake, and I. Reid. "Articulated Body Motion Capture by Annealed Particle Filtering." In *CVPR*, pp. 1144–1149, 2000.

[Devernay and Beardsley 10] Frédéric Devernay and Paul Beardsley. "Stereoscopic Cinema." In *Image and Geometry Processing for 3-D Cinematography*, Geometry and Computing, 5, edited by R. Ronfard and G. Taubin, 5, pp. 11–51. Springer Berlin Heidelberg, 2010.

[Devernay and Faugeras 01] F. Devernay and O. Faugeras. "Straight Lines Have to Be Straight: Automatic Calibration and Removal of Distortion from Scenes of Structured Environments." *Machine Vision and Applications* 13:1 (2001), 14–24.

[Doboš et al. 13] Jozef Doboš, Kristian Sons, Dmitri Rubinstein, Philipp Slusallek, and Anthony Steed. "XML3DRepo: A REST API for Version Controlled 3D Assets on the Web." In *Proceedings of the 18th International Conference on 3D Web Technology, Web3D '13*. ACM, 2013. To be published.

[Dodgson 02] Neil A. Dodgson. "Analysis of the Viewing Zone of Multi-View Autostereoscopic Displays." In *Stereoscopic Displays and Applications XIII*, pp. 254–265. Proc SPIE 4660, 2002.

[Donner and Jensen 06] Craig Donner and Henrik Wann Jensen. "A Spectral BSSRDF for Shading Human Skin." In *Eurographics Symposium on Rendering*, 2006.

[Donner et al. 08] Craig Donner, Tim Weyrich, Eugene d'Eon, Ravi Ramamoorthi, and Szymon Rusinkiewicz. "A Layered, Heterogeneous Reflectance Model for Acquiring and Rendering Human Skin." *ACM Transactions on Graphics (Proc. SIGGRAPH Asia 2008)* 27:5 (2008), 1–12.

[Dorrington et al. 11] A. Dorrington, J. Godbaz, M. Cree, A. Payne, and L. Streeter. "Separating True Range Measurements from Multi-Path and Scattering Interference in Commercial Range Cameras." In *Proc. IS&T/SPIE Electronic Imaging*, pp. 786404–786404, 2011.

[Drago and Myszkowski 01] Frédéric Drago and Karol Myszkowski. "Validation Proposal for Global Illumination and Rendering Techniques." *Computers & Graphics* 25:3 (2001), 511–518.

[Drettakis et al. 97] George Drettakis, Luc Robert, and Sylvain Bougnoux. "Interactive Common Illumination for Computer Augmented Reality." In *Rendering Techniques*, pp. 45–56, 1997.

[Dryden and Mardia 02] I. Dryden and K. Mardia. *Statistical Shape Analysis.* Wiley, 2002.

[Dubois 01] Eric Dubois. "A Projection Method to Generate Anaglyph Stereo Images." In *Proc. IEEE Int. Conf. Acoustics Speech Signal Processing*, pp. 1661–1664. IEEE Computer Society Press, 2001.

[Duda et al. 01] R. Duda, P. Hart, and D. Stork. *Pattern Classification, Second Edition.* John Wiley & Sons, Inc., 2001.

[Duncan 09] Jody Duncan. "The Unusual Birth of Benjamin Button." In *Cinefex 116*, 2009.

[Duncan 10] Jody Duncan. "The Seduction of Reality." In *Cinefex 120*, 2010.

[Duveau et al. 12] Estelle Duveau, Simon Courtemanche, Lionel Reveret, and Edmond Boyer. "Cage-based Motion Recovery using Manifold Learning." In *IEEE 3DIMPVT*, pp. 206–213, 2012.

[DŹmura 91] Michel DŹmura. "Shading Ambiguity: Reflectance and Illumination." *Computational Models of Visual Processing*, pp. 187–207.

[Eberhardt et al. 96] Bernhard Eberhardt, Andreas Weber, and Wolfgang Strasser. "A Fast, Flexible, Particle-System Model for Cloth Draping." *IEEE Comput. Graph. Appl.* 16:5 (1996), 52–59.

[Eden et al. 06] Ashley Eden, Matthew Uyttendaele, and Richard Szeliski. "Seamless Image Stitching of Scenes with Large Motions and Exposure Differences." In *Proc. Int. Conf. on Computer Vision and Pattern Recognition (CVPR)*, pp. 2498–2505. New York, USA, 2006.

[Efros and Freeman 01] Alexei A. Efros and William T. Freeman. "Image Quilting for Texture Synthesis and Transfer." In *Proceedings of the 28th*

Annual Conference on Computer Graphics and Interactive Techniques, pp. 341–346. ACM, 2001.

[Einarsson et al. 06] Per Einarsson, Charles-Felix Chabert, Andrew Jones, Wan-Chun Ma, Bruce Lamond, Tim Hawkins, Mark Bolas, Sebastian Sylwan, and Paul Debevec. "Relighting Human Locomotion with Flowed Reflectance Fields." In *Rendering Techniques 2006: 17th Eurographics Workshop on Rendering*, pp. 183–194, 2006.

[Eisemann et al. 07] Martin Eisemann, Anita Sellent, and Marcus Magnor. "Filtered Blending: A New, Minimal Reconstruction Filter for Ghosting-Free Projective Texturing with Multiple Images." In *Vision, Modeling and Visualization (VMV)*, pp. 119–126, 2007.

[Eisemann et al. 08] Martin Eisemann, Bert De Decker, Marcus Magnor, Philippe Bekaert, Edilson de Aguiar, Naveed Ahmed, Christian Theobalt, and Anita Sellent. "Floating Textures." *Computer Graphics Forum (Proc. Eurographics)* 27:2 (2008), 409–418.

[Eisert 03] Peter Eisert. "MPEG-4 Facial Animation in Video Analysis and Synthesis." *International Journal of Imaging Systems and Technology* 13:5 (2003), 245–256.

[Ekman and Friesen 78] Paul Ekman and Wallace Friesen. *Facial Action Coding System: A Technique for the Measurement of Facial Movement.* Palo Alto: Consulting Psychologists Press, 1978.

[Elhayek et al. 12] A. Elhayek, Carsten Stoll, Nils Hasler, Kwang In Kim, H. Seidel, and Christian Theobalt. "Spatio-Temporal Motion Tracking with Unsynchronized Cameras." In *Proc. CVPR*, pp. 1870–1877. IEEE, 2012.

[English and Bridson 08] Elliot English and Robert Bridson. "Animating Developable Surfaces Using Nonconforming Elements." *ACM Trans. Graph.* 27:3 (2008), 66:1–66:5.

[Etzmuss et al. 03] Olaf Etzmuss, Michael Keckeisen, and Wolfgang Strasser. "A Fast Finite Element Solution for Cloth Modelling." In *Proc. of the 11th Pacific Conference on Computer Graphics and Applications, PG '03*, pp. 244–251. Washington, DC, USA: IEEE Computer Society, 2003.

[Fairchild 13] Mark D. Fairchild. *Color Appearance Models.* John Wiley & Sons, 2013.

[Falie and Buzuloiu 08] Dragos Falie and Vasile Buzuloiu. "Distance Errors Correction for the Time of Flight (ToF) Cameras." In *Proc. European Conf. on Circuits and Systems for Communications*, pp. 193–196, 2008.

[Fan et al. 12] Shaojing Fan, Tian-Tsong Ng, Jonathan S. Herberg, Bryan L. Koenig, and Shiqing Xin. "Real or Fake?: Human Judgments about Photographs and Computer-Generated Images of Faces." In *SIGGRAPH Asia 2012 Technical Briefs, SA '12*, pp. 17:1–17:4. New York, NY, USA: ACM, 2012.

[Fechteler et al. 07] P. Fechteler, P. Eisert, and J. Rurainsky. "Fast and High Resolution 3D Face Scanning." In *ICIP07*, 3, 3, pp. 81–84, 2007.

[Fehn et al. 07] Christoph Fehn, Christoph Weissig, Ingo Feldmann, Markus Müller, Peter Eisert, Peter Kauff, and Hans Bloß. "Creation of High-Resolution Video Panoramas for Sport Events." *International Journal of Semantic Computing (IJSC)* 1:2 (2007), 493–505.

[Fehn 03] Christoph Fehn. "A 3D-TV Approach Using Depth-Image-Based Rendering (DIBR)." In *Proc. of VIIP*, 3, 3, 2003.

[Fehn 04] Christoph Fehn. "Depth-Image-Based Rendering (DIBR), Compression, and Transmission for a New Approach on 3D-TV." *Proc. SPIE* 5291:93 (2004), 93–104.

[Fender and Julesz 67] Derek Fender and Bela Julesz. "Extension of Panums Fusional Area in Binocularly Stabilized Vision." *JOSA* 57:6 (1967), 819–826.

[Feng et al. 10] Wei-Wen Feng, Yizhou Yu, and Byung-Uck Kim. "A Deformation Transformer for Real-Time Cloth Animation." *ACM Trans. Graph.* 29:4 (2010), 108:1–108:9.

[Ferwerda et al. 97] James A. Ferwerda, Peter Shirley, Sumanta N. Pattanaik, and Donald P. Greenberg. "A Model of Visual Masking for Computer Graphics." In *Proceedings of the 24th Annual Conference on Computer Graphics and Interactive Techniques, SIGGRAPH '97*, pp. 143–152. New York, NY, USA: ACM Press/Addison-Wesley Publishing Co., 1997.

[Ferwerda et al. 04] James A. Ferwerda, Stephen H. Westin, Randall C. Smith, and Richard Pawlicki. "Effects of Rendering on Shape Perception in Automobile Design." In *Proceedings of the 1st Symposium on Applied Perception in Graphics and Visualization, APGV '04*, pp. 107–114. New York, NY, USA: ACM, 2004.

[Ferwerda 08] James A. Ferwerda. "Psychophysics 101: How to Run Perception Experiments in Computer Graphics." In *ACM SIGGRAPH 2008 Classes, SIGGRAPH '08*, pp. 87:1–87:60. New York, NY, USA: ACM, 2008.

[Fischler and Bolles 81] Martin A. Fischler and Robert C. Bolles. "Random Sample Consensus: A Paradigm for Model Fitting with Applications to Image Analysis and Automated Cartography." *Communications of the ACM* 24:6 (1981), 381–395.

[Fleishman et al. 05] Shachar Fleishman, Daniel Cohen-Or, and Cláudio T. Silva. "Robust Moving Least-Squares Fitting with Sharp Features." *ACM Transactions on Graphics* 24:3 (2005), 544–552.

[Fleming 14] Roland W. Fleming. "Visual Perception of Materials and Their Properties." *Vision Research* 94 (2014), 62–75.

[Foote et al. 04] Jonathan Foote, Qiong Liu, Don Kimber, Patrick Chiu, and Frank Zhao. "Reach-Through-the-Screen: A New Metaphor for Remote Collaboration." In *Proc. 5th Pacific Rim Conference on Multimedia (PCM)*, pp. 73–80. Tokyo, Japan, 2004.

[Fordham 03] Joe Fordham. "Middle Earth Strikes Back." In *Cinefex 92*, 2003.

[Fordham 04] Joe Fordham. "Armed and Dangerous." In *Cinefex 99*, 2004.

[Fordham 09] Joe Fordham. "The Manhattan Project." In *Cinefex 117*, 2009.

[Fournier et al. 93] Alain Fournier, Atjeng S. Gunawan, and Chris Romanzin. "Common Illumination between Real and Computer Generated Scenes." In *Proceedings of Graphics Interface '93*, pp. 254–262, 1993.

[Frahm et al. 10] Jan-Michael Frahm, Pierre Fite-Georgel, David Gallup, Tim Johnson, Rahul Raguram, Changchang Wu, Yi-Hung Jen, Enrique Dunn, Brian Clipp, Svetlana Lazebnik, and Marc Pollefeys. "Building Rome on a Cloudless Day." In *Proceedings of the 11th European Conference on Computer Vision: Part IV, ECCV'10*, pp. 368–381, 2010.

[Franco and Boyer 09] Jean-Sébastien Franco and Edmond Boyer. "Efficient Polyhedral Modeling from Silhouettes." *IEEE Transactions on PAMI* 31:3 (2009), 414–427.

[Fuchs and May 08] Stefan Fuchs and Stefan May. "Calibration and Registration for Precise Surface Reconstruction with Time-Of-Flight Cameras." *Int. J. on Intell. Systems Techn. and App., Issue on Dynamic 3D Imaging* 5:3/4 (2008), 278–284.

[Fuchs et al. 05] Martin Fuchs, Volker Blanz, and Hans-Peter Seidel. "Bayesian Relighting." In *Eurographics Symposium on Rendering (EGSR)*, pp. 157–164, 2005.

[Fuchs et al. 07] Martin Fuchs, Volker Blanz, Hendrik P.A. Lensch, and Hans-Peter Seidel. "Adaptive Sampling of Reflectance Fields." *ACM Transactions on Graphics* 26:2 (2007), 10.

[Furukawa and Ponce 08] Yasutaka Furukawa and Jean Ponce. "Dense 3D Motion Capture from Synchronized Video Streams." In *CVPR*, pp. 193–211. IEEE Computer Society, 2008.

[Furukawa and Ponce 10] Yasutaka Furukawa and Jean Ponce. "Accurate, Dense, and Robust Multiview Stereopsis." *IEEE Transactions on Pattern Analysis and Machine Intelligence (PAMI)* 32:8 (2010), 1362–1376.

[Fusiello et al. 97] Aandrea Fusiello, Vito Roberto, and Emanuele Trucco. "Efficient Stereo with Multiple Windowing." In *Conference on Computer Vision and Pattern Recognition*, pp. 858–863, 1997.

[Fusiello 07] A. Fusiello. "Specifying Virtual Cameras in Uncalibrated View Synthesis." *Circuits and Systems for Video Technology, IEEE Transactions on* 17:5 (2007), 604 –611.

[Gagvani and Silver 99] N. Gagvani and D. Silver. "Parameter Controlled Volume Thinning." *Graphical Models and Image Processing* 61:3 (1999), 149–164.

[Gal et al. 10] R. Gal, Y. Wexler, E. Ofek, H. Hoppe, and D. Cohen-Or. "Seamless Montage for Texturing Models." In *Computer Graphics Forum*, 29, pp. 479–486. Wiley Online Library, 2010.

[Gall et al. 09] Juergen Gall, Carsten Stoll, Edilson De Aguiar, Christian Theobalt, Bodo Rosenhahn, and Hans-Peter Seidel. "Motion Capture Using Joint Skeleton Tracking and Surface Estimation." In *IEEE Computer Vision and Patttern Recognition (CVPR)*, pp. 1746–1753, 2009.

[Gall et al. 10] Juergen Gall, Bodo Rosenhahn, Thomas Brox, and Hans-Peter Seidel. "Optimization and Filtering for Human Motion Capture." *International Journal of Computer Vision* 87:1-2 (2010), 75–92.

[Gallup et al. 07] David Gallup, Jan-Michael Frahm, Philippos Mordohai, Qingxiong Yang, and Marc Pollefeys. "Real-Time Plane-Sweeping Stereo with Multiple Sweeping Directions." In *Conference on Computer Vision and Pattern Recognition*, 2007.

[Ganapathi et al. 10] Varun Ganapathi, Christian Plagemann, Daphne Koller, and Sebastian Thrun. "Real-Time Motion Capture Using a Single Time-of-Flight Camera." In *CVPR*, pp. 755–762, 2010.

[Garg et al. 06] Gaurav Garg, Eino-Ville Talvala, Marc Levoy, and Hendrik P. A. Lensch. "Symmetric Photography: Exploiting Data-Sparseness in Reflectance Fields." In *Rendering Techniques*, pp. 251–262, 2006.

[Gargallo and Sturm 05] Pau Gargallo and Peter Sturm. "Bayesian 3D Modeling from Images Using Multiple Depth Maps." In *Proceedings of the 2005 IEEE Computer Society Conference on Computer Vision and Pattern Recognition (CVPR'05) Volume 02*, pp. 885–891, 2005.

[Garland and Heckbert 97] M. Garland and P.S. Heckbert. "Surface Simplification Using Quadric Error Metrics." In *ACM SIGGRAPH*, pp. 209–216, 1997.

[Gehrig et al. 12] Stefan K. Gehrig, Hernán Badino, and Uwe Franke. "Improving Sub-Pixel Accuracy for Long Range Stereo." *Computer Vision and Image Understanding* 116:1 (2012), 16–24.

[Geiger 12] Andreas Geiger. "Are We Ready for Autonomous Driving? The KITTI Vision Benchmark Suite." In *Computer Vision and Pattern Recognition*, pp. 3354–3361, 2012.

[Georgiev et al. 11] T. Georgiev, A. Lumsdaine, and G. Chunev. "Using Focused Plenoptic Cameras for Rich Image Capture." *CGA* 31:1 (2011), 62–73.

[Georgiev et al. 12] Iliyan Georgiev, Jaroslav Krivanek, Tomas Davidovic, and Philipp Slusallek. "Light Transport Simulation with Vertex Connection and Merging." *ACM Transactions on Graphics (Proc. SIGGRAPH Asia* 31:6 (2012), 192:1–192:10.

[Germann et al. 10] Marcel Germann, Alexander Hornung, Richard Keiser, Remo Ziegler, Stephan Würmlin, and Markus Gross. "Articulated Billboards for Video-Based Rendering." *Comput. Graphics Forum (Proc. Eurographics)* 29:2 (2010), 585–594.

[Germann et al. 12] Marcel Germann, Tiberiu Popa, Richard Keiser, Remo Ziegler, and Markus Gross. "Novel-View Synthesis of Outdoor Sport

Events Using an Adaptive View-Dependent Geometry." *Comput. Graphics Forum (Proc. Eurographics)* 31:2 (2012), 325–333.

[Germann 12] Marcel Germann. "Video-Based Rendering Techniques." Ph.D. thesis, Zürich, 2012. Diss., Eidgenssische Technische Hochschule ETH Zürich, Nr. 20290, 2012.

[Gershun 36] A. Gershun. "The Light Field." *J. Math. and Physics* 18 (1936), 51–151.

[Ghosh et al. 10] Abhijeet Ghosh, Wolfgang Heidrich, Shruthi Achutha, and Matthew O'Toole. "A Basis Illumination Approach to BRDF Measurement." *International Journal on Computer Vision* 90:2 (2010), 183–197.

[Ghosh et al. 11] Abhijeet Ghosh, Graham Fyffee, Borom Tunwattanapong, Jay Busch, Xueming Yu, and Paul Debevec. "Multiview Face Capture Using Polarized Spherical Gradient Illumination." In *ACM Transactions on Graphics (Proc. SIGGRAPH Asia 2011)*, 2011.

[Giachetti et al. 98] Andrea Giachetti, Marco Campani, and Vincent Torre. "The Use of Optical Flow for Road Navigation." *IEEE T. Robotics and Automation* 14:1 (1998), 34–48.

[Gibson and Mirtich 97] Sarah F. F. Gibson and Brian Mirtich. "A Survey of Deformable Modeling in Computer Graphics, Technical Report TR-97-19, MERL." Technical report, Cambridge, MA, 1997.

[Giesen et al. 09] J. Giesen, B. Miklos, M. Pauly, and C. Wormser. "The Scale Axis Transform." In *Proceedings of the Annual Symposium on Computational Geometry*, pp. 106–115, 2009.

[Giraudot et al. 13] Simon Giraudot, David Cohen-Steiner, and Pierre Alliez. "Noise-Adaptive Shape Reconstruction from Raw Point Sets." *Computer Graphics Forum* 32:5 (2013), 229–238.

[Glondu et al. 12] Loeiz Glondu, Lien Muguercia, Maud Marchal, Carles Bosch, Holly Rushmeier, Georges Dumont, and George Drettakis. "Example-Based Fractured Appearance." *Computer Graphics Forum* 31 (2012), 1547–1556. Available online (http://onlinelibrary.wiley.com/doi/10.1111/j.1467-8659.2012.03151.x/full).

[Goesele et al. 00] Michael Goesele, Wolfgang Heidrich, Hendrik P. A. Lensch, and Hans-Peter Seidel. "Building a Photo Studio for Measurement Purposes." In *VMV*, pp. 231–238, 2000.

[Goesele et al. 03] Michael Goesele, Xavier Granier, Wolfgang Heidrich, and Hans-Peter Seidel. "Accurate Light Source Acquisition and Rendering." *ACM Trans. Graph.* 22:3 (2003), 621–630.

[Goesele et al. 04] Michael Goesele, Hendrik P. A. Lensch, Jochen Lang, Christian Fuchs, and Hans-Peter Seidel. "DISCO: Acquisition of Translucent Objects." *ACM Trans. Graph.* 23:3 (2004), 835–844.

[Goesele et al. 06] Michael Goesele, Brian Curless, and Steven M. Seitz. "Multi-View Stereo Revisited." In *Computer Vision and Pattern Recognition*, CVPR '06, 2, CVPR '06, 2, pp. 2402–2409, 2006.

[Goesele et al. 07] Michael Goesele, Noah Snavely, Brian Curless, Hugues Hoppe, and Steven M. Seitz. "Multi-View Stereo for Community Photo Collections." In *International Conference on Computer Vision*, pp. 1–8, 2007.

[Goesele et al. 10] Michael Goesele, Jens Ackermann, Simon Fuhrmann, Carsten Haubold, Ronny Klowsky, and T.U. Darmstadt. "Ambient Point Clouds for View Interpolation." *ACM Trans. Graph.* 29 (2010), 95:1–95:6.

[Goldlücke and Magnor 03] B. Goldlücke and M. Magnor. "Real-time Microfacet Billboarding for Free-Viewpoint Video Rendering." In *IEEE International Conference on Image Processing (ICIP)*, 3, pp. 713–716. IEEE, 2003.

[Goldlücke and Magnor 04] B. Goldlücke and M. Magnor. "Space-time Isosurface Evolution for Temporally Coherent 3D Reconstruction." In *IEEE Computer Vision and Pattern Recognition (CVPR)*, 1, pp. 350–355, 2004.

[Goldlücke et al. 02] Bastian Goldlücke, Marcus Magnor, and Bennett Wilburn. "Hardware-Accelerated Dynamic Light Field Rendering." In *Vision, Modeling, and Visualization (VMV)*, pp. 455–462, 2002.

[Goldlücke and Magnor 04] Bastian Goldlücke and Marcus Magnor. "Weighted Minimal Hypersurfaces and Their Applications in Computer Vision." In *European Conference on Computer Vision (ECCV)*, pp. 366–378, 2004.

[Goldlücke and Wanner 13] B. Goldlücke and S. Wanner. "The Variational Structure of Disparity and Regularization of 4D Light Fields." In *Proc. International Conference on Computer Vision and Pattern Recognition*, 2013.

[google 12] google. "Google Sketchup." http://sketchup.google.com, 2012.

[Goorts et al. 13] Patrik Goorts, Cosmin Ancuti, Maarten Dumont, Sammy Rogmans, and Philippe Bekaert. "Real-Time Video-Based View Interpolation of Soccer Events Using Depth-Selective Plane Sweeping."

[Gortler et al. 96] S.J. Gortler, R. Grzeszczuk, R. Szeliski, and M.F. Cohen. "The Lumigraph." In *Proc. SIGGRAPH*, pp. 43–54, 1996.

[Granados et al. 10] Miguel Granados, Boris Ajdin, Michael Wand, Christian Theobalt, H.-P. Seidel, and H. Lensch. "Optimal HDR Reconstruction with Linear Digital Cameras." In *Computer Vision and Pattern Recognition (CVPR), 2010 IEEE Conference on*, pp. 215–222. IEEE, 2010.

[Grau and Easterbrook 08] Oliver Grau and Jim Easterbrook. "Effects of Camera Aperture Correction on Keying of Broadcast Video."

[Grau et al. 04] Oliver Grau, Tim Pullen, and Graham A. Thomas. "A Combined Studio Production System for 3-D Capturing of Live Action and Immersive Actor Feedback." *IEEE Transactions on Circuits and Systems for Video Technology* 14:3 (2004), 370–380.

[Grau 05] Oliver Grau. "A 3D Production Pipeline for Special Effects in TV and Film." In *Mirage 2005, Computer Vision/Computer Graphics Collaboration Techniques and Applications*, 2005.

[Grosch et al. 07] Thorsten Grosch, Tobias Eble, and Stefan Mueller. "Consistent Interactive Augmentation of Live Camera Images with Correct Near-Field Illumination." In *Proceedings of the 2007 ACM Symposium on Virtual Reality Software and Technology*, pp. 125–132, 2007.

[Grosch 05] Thorsten Grosch. "Differential Photon Mapping—Consistent Augmentation of Photographs with Correction of all Light Paths." In *EG Short Presentations*, pp. 53–56, 2005.

[Grossberg and Nayar 04] Michael D. Grossberg and Shree K. Nayar. "Modeling the Space of Camera Response Functions." *IEEE Trans. on Pattern Analysis Machine Intelligence* 26:10 (2004), 1272–1282.

[Grossman and Balakrishnan 06] Tovi Grossman and Ravin Balakrishnan. "An Evaluation of Depth Perception on Volumetric Displays." In *Proceedings of the Working Conference on Advanced Visual Interfaces*, pp. 193–200. ACM, 2006.

[Guan et al. 09] P. Guan, A. Weiss, A. Bălan, and M. Black. "Estimating Human Shape and Pose from a Single Image." In *International Conference on Computer Vision*, 2009.

[Guan et al. 12] Peng Guan, Loretta Reiss, David Hirshberg, Alex Weiss, and Micheal J. Black. "DRAPE: DRessing Any PErson." *ACM Trans. Graph.* 31:3 (2012), 35:1–35:10.

[Gudmundsson et al. 08] Sigurjón Árni Gudmundsson, Henrik Aanæs, and Rasmus Larsen. "Fusion of Stereo Vision and Time-of-Flight Imaging for Improved 3D Estimation." *IJISTA* 5:3/4 (2008), 425–433.

[Guennebaud and Gross 07] Gaël Guennebaud and Markus Gross. "Algebraic Point Set Surfaces." *ACM Transactions on Graphics* 26:3.

[Guillemaut et al. 10] J.-Y. Guillemaut, M. Sarim, and A. Hilton. "Stereoscopic Content Production of Complex Dynamic Scenes Using a Wide-Baseline Monoscopic Camera Set-Up." In *Proc. International Conference on Image Processing (ICIP 2010), Special Session on Image Processing for Stereo Digital Cinema Production*, pp. 9–12, 2010.

[Guillemot et al. 12] Thiery Guillemot, Andrés Almansa, and Tamy Boubekeur. "Non-Local Point Set Surfaces." In *Proceedings of 3DIMPVT 2012*, pp. 61–70, 2012.

[Guomundsson and Sveinsson 11] SA Guomundsson and Johannes R Sveinsson. "TOF-CCD Image Fusion Using Complex Wavelets." In *Proc. IEEE Int. Conf. Acoustics, Speech and Signal Processing (ICASSP)*, pp. 1557–1560, 2011.

[Guskov et al. 03] Igor Guskov, Sergey Klibanov, and Benjamin Bryant. "Trackable Surfaces." In *Proc. Eurographics Symp. on Computer Animation (SCA)*, pp. 251–257, 2003.

[Gvili et al. 03] Ronen Gvili, Amir Kaplan, Eyal Ofek, and Giora Yahav. "Depth Keying." In *Proc SPIE, Video-Based Image Techniques and Emerging Work*, 5006, 5006, 2003. DOI: 10.1117/12.474052.

[Habbecke and Kobbelt 06] Martin Habbecke and Leif Kobbelt. "Iterative Multi-View Plane Fitting." In *Vision, Modelling and Visualization*, pp. 73–80, 2006.

[Habbecke and Kobbelt 07] Martin Habbecke and Leif Kobbelt. "A Surface-Growing Approach to Multi-View Stereo Reconstruction." In *Computer Vision and Pattern Recognition, 2007. CVPR '07. IEEE, Conference on*, pp. 1–8, 2007.

[Haber et al. 09] Tom Haber, Christian Fuchs, Philippe Bekaert, Hans-Peter Seidel, Michael Goesele, and Hendrik P. A. Lensch. "Relighting Objects from Image Collections." In *Computer Vision and Pattern Recognition*, pp. 627–634, 2009.

[Hachisuka et al. 08] Toshiya Hachisuka, Shinji Ogaki, and Henrik Wann Jensen. "Progressive Photon Mapping." *ACM Transactions on Graphics* 27:5 (2008), 130:1–130:8.

[Hachisuka et al. 12] Toshiya Hachisuka, Jacopo Pantaleoni, and Henrik Wann Jensen. "A Path Space Extension for Robust Light Transport Simulation." *ACM Transactions on Graphics (Proc. SIGGRAPH Asia)* 31:6 (2012), 191:1–191:10.

[Hackbusch 99] Wolfgang Hackbusch. "A Sparse Matrix Arithmetic Based on H-matrices. Part I: Introduction to H-matrices." *Computing* 62:2 (1999), 89–108. Available online (http://dx.doi.org/10.1007/s006070050015).

[Hadap et al. 99] Sunil Hadap, Endre Bangerter, Pascal Volino, and Nadia Magnenat-Thakmann. "Animating Wrinkles on Cloth." In *Proc. of IEEE Conf. on Visualization*, pp. 175–182, 1999.

[Hahne and Alexa 09] Uwe Hahne and Marc Alexa. "Depth Imaging by Combining Time-of-Flight and On-Demand Stereo." In *Dyn3D*, pp. 70–83, 2009.

[Hall-Holt and Rusinkiewicz 01] Olaf Hall-Holt and Szymon Rusinkiewicz. "Stripe Boundary Codes for Real-Time Structured-light Range Scanning of Moving Objects." In *Proceedings of the Eighth IEEE International Conference on Computer Vision, 2001. ICCV 2001*, 2, 2, pp. 359–366, 2001.

[Hanika and Dachsbacher 14] Johannes Hanika and Carsten Dachsbacher. "Efficient Monte Carlo Rendering with Realistic Lenses." *Computer Graphics Forum (Proc. of Eurographics)* 33:2, 2014.

[Hao and Varshney 01] Xuejun Hao and Amitabh Varshney. "Variable-Precision Rendering." In *Proc. I3D*, pp. 149–158, 2001.

[Hara et al. 08] Kenji Hara, Ko Nishino, and Katsushi Ikeuchi. "Mixture of Spherical Distributions for Single-View Relighting." *IEEE Trans. Pattern Anal. Mach. Intell.* 30:1 (2008), 25–35.

[Haralick et al. 94] Robert M. Haralick, Chung-Nan Lee, Karsten Ottenberg, and Michael Nölle. "Review and Analysis of Solutions of the Three Point Perspective Pose Estimation Problem." *Int. J. Comput. Vision* 13:3 (1994), 331–356.

[Harker and O'Leary 08] Matthew Harker and Paul O'Leary. "Least Squares Surface Reconstruction from Measured Gradient Fields." In *CVPR*, 2008.

[Harris and Stephens 88] Chris Harris and Mike Stephens. "A Combined Corner and Edge Detector." In *Alvey Vision Conference*, 15, p. 50. Manchester, UK, 1988.

[Hartley and Kang 05] R. Hartley and S. B. Kang. "Parameter Free Radial Distortion Correction with Center of Distortion Estimation." Technical Report TR=2005-42, Microsoft Research, 2005.

[Hartley and Sturm 97] Richard Hartley and Peter Sturm. "Triangulation." *Computer Vision and Image Understanding* 68:2 (1997), 146–157.

[Hartley and Zisserman 03] Richard Hartley and Andrew Zisserman. *Multiple View Geometry in Computer Vision*, Second edition. Cambridge University Press, 2003.

[Hartley 93] Richard Hartley. "Extraction of Focal Lengths from the Fundamental Matrix." Technical Report, G.E. CRD, Schenectady, NY.

[Hartley 97] Richard Hartley. "In Defense of the Eight-Point Algorithm." *Pattern Analysis and Machine Intelligence, IEEE Transactions on* 19:6 (1997), 580–593.

[Hasinoff et al. 10] Samuel W. Hasinoff, Fredo Durand, and William T. Freeman. "Noise-Optimal Capture for High Dynamic Range Photography." *2013 IEEE Conference on Computer Vision and Pattern Recognition* (2010), 553–560.

[Hasler et al. 06] Nils Hasler, Mark Asbach, Bodo Rosenhahn, Jens-Rainer Ohm, and Hans-Peter Seidel. "Physically-Based Tracking of Cloth." In *Proc. of the Int. Workshop on Vision, Modeling, and Visualization, VMV*, pp. 49–56, 2006.

[Hasler et al. 09a] N. Hasler, B. Rosenhahn, T. Thormählen, M. Wand, J. Gall, and H.-P. Seidel. "Markerless Motion Capture with Unsynchronized Moving Cameras." In *Computer Vision and Pattern Recognition (CVPR)*, pp. 224–231, 2009.

[Hasler et al. 09b] N. Hasler, C. Stoll, B. Rosenhahn, T. Thormählen, and H.-P. Seidel. "Estimating Body Shape of Dressed Humans." *Computers and Graphics* 33:3 (2009), 211–216. Proceedings of Shape Modeling International.

[Hasler et al. 09c] N. Hasler, C. Stoll, M. Sunkel, B. Rosenhahn, and H.-P. Seidel. "A Statistical Model of Human Pose and Body Shape." *Computer Graphics Forum* 2:28 (2009), 337–346. Proceedings of Eurographics.

[Hasler et al. 10a] N. Hasler, H. Ackermann, B. Rosenhahn, T. Thormählen, and H.-P. Seidel. "Multilinear Pose and Body Shape Estimation of Dressed Subjects from Image Sets." In *Conference on Computer Vision and Pattern Recognition*, pp. 1823–1830, 2010.

[Hasler et al. 10b] N. Hasler, T. Thormahlen, B. Rosenhahn, and H.-P. Seidel. "Learning Skeletons for Shape and Pose." In *I3D10: Proc. of Symp. on Interactive 3D Graphics and Games*, pp. 23–30, 2010.

[Hauswiesner et al. 11] Stefan Hauswiesner, Matthias Straka, and Gerhard Reitmayr. "Image-Based Clothes Transfer." In *Proc. Int. Symp.on Mixed and Augmented Reality (ISMAR)*, pp. 169–172. IEEE Computer Society, 2011.

[Hauswiesner et al. 13] Stefan Hauswiesner, Matthias Straka, and Gerhard Reitmayr. "Virtual Try-On through Image-Based Rendering." *IEEE Trans. on Visualization and Computer Graphics* 19:9 (2013), 1552–1565.

[Havaldar et al. 06] Parag Havaldar, Fred Pighin, and John Peter Lewis. "Performance Driven Facial Animation." In *SIGGRAPH Course Notes 2006*, 2006.

[Hawkins et al. 07] Tim Hawkins, Jonathan Cohen, Chris Tchou, and Paul Debevec. "Light Stage 2.0." In *SIGGRAPH 2001 Sketch*, p. 229, 2007.

[He et al. 10] Kaiming He, Jian Sun, and Xiaoou Tang. "Guided Image Filtering." In *Computer Vision–ECCV 2010*, pp. 1–14. Springer, 2010.

[Heck and Gleicher 07] R. Heck and M. Gleicher. "Parametric Motion Graphs." In *ACM Symposium on Interactive 3D Graphics and Graphics*, 2007.

[Heer and Bostock 10] Jeffrey Heer and Michael Bostock. "Crowdsourcing Graphical Perception: Using Mechanical Turk to Assess Visualization Design." In *Proceedings of the SIGCHI Conference on Human Factors in Computing Systems*, pp. 203–212. ACM, 2010.

[Heide et al. 13] Felix Heide, Matthias B. Hullin, James Gregson, and Wolfgang Heidrich. "Low-Budget Transient Imaging Using Photonic Mixer Devices." *ACM Trans. Graph.* 32:4 (2013), 45:1–45:10.

[Heidrich et al. 99] Wolfgang Heidrich, Hartmut Schirmacher, Hendrik Kück, and Hans-Peter Seidel. "A Warping-Based Refinement of Lumigraphs." In *Proc. WSCG*, 99, 99, pp. 102–109, 1999.

[Heikkila and Silven 97] J. Heikkila and O. Silven. "A Four-step Camera Calibration Procedure with Implicit Image Correction." In *IEEE CVPR*, 1997.

[Heinly et al. 12] Jared Heinly, Enrique Dunn, and Jan-Michael Frahm. "Comparative Evaluation of Binary Features." In *Computer Vision–ECCV 2012*, pp. 759–773. Springer, 2012.

[Helten et al. 13] T. Helten, A. Baak, G. Bharai, M. Müller, H.-P. Seidel, and C. Theobalt. "Personalization and Evaluation of a Real-Time Depth-Based Full Body Scanner." In *3D Vision*, 2013.

[Helzle et al. 04] Volker Helzle, Christoph Biehn, Thomas Schlömer, and Florian Linner. "Adaptable Setup for Performance Driven Facial Animation." *ACM Transactions on Graphics (Proc. SIGGRAPH 2004)*.

[Herbst et al. 09] Evan Herbst, Steve Seitz, and Simon Baker. "Occlusion Reasoning for Temporal Interpolation Using Optical Flow." Technical report, Microsoft Research Technical Report, MSR-TR-2009-2014, 2009. No. MSR-TR-2009-2014.

[Herrera C. et al. 12] D. Herrera C., J. Kannala, and J. Heikkilä. "Joint Depth and Color Camera Calibration with Distortion Correction." *IEEE Trans. Pattern Anal. Mach. Intell.* 34:10 (2012), 2058–2064.

[Herzog et al. 12] Robert Herzog, Martin Cadik, Tunc O. Aydin, Kwang In Kim, Karol Myszkowski, and Hans-P. Seidel. "NoRM: No-Reference Image Quality Metric for Realistic Image Synthesis." *Comp. Graph. Forum* 31:2pt4 (2012), 545–554.

[Hildebrandt et al. 12] K. Hildebrandt, C. Schulz, C. von Tycowicz, and K. Polthier. "Interactive Space-Time Control of Deformable Objects." *ACM Transactions on Graphics (Proceedings of SIGGRAPH)* 31:4.

[Hill 53] A. J. Hill. "A Mathematical and Experimental Foundation for Stereoscopic Photography." In *Journal of SMPTE*, 61, pp. 461–486, 1953.

[Hilsmann et al. 10] Anna Hilsmann, David C. Schneider, and Peter Eisert. "Realistic Cloth Augmentation in Single View Video under Occlusions." *Computers & Graphics* 34:5 (2010), 567–574.

[Hilsmann et al. 11] Anna Hilsmann, David C. Schneider, and Peter Eisert. "Warp-Based Near-Regular Texture Analysis for Image-Based Texture Overlay." In *Proc. Int. Workshop on Vision, Modeling, and Visualization (VMV)*, pp. 73–80. Eurographics Association, 2011.

[Hilsmann et al. 13] Anna Hilsmann, Philipp Fechteler, and Peter Eisert. "Pose Space Image Based Rendering." *Comput. Graph. Forum (Proc. Eurographics)* 32:2 (2013), 265–274.

[Hilton et al. 10] Adrian Hilton, Jean-Yves Guillemaut, Joe Kilner, Oliver Grau, and Graham Thomas. "Free-Viewpoint Video for TV Sport Production." In *Image and Geometry Processing for 3-D Cinematography*, pp. 77–106. Springer, 2010.

[Hilton et al. 11] A. Hilton, J.-Y. Guillemaut, J. Kilner, O. Grau, and G. Thomas. "3D-TV Production from Conventional Cameras for Sports Broadcast." *IEEE Transactions on Broadcasting* 57:2 (2011), 462–476.

[Hirschmüller et al. 02] Heiko Hirschmüller, Peter R. Innocent, and Jon M. Garibaldi. "Real-Time Correlation-Based Stereo Vision with Reduced Border Errors." *International Journal of Computer Vision* 47:1-3 (2002), 229–246.

[Hirschmüller 05] Heiko Hirschmüller. "Accurate and Efficient Stereo Processing by Semi-Global Matching and Mutual Information." In *Computer Vision and Pattern Recognition*, 2, pp. 807–814, 2005.

[Hodgins et al. 98] Jessica K Hodgins, James F O'Brien, and Jack Tumblin. "Perception of Human Motion with Different Geometric Models." *Visualization and Computer Graphics, IEEE Transactions on* 4:4 (1998), 307–316.

[Högg et al. 13] T. Högg, D. Lefloch, and A. Kolb. "Real-Time Motion Artifact Compensation for PMD-ToF Images." In *Proc. Workshop Imaging New Modalities, German Conference of Pattern Recognition (GCPR)*, LNCS, 8200, LNCS, 8200, pp. 273–288. Springer, 2013.

[Hong and Chen 04] Li Hong and George Chen. "Segment-Based Stereo Matching Using Graph Cuts." In *Conference on Computer Vision and Pattern Recognition*, pp. 74–81, 2004.

[Hong et al. 10] Q. Y. Hong, S. I. Park, and J. K. Hodgins. "A Data-Driven Segmentation for the Shoulder Complex." *Computer Graphics Forum.* 29:2 (2010), 537–544.

[Hoppe et al. 92] Hugues Hoppe, Tony DeRose, Tom Duchamp, John McDonald, and Werner Stuetzle. "Surface Reconstruction from Unorganized Points." In *Proceedings of ACM SIGGRAPH 1992*, pp. 71–78, 1992.

[Hoppe 96] Hugues Hoppe. "Progressive Meshes." In *Proc. SIGGRAPH*, pp. 99–108, 1996.

[Hoppe 98] Hugues Hoppe. "Efficient Implementation of Progressive Meshes." *Computers & Graphics* 22 (1998), 27–36.

[Horaud et al. 09] Radu Horaud, Matti Niskanen, Guillaume Dewaele, and Edmond Boyer. "Human Motion Tracking by Registering an Articulated Surface to 3-D Points and Normals." *IEEE Transactions on PAMI* 31:1 (2009), 158–163.

[Horn and Brooks 86] Berthold K. P. Horn and Michael J. Brooks. "The Variational Approach to Shape from Shading." *Computer Vision, Graphics, and Image Processing* 33:2 (1986), 174–208.

[Horn and Schunck 81] Berthold K. P. Horn and Brian G. Schunck. "Determining Optical Flow." *Artificial Intelligence* 17 (1981), 185–203.

[Hornung and Kobbelt 06] Alexander Hornung and Leif Kobbelt. "Robust Reconstruction of Watertight 3D Models from Non-uniformly Sampled Point Clouds without Normal Information." In *Proceedings of the Fourth Eurographics Symposium on Geometry Processing*, pp. 41–50, 2006.

[Hornung and Kobbelt 09] Alexander Hornung and Leif Kobbelt. "Interactive Pixel-Accurate Free Viewpoint Rendering from Images with Silhouette Aware Sampling." *Computer Graphics Forum* 28:8 (2009), 2090–2103.

[Hosni et al. 09] Asmaa Hosni, Michael Bleyer, Margrit Gelautz, and Christoph Rhemann. "Local Stereo Matching Using Geodesic Support Weights." In *International Conference on Image Processing*, pp. 2069–2072, 2009.

[Hosni et al. 11a] Asmaa Hosni, Michael Bleyer, Christoph Rhemann, Margrit Gelautz, and Carsten Rother. "Real-Time Local Stereo Matching Using Guided Image Filtering." In *Multimedia and Expo (ICME), 2011 IEEE International Conference on*, pp. 1–6. IEEE, 2011.

[Hosni et al. 11b] Asmaa Hosni, Christoph Rhemann, Michael Bleyer, Carsten Rother, and Margrit Gelautz. "Fast Cost-Volume Filtering for Visual Correspondence and Beyond." In *Computer Vision and Pattern Recognition*, pp. 3017–3024, 2011.

[Hosni et al. 13] Asmaa Hosni, Michael Bleyer, and Margrit Gelautz. "Secrets of Adaptive Support Weight Techniques for Local Stereo Matching." *Comput. Vis. Image Underst.* 117:6 (2013), 620–632.

[Hostica et al. 06] B. Hostica, P. Seitz, and A. Simoni. *Encyclopedia of Sensors*, 7, Chapter Optical Time-of-Flight Sensors for Solid-State 3D-vision, pp. 259–289. American Scientific Pub, 2006.

[House and Breen 00] Donald H. House and David E. Breen, editors. *Cloth Modeling and Animation.* Natick, MA, USA: A. K. Peters, Ltd., 2000.

[Howlett et al. 05] Sarah Howlett, John Hamill, and Carol O'Sullivan. "Predicting and Evaluating Saliency for Simplified Polygonal Models." *ACM Trans. Appl. Percept.* 2:3 (2005), 286–308.

[Hua et al. 07] Hong Hua, Narendra Ahuja, and Chunyu Gao. "Design Analysis of a High-Resolution Panoramic Camera Using Conventional Imagers and a Mirror Pyramid." *IEEE Trans. on Pattern Analysis and Machine Intelligence* 29:2 (2007), 356–945.

[Huang et al. 09] P. Huang, A. Hilton, and J. Starck. "Human Motion Synthesis from 3D Video." In *Proceedings of IEEE Conference on Computer Vision and Pattern Recognition*, pp. 1478–1485, 2009.

[Huang et al. 10] Peng Huang, Adrian Hilton, and Jonathan Starck. "Shape Similarity for 3D Video Sequences of People." *International Journal of Computer Vision* 89:2-3 (2010), 362–381.

[Huang et al. 13] Chun-Hao Huang, Edmond Boyer, and Slobodan Ilic. "Robust Human Body Shape and Pose Tracking." In *3DV*, 2013.

[Hullin et al. 11] Matthias B. Hullin, Elmar Eisemann, Hans-Peter Seidel, and Sungkil Lee. "Physically-Based Real-Time Lens Flare Rendering." *ACM Transactions on Graphics (Proc. SIGGRAPH)* 30:4 (2011), 108:1–108:9.

[Hypr3D Development Team 12] Hypr3D Development Team. "hypr3D." http://www.agisoft.ru/, 2012.

[Iddan and Yahav 01] G. J. Iddan and G. Yahav. "3D Imaging in the Studio." In *Proc. of SPIE*, 4298, 4298, pp. 48–56, 2001.

[Ihrke et al. 08] I. Ihrke, T. Stich, H. Gottschlich, M. Magnor, and H. Seidel. "Fast Incident Light Field Acquisition and Rendering." *Journal of WSCG*, pp. 177–184.

[Ihrke et al. 10] Ivo Ihrke, Kiriakos N Kutulakos, Hendrik Lensch, Marcus Magnor, and Wolfgang Heidrich. "Transparent and Specular Object Reconstruction." *Computer Graphics Forum (CGF)* 29:8 (2010), 2400–2426.

[Imre and Hilton 12] Evren Imre and Adrian Hilton. "Through-the-Lens Synchronisation for Heterogeneous Camera Networks." In *Proceedings of the British Machine Vision Conference*, pp. 97.1–97.11. BMVA Press, 2012.

[Imre et al. 12] Evren Imre, Jean-Yves Guillemaut, and Adrian Hilton. "Through-the-Lens Multi-camera Synchronisation and Frame-Drop Detection for 3D Reconstruction." In *3D Imaging, Modeling, Processing, Visualization and Transmission (3DIMPVT), 2012 Second International Conference on*, pp. 395–402, 2012.

[Ince and Konrad 08] Serdar Ince and Janusz Konrad. "Occlusion-Aware View Interpolation." *EURASIP Journal on Image and Video Processing* (2008), 1–15.

[Irani and Peleg 91] Michael Irani and Shmuel Peleg. "Improving Resolution by Image Registration." *CVGIP: Graph. Models Image Process.* 53:3 (1991), 231–239.

[Irawan and Marschner 12] Piti Irawan and Steve Marschner. "Specular Reflection from Woven Cloth." *ACM Transactions on Graphics (TOG)* 31:1 (2012), 11.

[Irawan 08] Piti Irawan. "Appearance of Woven Cloth." Ph.D. thesis, Cornell University, 2008.

[Ismael et al. 14] Muhannad Ismael, Stéphanie Prévost, Céline Loscos, and Yannick Rémion. "Materiality Maps: A Novel Scene-Based Framework for Direct Multi-View Stereovision Reconstruction." In *IEEE International Conference on Image Processing*, p. to appear. IEEE, 2014.

[Ives 03] F. Ives. "Parallax Stereogram and Process of Making Same." 1903.

[Izadi et al. 11] S. Izadi, D. Kim, O. Hilliges, D. Molyneaux, R. Newcombe, P. Kohli, J. Shotton, S. Hodges, D. Freeman, A. Davison, and A. Fitzgibbon. "KinectFusion: Real-Time 3D Reconstruction and Interaction Using a Moving Depth Camera." In *Proc. ACM Symp. User Interface Softw. & Tech.*, pp. 559–568, 2011.

[Jacobson et al. 12] A. Jacobson, I. Baran, L. Kavan, J. Popović, and O. Sorkine. "Fast Automatic Skinning Transformations." *ACM Transactions on Graphics (Proceedings of SIGGRAPH Asia)* 31:4.

[Jain et al. 10] A. Jain, T. Thormählen, H.-P. Seidel, and C. Theobalt. "MovieReshape: Tracking and Reshaping of Humans in Videos." *ACM Transactions on Graphics* 29 (2010), 148:1–10. Proceedings of SIGGRAPH Asia.

[Jakob and Marschner 12] Wenzel Jakob and Steve Marschner. "Manifold Exploration: A Markov Chain Monte Carlo Technique for Rendering Scenes with Difficult Specular Transport." *ACM Transactions on Graphics (Proc. of SIGGRAPH)* 31:4 (2012), 58:1–58:13.

[James and Twigg 05] D. James and C. Twigg. "Skinning Mesh Animations." *ACM Transactions on Graphics (Proceedings of SIGGRAPH)* 24:3, 2005.

[Jarabo et al. 14] Adrian Jarabo, Hongzhi Wu, Julie Dorsey, Holly Rushmeier, and Diego Gutierrez. "Effects of Approximate Filtering on the Appearance of Bidirectional Texture Functions." *IEEE Transactions on Visualization and Computer Graphics.*

[Jensen et al. 01] Henrik Wann Jensen, Stephen R. Marschner, Marc Levoy, and Pat Hanrahan. "A Practical Model for Subsurface Light Transport." *ACM Transactions on Graphics (Proc. SIGGRAPH 2001).*

[Jensen 96] Henrik Wann Jensen. "Global Illumination Using Photon Maps." In *Proc. Eurographics Workshop on Rendering*, pp. 21–30, 1996.

[Jian and Vemuri 11] Bing Jian and Baba C. Vemuri. "Robust Point Set Registration Using Gaussian Mixture Models." *IEEE Transactions on PAMI* 33:8 (2011), 1633–1645.

[Jimenez et al. 12] Jorge Jimenez, Adrian Jarabo, and Diego Gutierrez. "Separable Subsurface Scattering and Photorealistic Eyes Rendering." In *SIGGRAPH Course Notes 2012*, 2012.

[Johannsen et al. 13] O. Johannsen, C. Heinze, B. Goldlücke, and C. Perwass. "On the Calibration of Focused Plenoptic Cameras." In *GCPR Workshop on Imaging New Modalities*, 2013.

[Johnson and Hebert 97] A. Johnson and M. Hebert. "Recognizing Objects by Matching Oriented Points." In *Conference on Computer Vision and Pattern Recognition*, pp. 684–692, 1997.

[Johnson 02] T. Johnson. "Methods for Characterizing Colour Scanners and Digital Cameras." In *Colour Engineering: Achieving Device Independent Colour*, pp. 165–178. John Wiley & Sons Inc., 2002.

[Jones et al. 01] Graham Jones, Delman Lee, Nicolas Holliman, and David Ezra. "Controlling Perceived Depth in Stereoscopic Images." 4297, pp. 42–53, 2001.

[Ju and Kang 09] Myung-Ho Ju and Hang-Bong Kang. "Constant Time Stereo Matching." *International Machine Vision and Image Processing Conference* 0 (2009), 13–17.

[Ju et al. 02] Tao Ju, Frank Losasso, Scott Schaefer, and Joe Warren. "Dual Contouring of Hermite Data." *ACM Transactions on Graphics* 21:3 (2002), 339–346.

[Jung et al. 12] Yvonne Jung, Johannes Behr, Timm Drevensek, and Sebastian Wagner. "Declarative 3D Approaches for Distributed Web-Based Scientific Visualization Services." In *Dec3D*, 869, edited by Johannes Behr,

Donald P. Brutzman, Ivan Herman, Jacek Jankowski, and Kristian Sons, 869, 2012.

[Jung et al. 13] Yvonne Jung, Max Limper, Pasquale Herzig, Karsten Schwenk, and Johannes Behr. "Fast and Efficient Vertex Data Representations for the Web." In *IVAPP*, pp. 77–86, 2013.

[Kahlmann et al. 07] T. Kahlmann, F. Remondino, and S. Guillaume. "Range Imaging Technology: New Developments and Applications for People Identification and Tracking." In *Proc. of Videometrics IX - SPIE-IS&T Electronic Imaging*, 6491, 6491, 2007. DOI: 10.1117/12.702512.

[Kajiya 86] James T. Kajiya. "The Rendering Equation." *Computer Graphics (Proc. of SIGGRAPH)*, pp. 143–150.

[Kaldor et al. 08] Jonathan M. Kaldor, Doug L. James, and Steve Marschner. "Simulating Knitted Cloth at the Yarn Level." *ACM Trans. Graph.* 27:3 (2008), 65.

[Kaldor et al. 10] Jonathan M. Kaldor, Doug L. James, and Steve Marschner. "Efficient Yarn-Based Cloth with Adaptive Contact Linearization." *ACM Trans. Graph.* 29:4 (2010), 105.

[Kán 12] Hannes Kán, Peter and Kaufmann. "Physically-Based Depth of Field in Augmented Reality." In *Proceedings of EUROGRAPHICS 2012*, pp. 89–92, 2012.

[Kanatani et al. 08] Kenichi Kanatani, Yasuyuki Sugaya, and Hirotaka Niitsuma. "Triangulation from Two Views Revisited: Hartley-Sturm vs. Optimal Correction." *In Practice* 4 (2008), 5.

[Kang and Szeliski 04] Sing Bing Kang and Richard Szeliski. "Extracting View-Dependent Depth Maps from a Collection of Images." *International Journal of Computer Vision* 58 (2004), 139–163.

[Kang et al. 04] Sing Bing Kang, Richard Szeliski, and Matthew Uyttendaele. "Seamless Stitching Using Multi-Perspective Plane Sweep." Technical Report MSR- TR-2004-48, June 2004.

[Kaplanyan and Dachsbacher 13a] Anton S. Kaplanyan and Carsten Dachsbacher. "Adaptive Progressive Photon Mapping." *ACM Transactions on Graphics* 32:2, 2013.

[Kaplanyan and Dachsbacher 13b] Anton S. Kaplanyan and Carsten Dachsbacher. "Path Space Regularization for Holistic and Robust Light Transport." *Computer Graphics Forum (Proc. of Eurographics)* 32:2, 2013.

[Kaplanyan et al. 14] Anton S. Kaplanyan, Johannes Hanika, and Carsten Dachsbacher. "The Natural-Constraint Representation of the Path Space for Efficient Light Transport Simulation." *ACM Transactions on Graphics (Proc. SIGGRAPH)* 33:4, 2014.

[Karsch et al. 11] Kevin Karsch, Varsha Hedau, David Forsyth, and Derek Hoiem. "Rendering Synthetic Objects into Legacy Photographs." *ACM Trans. Graph.* 30:6 (2011), 157:1–157:12.

[Kastrinaki et al. 03] V. Kastrinaki, Michael E. Zervakis, and Kostas Kalaitzakis. "A Survey of Video Processing Techniques for Traffic Applications." *Image Vision Comput.* 21:4 (2003), 359–381.

[Katz and Tal 03] S. Katz and A. Tal. "Hierarchical Mesh Decomposition Using Fuzzy Clustering and Cuts." *ACM Transactions on Graphics* 22:3 (2003), 954–961.

[Kavan et al. 10] L. Kavan, P-P. Sloan, and C. O'Sullivan. "Fast and Efficient Skinning of Animated Meshes." *Computer Graphics Forum* 29:2 (2010), 327–336.

[Kavan et al. 11] Ladislav Kavan, Dan Gerszewski, Adam W. Bargteil, and Peter-Pike Sloan. "Physics-Inspired Upsampling for Cloth Simulation in Games." *ACM Trans. Graph.* 30:4 (2011), 93:1–93:10.

[Kazhdan et al. 06] Michael Kazhdan, Matthew Bolitho, and Hugues Hoppe. "Poisson Surface Reconstruction." In *Proceedings of the Fourth Eurographics Symposium on Geometry Processing, SGP '06*, pp. 61–70. Aire-la-Ville, Switzerland, Switzerland: Eurographics Association, 2006.

[Keller et al. 13] M. Keller, D. Lefloch, M. Lambers, S. Izadi, T. Weyrich, and A. Kolb. "Real-Time 3D Reconstruction in Dynamic Scenes Using Point-Based Fusion." In *Proc. Conf. 3D Vision (3DV)*, 2013. DOI:10.1109/3DV.2013.9.

[Khan et al. 06] E. A. Khan, A. O. Akyz, and E. Reinhard. "Ghost Removal in High Dynamic Range Images." In *IEEE International Conference on Image Processing*, pp. 2005–2008, 2006.

[Kholgade et al. 14] Natasha Kholgade, Tomas Simon, Alexei Efros, and Yaser Sheikh. "3D Object Manipulation in a Single Photograph Using Stock 3D Models." *ACM Transactions on Computer Graphics* 33:4, 2014.

[Khoshelham and Elberink 12] Kourosh Khoshelham and Sander Oude Elberink. "Accuracy and Resolution of Kinect Depth Data for Indoor Mapping Applications." *Sensors* 12:2 (2012), 1437–1454.

[Kilner et al. 06] J.J. Kilner, J.R. Starck, and A. Hilton. "A Comparative Study of Free Viewpoint Video Techniques for Sports Events." In *IET European Conference on Visual Media Production*, pp. 87–96. IET, 2006.

[Kilner et al. 09] J. Kilner, J.-Y. Guillemaut, and A. Hilton. "3D Action Matching with Key-Pose Detection." In *Computer Vision Workshops (ICCV Workshops), 2009 IEEE 12th International Conference on*, pp. 1–8. IEEE, 2009.

[Kim and Vendrovsky 08] Tae-Yong Kim and Eugene Vendrovsky. "DrivenShape: A Data-driven Approach for Shape Deformation." In *Proc. of the 2008 ACM SIGGRAPH/Eurographics Symposium on Computer Animation, SCA '08*, pp. 49–55. Aire-la-Ville, Switzerland, Switzerland: Eurographics Association, 2008.

[Kim et al. 07] Seon Joo Kim, Jan-Michael Frahm, and Marc Pollefeys. "Joint Feature Tracking and Radiometric Calibration from Auto-Exposure

Video." In *Computer Vision, 2007. ICCV 2007. IEEE 11th International Conference on*, pp. 1–8. IEEE, 2007.

[Kim et al. 09] Y. M. Kim, Christian Theobalt, J. Diebel, J. Kosecka, B. Micusik, and S. Thrun. "Multi-View Image and ToF Sensor Fusion for Dense 3D Reconstruction." In *IEEE Workshop on 3-D Digital Imaging and Modeling (3DIM)*, edited by Adrian Hilton, Takeshi Masuda, and Chang Shu, pp. 1542–1549. Kyoto, Japan: IEEE, 2009.

[Kim et al. 10] Youngmin Kim, Amitabh Varshney, David W. Jacobs, and François Guimbretière. "Mesh Saliency and Human Eye Fixations." *ACM Trans. Appl. Percept.* 7:2 (2010), 12:1–12:13. Available online (http://doi.acm.org/10.1145/1670671.1670676).

[Kim et al. 12] Min H. Kim, Todd Alan Harvey, David S. Kittle, Holly Rushmeier, Julie Dorsey, Richard O. Prum, and David J. Brady. "3D Imaging Spectroscopy for Measuring Hyperspectral Patterns on Solid Objects." *ACM Trans. Graph.* 31:4 (2012), 38:1–38:11. Available online (http://doi.acm.org/10.1145/2185520.2185534).

[Kim et al. 13] C. Kim, H. Zimmer, Y. Pritch, A. Sorkine-Hornung, and M. Gross. "Scene Reconstruction from High Spatio-Angular Resolution Light Fields." In *ACM Transactions on Graphics (Proc. SIGGRAPH)*, 32, 32, pp. 73:1–73:12, 2013.

[Kircher and Garland 06] S. Kircher and M. Garland. "Editing Arbitrarily Deforming Surface Animations." *ACM Transactions on Graphics (Proceedings of SIGGRAPH)*, pp. 1098–1107, 2006.

[Klaudiny et al. 12] Martin Klaudiny, Chris Budd, and Adrian Hilton. "Towards Optimal Non-Rigid Surface Tracking." In *ECCV*, pp. 743–756, 2012.

[Klose et al. 11] Felix Klose, Kai Ruhl, Christian Lipski, and Marcus Magnor. "Flowlab—An Interactive Tool for Editing Dense Image Correspondences." In *European Conference on Visual Media Production (CVMP)*, pp. 59–66, 2011.

[Kobbelt et al. 01] Leif P. Kobbelt, Mario Botsch, Ulrich Schwanecke, and Hans-Peter Seidel. "Feature Sensitive Surface Extraction from Volume Data." In *Proceedings of ACM SIGGRAPH 2001*, pp. 57–66, 2001.

[Kolb et al. 10] A. Kolb, E. Barth, R. Koch, and R. Larsen. "Time-of-Flight Cameras in Computer Graphics." *J. Computer Graphics Forum* 29:1 (2010), 141–159.

[Kolmogorov and Zabih 01] Vladimir Kolmogorov and Ramin Zabih. "Computing Visual Correspondence with Occlusions Using Graph Cuts." In *International Conference on Computer Vision*, 2, pp. 508–515 vol. 2, 2001.

[Kolmogorov and Zabih 02] Vladimir Kolmogorov and Ramin Zabih. "Multi-Camera Scene Reconstruction via Graph Cuts." In *Proceedings of the 7th European Conference on Computer Vision-Part III*, pp. 82–96, 2002.

[Kovar et al. 02] L. Kovar, M. Gleicher, and F. Pighin. "Motion Graphs." In *Proc. ACM SIGGRAPH*, pp. 473–482, 2002.

[Kriegel et al. 09] Hans-Peter Kriegel, Peer Kröger, Erich Schubert, and Arthur Zimek. "LoOP: Local Outlier Probabilities." In *Proceedings of the 18th ACM Conference on Information and Knowledge Management*, pp. 1649–1652, 2009.

[Krishnan and Nayar 08] Gurunandan Krishnan and Shree Nayar. "Cata-Fisheye Camera for Panoramic Imaging." In *Proc. IEEE Workshop on Application of Computer Vision (WACV)*, 2008.

[Křivánek et al. 13] Jaroslav Křivánek, Iliyan Georgiev, Anton Kaplanyan, and Juan Canada. "Recent Advances in Light Transport Simulation: Theory and Practice." In *ACM SIGGRAPH Courses*, 2013.

[Kry et al. 02] P. G. Kry, D. L. James, and D. K. Pai. "EigenSkin: Real-Time Large Deformation Character Skinning in Hardware." In *Proceedings of Symposium on Computer Animation (SCA)*, pp. 153–160, 2002.

[Kuhnert and Stommel 06] Klaus-Dieter Kuhnert and Martin Stommel. "Fusion of Stereo-Camera and PMD-Camera Data for Real-Time Suited Precise 3D Environment Reconstruction." In *IROS*, pp. 4780–4785, 2006.

[Kunitomo et al. 10] Shoji Kunitomo, Shinsuke Nakamura, and Shigeo Morishima. "Optimization of Cloth Simulation Parameters by Considering Static and Dynamic Features." In *ACM SIGGRAPH 2010 Posters, SIGGRAPH '10*, pp. 15:1–15:1. New York, NY, USA: ACM, 2010.

[Kuster et al. 11] C. Kuster, T. Popa, C. Zach, C. Gotsman, and M. Gross. "FreeCam: A Hybrid Camera System for Interactive Free-Viewpoint Video." In *Proceedings of Vision, Modeling, and Visualization (VMV)*, pp. 17–24, 2011.

[Kutulakos and Seitz 00] Kiriakos N. Kutulakos and Steven M. Seitz. "A Theory of Shape by Space Carving." *Int. J. Comput. Vision* 38:3 (2000), 199–218.

[Kwatra et al. 03] Vivek Kwatra, Arno Schödl, Irfan Essa, Greg Turk, and Aaron Bobick. "Graphcut Textures: Image and Video Synthesis Using Graph Cuts." In *ACM Transactions on Graphics (ToG)*, 22, 22, pp. 277–286. ACM, 2003.

[Lafortune and Willems 93] Eric P. Lafortune and Yves D. Willems. "Bi-Directional Path Tracing." In *Compugraphics '93*, pp. 145–153, 1993.

[Lalonde et al. 09] Jean-Francois Lalonde, Alexei A. Efros, and Srinivasa G. Narasimhan. "Estimating Natural Illumination from a Single Outdoor Image." In *IEEE International Conference on Computer Vision*, 2009.

[Lamond et al. 09] Bruce Lamond, Pieter Peers, Abhijeet Ghosh, and Paul Debevec. "Image-based Separation of Diffuse and Specular Reflections Using Environmental Structured Illumination." In *IEEE International Conference on Computational Photography*, 2009.

[Lang et al. 10] Manuel Lang, Alexander Hornung, Oliver Wang, Steven Poulakos, Aljoscha Smolic, and Markus Gross. "Nonlinear Disparity Mapping for Stereoscopic 3D." *ACM Transactions on Graphics (TOG)* 29:4 (2010), 75.

[Lanman et al. 08] D. Lanman, R. Raskar, A. Agrawal, and G. Taubin. "Shield Fields: Modeling and Capturing 3D Occluders." *Proc. SIGGRAPH Asia* 27:5 (2008), 1–10.

[Lantz 07] Ed Lantz. "A Survey of Large-scale Immersive Displays." In *Proceedings of the 2007 Workshop on Emerging Displays Technologies: Images and Beyond: The Future of Displays and Interacton*, p. 1. ACM, 2007.

[Laurentini 94] Aldo Laurentini. "The Visual Hull Concept for Silhouette-Based Image Understanding." *IEEE Transactions on Pattern Analysis and Machine Intelligence* 16:2 (1994), 150–162.

[Lavoué et al. 13] Guillaume Lavoué, Laurent Chevalier, and Florent Dupont. "Streaming Compressed 3D Data on the Web Using JavaScript and WebGL." In *Proceedings of the 18th International Conference on 3D Web Technology, Web3D '13*, pp. 19–27, 2013.

[Lee et al. 98] S. Lee, G. Wolberg, and S. Shin. "Polymorph: Morphing among Multiple Images." *IEEE Computer Graphics and Applications*, pp. 58–71.

[Lee et al. 02] J. Lee, J. Chai, P.S.A. Reitsma, J.K. Hodgins, and N.S. Pollard. "Interactive Control of Avatars Animated with Human Motion Data." In *Proc. ACM SIGGRAPH 2002*, pp. 491–500, 2002.

[Lee et al. 09] S.-H. Lee, E. Sifakis, and D. Terzopoulos. "Comprehensive Biomechanical Modeling and Simulation of the Upper Body." *ACM Transactions on Graphics* 28:4 (2009), 99:1–99:17.

[Lee et al. 10a] Jongseok Lee, Sungyul Choe, and Seungyong Lee. "Mesh Geometry Compression for Mobile Graphics." In *Proc. CCNC*, pp. 301–305, 2010.

[Lee et al. 10b] Sungkil Lee, Elmar Eisemann, and Hans-Peter Seidel. "Real-Time Lens Blur Effects and Focus Control." *ACM Transaction on Graphics (Proc. SIGGRAPH Asia)* 29:4.

[Lefloch et al. 13] D. Lefloch, T. Hoegg, and A. Kolb. "Real-Time Motion Artifacts Compensation of ToF Sensors Data on GPU." In *Proc. SPIE Defense, Security—Three-Dimensional Imaging, Visualization, and Display*, pp. 87380U–87380U–7, 2013.

[Lempitsky and Ivanov 07] Victor Lempitsky and Denis Ivanov. "Seamless Mosaicing of Image-Based Texture Maps." In *Computer Vision and Pattern Recognition, 2007. CVPR'07. IEEE Conference on*, pp. 1–6. IEEE, 2007.

[Lengyel 98] J. Lengyel. "The Convergence of Graphics and Vision." *Computer* 31:7 (1998), 46–53.

[Lensch et al. 03] Hendrik P. A. Lensch, Jan Kautz, Michael Goesele, Wolfgang Heidrich, and Hans-Peter Seidel. "Image-Based Reconstruction of Spatial Appearance and Geometric Detail." *ACM Trans. Graph.* 22:2 (2003), 234–257.

[Leutenegger et al. 11] Stefan Leutenegger, Margarita Chli, and Roland Yves Siegwart. "BRISK: Binary Robust Invariant Scalable Keypoints." In *Computer Vision (ICCV), 2011 IEEE International Conference on*, pp. 2548–2555. IEEE, 2011.

[Levin 98] David Levin. "The Approximation Power of Moving Least-Squares." *Mathematics of Computation* 67 (1998), 1517–1531.

[Levin 03] David Levin. "Mesh-Independent Surface Interpolation." *Geometric Modeling for Scientific Visualization* 3 (2003), 37–49.

[Levoy and Hanrahan 96] Marc Levoy and Pat Hanrahan. "Light Field Rendering." In *Proceedings of the 23rd Annual Conference on Computer Graphics and Interactive Techniques*, pp. 31–42. New York: ACM, 1996.

[Levoy et al. 00] M. Levoy, K. Pulli, B. Curless, S. Rusinkiewicz, D. Koller, L. Pereira, M. Ginzton, S. Anderson, J. Davis, and J. Ginsberg. "The Digital Michelangelo Project: 3D Scanning of Large Statues." In *Proc. Conf. Computer Graphics and Interactive Techniques (SIGGRAPH)*, pp. 131–144, 2000.

[Levoy 06] M. Levoy. "Light Fields and Computational Imaging." *Computer* 39:8 (2006), 46–55.

[Lewis and Anjyo 10] J. P. Lewis and K. Anjyo. "Direct Manipulation Blendshapes." In *Proceeedings of IEEE CGAA*, 2010.

[Lewis et al. 00] J. P. Lewis, Matt Cordner, and Nickson Fong. "Pose Space Deformation: A Unified Approach to Shape Interpolation and Skeleton-Driven Deformation." In *ACM SIGGRAPH*, pp. 165–172, 2000.

[Lhuillier and Quan 05] Maxime Lhuillier and Long Quan. "A Quasi-Dense Approach to Surface Reconstruction from Uncalibrated Images." *Pattern Analysis and Machine Intelligence, IEEE Transactions on* 27:3 (2005), 418–433.

[Li et al. 03] Ming Li, Marcus A. Magnor, and Hans-Peter Seidel. "Hardware-Accelerated Visual Hull Reconstruction and Rendering." In *Graphics Interface (GI)*, pp. 65–71, 2003.

[Li et al. 08] Xin Li, Bahadir Gunturk, and Lei Zhang. "Image Demosaicing: A Systematic Survey." In *Electronic Imaging 2008*, pp. 68221J–68221J. International Society for Optics and Photonics, 2008.

[Li et al. 10] Hao Li, Thibaut Weise, and Mark Pauly. "Example-Based Facial Rigging." *ACM Transactions on Graphics (Proc. SIGGRAPH 2010)* 29:4.

[Li et al. 12a] Hao Li, Linjie Luo, Daniel Vlasic, Pieter Peers, Jovan Popović, Mark Pauly, and Szymon Rusinkiewicz. "Temporally Coherent Completion of Dynamic Shapes." *ACM Trans. Graph.* 31:1 (2012), 2:1–2:11.

[Li et al. 12b] K. Li, Q. Dai, W. Xu, and J. Yang. "Temporal-Dense Dynamic 3-D Reconstruction with Low Frame Rate Cameras." *Selected Topics in Signal Processing, IEEE Journal of* 6:5 (2012), 447–459.

[Li et al. 13] Guannan Li, Chenglei Wu, Carsten Stoll, Yebin Liu, Kiran Varanasi, Qionghai Dai, and Christian Theobalt. "Capturing Relightable Human Performances under General Uncontrolled Illumination." In *Computer Graphics Forum*, 32, 32, pp. 275–284, 2013.

[Liang et al. 08] C.-K. Liang, T.-H. Lin, B.-Y. Wong, C. Liu, and H. Chen. "Programmable Aperture Photography: Multiplexed Light Field Acquisition." *ACM Transactions on Graphics (Proc. SIGGRAPH)* 27:3 (2008), 1–10.

[Lien et al. 06] J. M. Lien, J. Keyser, and N. M. Amato. "Simultaneous Shape Decomposition and Skeletonization." In *Proceedings of SPM*, pp. 219–228, 2006.

[Lim et al. 05] SukHwan Lim, Jon G. Apostolopoulos, and Abbas El Gamal. "Optical Flow Estimation Using Temporally Oversampled Video." *IEEE Transactions on Image Processing* 14:8 (2005), 1074–1087.

[Limper et al. 13a] Max Limper, Yvonne Jung, Johannes Behr, and Marc Alexa. "The POP Buffer: Rapid Progressive Clustering by Geometry Quantization." *Computer Graphics Forum* 32:7 (2013), 197–206.

[Limper et al. 13b] Max Limper, Stefan Wagner, Christian Stein, Yvonne Jung, and André Stork. "Fast Delivery of 3D Web Content: A Case Study." In *Proceedings of the 18th International Conference on 3D Web Technology, Web3D '13*, pp. 11–17, 2013.

[Lin and Tomasi 04] Micahel H. Lin and Carlo Tomasi. "Surfaces with Occlusions from Layered Stereo." *IEEE Trans. Pattern Anal. Mach. Intell.* 26 (2004), 1073–1078.

[Lindemann et al. 11] Lea Lindemann, Stephan Wenger, and Marcus Magnor. "Evaluation of Video Artifact Perception Using Event-Related Potentials." In *ACM Symposium on Applied Perception in Graphics and Visualization (APGV)*, pp. 53–58, 2011.

[Lindner and Kolb 06] M. Lindner and A. Kolb. "Lateral and Depth Calibration of PMD-Distance Sensors." In *Proc. Int. Symp. on Visual Computing, LNCS*, pp. 524–533. Springer, 2006.

[Lindner and Kolb 07] M. Lindner and A. Kolb. "Calibration of the Intensity-Related Distance Error of the PMD ToF-Camera." In *Proc. SPIE, Intelligent Robots and Computer Vision*, 6764, pp. 6764–35, 2007.

[Lindner and Kolb 09] M. Lindner and A. Kolb. "Compensation of Motion Artifacts for Time-of-Flight Cameras." In *Proc. Dynamic 3D Imaging*, LNCS, 5742, pp. 16–27. Springer, 2009.

[Lindner et al. 08] M. Lindner, M. Lambers, and A. Kolb. "Sub-Pixel Data Fusion and Edge-Enhanced Distance Refinement for 2D/3D Images." *Int. J. on Intell. Systems and Techn. and App., Special Issue on Dynamic 3D Imaging* 5:3/4 (2008), 344–354.

[Lindner et al. 10] M. Lindner, I. Schiller, A. Kolb, and R. Koch. "Time-of-Flight Sensor Calibration for Accurate Range Sensing." *Computer Vision and Image Understanding* 114:12 (2010), 1318–1328.

[Lindstrom 10] Peter Lindstrom. "Triangulation Made Easy." In *Computer Vision and Pattern Recognition (CVPR), 2010 IEEE Conference on*, pp. 1554–1561. IEEE, 2010.

[Linz et al. 10a] Christian Linz, Christian Lipski, and Marcus Magnor. "Multi-Image Interpolation Based on Graph-Cuts and Symmetric Optic Flow." In *Vision, Modeling, and Visualization (VMV)*, pp. 115–122, 2010.

[Linz et al. 10b] Christian Linz, Christian Lipski, Lorenz Rogge, Christian Theobalt, and Marcus Magnor. "Space-Time Visual Effects as a Post-Production Process." In *ACM International Workshop on 3D Video Processing (3DVP)*, pp. 1–6, 2010.

[Linz 11] Christian Linz. "Correspondence Estimation and Image Interpolation for Photo-Realistic Rendering." Ph.D. thesis, Braunschweig, 2011.

[Lippmann 08] G. Lippmann. "Épreuves réversibles donnant la sensation du relief." *J. Phys. Theor. Appl.* 7:1 (1908), 821–825.

[Lipski et al. 10a] Christian Lipski, Christian Linz, Kai Berger, Anita Sellent, and Marcus Magnor. "Virtual Video Camera: Image-Based Viewpoint Navigation through Space and Time." *Computer Graphics Forum* 29:8 (2010), 2555–2568.

[Lipski et al. 10b] Christian Lipski, Christian Linz, Thomas Neumann, Markus Wacker, and Marcus Magnor. "High Resolution Image Correspondences for Video Post-Production." In *European Conference on Visual Media Production (CVMP)*, pp. 33–39, 2010.

[Lipski et al. 12] Christian Lipski, Christian Linz, Thomas Neumann, Markus Wacker, and Marcus Magnor. "High Resolution Image Correspondences for Video Post-Production." *Journal of Virtual Reality and Broadcasting (JVRB)* 9:8 (2012), 1–12.

[Lipski et al. 14] Christian Lipski, Felix Klose, and Marcus Magnor. "Correspondence and Depth-Image Based Rendering: A Hybrid Approach for Free-Viewpoint Video." *IEEE Trans. Circuits and Systems for Video Technology (CSVT)* 24:6 (2014), 942–951.

[Lipski 12] Christian Lipski. "Making of 3D-Gauss, Vision, Modeling, and Visualization Workshop 2009." http://vmv09.tu-bs.de/index.php?page=gauss, 2012.

[Liu et al. 03] P. C. Liu, F. C. Wu, W. C. Ma, R. H Liang, and M. Ouhyoung. "Automatic Animation Skeleton Using Repulsive Force Field." In *Proceedings of the 11th Pacific Conference on Computer Graphics and Applications*, pp. 409–413, 2003.

[Liu et al. 08] Ce Liu, Jenny Yuen, Antonio Torralba, Josef Sivic, and William T. Freeman. "SIFT Flow: Dense Correspondence Across Different Scenes." In *European Conference on Computer Vision*, pp. 28–42, 2008.

[Liu et al. 10a] Kai Liu, Yongchang Wang, Daniel L. Lau, Qi Hao, and Laurence G. Hassebrook. "Dual-Frequency Pattern Scheme for High-Speed 3-D Shape Measurement." *Optics Express* 18:5 (2010), 5229–5244.

[Liu et al. 10b] Yebin Liu, Qionghai Dai, and Wenli Xu. "A Point-Cloud-Based Multiview Stereo Algorithm for Free-Viewpoint Video." *IEEE Transactions on Visualization and Computer Graphics* 16:3 (2010), 407–418.

[Liu et al. 11] Yebin Liu, Carsten Stoll, Juergen Gall, Hans-Peter Seidel, and Christian Theobalt. "Markerless Motion Capture of Interacting Characters Using Multi-View Image Segmentation." In *CVPR*, pp. 1249–1256, 2011.

[Liu et al. 13] Yebin Liu, Juergen Gall, Carsten Stoll, Qionghai Dai, Hans-Peter Seidel, and Christian Theobalt. "Markerless Motion Capture of Multiple Characters Using Multiview Image Segmentation." *IEEE Transactions on Pattern Analysis Machine Intelligence (PAMI)* 35:11 (2013), 2720–2735.

[Lokovic and Veach 00] Tom Lokovic and Eric Veach. "Deep Shadow Maps." *ACM Transactions on Graphics (Proc. SIGGRAPH 2000)*, pp. 385–392.

[Longere et al. 02] P. Longere, Xuemei Zhang, P.B. Delahunt, and D.H. Brainard. "Perceptual Assessment of Demosaicing Algorithm Performance." *Proceedings of the IEEE* 90:1 (2002), 123–132.

[Lorensen and Cline 87] William E. Lorensen and Harvey E. Cline. "Marching Cubes: A High Resolution 3D Surface Construction Algorithm." In *Proceedings of ACM SIGGRAPH 1987*, pp. 163–169, 1987.

[Loscos and Jacobs 10] C. Loscos and K. Jacobs. "High-Dynamic Range Imaging for Dynamic Scenes." In *Computational Photography, Methods and Applications*, p. 259–281. CRC Press, 2010.

[Loscos et al. 99] Cline Loscos, Marie-Claude Frasson, George Drettakis, Bruce Walter, Xavier Granier, and Pierre Poulin. "Interactive Virtual Relighting and Remodeling of Real Scenes." In *Rendering Techniques*, pp. 329–340, 1999.

[Lowe 04] David G. Lowe. "Distinctive Image Features from Scale-Invariant Keypoints." *International Journal of Computer Vision* 60:2 (2004), 91–110.

[Lucas and Kanade 81] Bruce D. Lucas and Takeo Kanade. "An Iterative Image Registration Technique with an Application to Stereo Vision." In *International Joint Conference on Artificial Intelligence*, 2, pp. 674–679, 1981.

[Lucas et al. 13] Laurent Lucas, Philippe Souchet, Muhannad Ismael, Olivier Nocent, Cédric Niquin, Céline Loscos, Ludovic Blache, Stéphanie Prévost, and Yannick Remion. "RECOVER3D: A Hybrid Multi-View System for 4D Reconstruction of Moving Actors." In *4th International Conference and Exhibition on 3D Body Scanning Technologies*, pp. 219–230. Hometrica Consulting, 2013.

[Luenberger 73] D. G. Luenberger. "Introduction to Linear and Non-Linear Programming." Reading MA: Addison-Wesley, 1973.

[Lumsdaine and Georgiev 09] Andrew Lumsdaine and Todor Georgiev. "The Focused Plenoptic Camera." In *Proc. IEEE International Conference on Computational Photography*, pp. 1–8, 2009.

[Lytro, Inc. 12] Lytro, Inc. "The Lytro Camera." https://www.lytro.com/, 2012.

[Ma et al. 07] Wan-Chun Ma, Tim Hawkins, Pieter Peers, Charles-Felix Chabert, Malte Weiss, and Paul Debevec. "Rapid Acquisition of Specular and Diffuse Normal Maps from Polarized Spherical Gradient Illumination." In *Eurographics Symposium on Rendering*, pp. 183–194, 2007.

[Magnenat-Thalmann and Volino 05] Nadia Magnenat-Thalmann and Pascal Volino. "From Early Draping to Haute Couture Models: 20 Years of Research." *The Visual Computer* 21:8-10 (2005), 506–519.

[Magnenat-Thalmann et al. 88] N. Magnenat-Thalmann, R. Laperrière, and D. Thalmann. "Joint-Dependent Local Deformations for Hand Animation and Object Grasping." In *Proceedings on Graphics Interface '88*, pp. 26–33. Canadian Information Processing Society, 1988.

[Magnenat-Thalmann et al. 04] Nadia Magnenat-Thalmann, Frederic Cordier, Michael Keckeisen, Stefan Kimmerle, Reinhard Klein, and Jan Meseth. "Simulation of Clothes for Real-Time Applications." In *Eurographics 2004, Tutorials 1: Simulation of Clothes for Real-time Applications*. Eurographics Association, 2004.

[Magnor 05] M. Magnor. *Video-Based Rendering*. A. K. Peters, 2005.

[Malvar et al. 04] Henrique S. Malvar, Li-wei He, and Ross Cutler. "High-Quality Linear Interpolation for Demosaicing of Bayer-Patterned Color Images." In *Proc. IEEE Int. Conf. Acoustics, Speech, and Signal Processing*, 3, pp. 485–488, 2004.

[Mamou et al. 09] Khaled Mamou, Titus Zaharia, and Françoise Prêteux. "TFAN: A Low Complexity 3D Mesh Compression Algorithm." *Comput. Animat. Virtual Worlds* 20:3 (2009), 343–354.

[Manakov et al. 13] Alkhazur Manakov, John F. Restrepo, Oliver Klehm, Ramon Hegedüs, Elmar Eisemann, Hans-Peter Seidel, and Ivo Ihrke. "A Reconfigurable Camera Add-on for High Dynamic Range, Multi-Spectral, Polarization, and Light-Field Imaging." *ACM Transactions on Graphics (Proc. SIGGRAPH)* 32:4 (2013), 47:1–47:14.

[Mann and Picard 95] Steve Mann and Rosalind W. Picard. "Being Undigital with Digital Cameras: Extending Dynamic Range by Combining Differently Exposed Pictures." In *IST 46th Annual Conference*, pp. 422–428, 1995.

[Marr and Poggio 76] D. Marr and T. Poggio. "Cooperative Computation of Stereo Disparity." *Science* 194 (1976), 283–287.

[Marschner et al. 03] Stephen Marschner, Henrik Wann Jensen, Mike Cammarano, Steve Worley, and Pat Hanrahan. "Light Scattering from Human Hair Fibers." *ACM Transactions on Graphics (Proc. SIGGRAPH 2003)*.

[Martin and Crowley 95] Jerome Martin and James L. Crowley. "Comparison of Correlation Techniques." In *Intelligent Autonomous Systems*, pp. 86–93, 1995.

[Martínez et al. 14] Leonardo Sánchez-Mesa Martínez, Francisco Javier Melero Rus, and Jorge Revelles Moreno. "Portal Virtual de Patrimonio de las Universidades Andaluzas." www.http://patrimonio3d.ugr.es, February 28, 2014.

[Marwah et al. 13] K. Marwah, G. Wetzstein, Y. Bando, and R. Raskar. "Compressive Light Field Photography Using Overcomplete Dictionaries and Optimized Projections." *ACM Transactions on Graphics (Proc. SIGGRAPH)* 32:4 (2013), 46:1–46:11.

[Masselus et al. 02] Vincent Masselus, Philip Dutré, and Frederik Anrys. "The Free-Form Light Stage." In *Proceedings of the 13th Eurographics Workshop on Rendering, EGRW '02*, pp. 247–256. Aire-la-Ville, Switzerland, Switzerland: Eurographics Association, 2002.

[Masselus et al. 03] Vincent Masselus, Pieter Peers, Philip Dutré, and Yves D. Willems. "Relighting with 4D Incident Light Fields." *ACM Transactions on Graphics (Proc. SIGGRAPH 2003)* 22:3 (2003), 613–620.

[Matas et al. 04] Jiri Matas, Ondrej Chum, Martin Urban, and Tomás Pajdla. "Robust Wide-Baseline Stereo from Maximally Stable Extremal Regions." *Image and Vision Computing* 22:10 (2004), 761–767.

[Matsushita et al. 06] Yasuyuki Matsushita, Eyal Ofek, Weina Ge, Xiaoou Tang, and Heung-Yeung Shum. "Full-Frame Video Stabilization with Motion Inpainting." *IEEE Trans. Pattern Anal. Mach. Intell.* 28:7 (2006), 1150–1163.

[Matsuyama et al. 04] T. Matsuyama, X. Wu, T. Takai, and S. Nobuhara. "Real-Time 3D Shape Reconstruction, Dynamic 3D Mesh Deformation, and High Fidelity Visualization for 3D Video." *Computer Vision and Image Understanding* 96:3 (2004), 393–434.

[Matusik et al. 00] Wojciech Matusik, Chris Buehler, Ramesh Raskar, Steven J. Gortler, and Leonard McMillan. "Image-Based Visual Hulls." In *Proceedings of the 27th Annual Conference on Computer Graphics and Interactive Techniques, SIGGRAPH '00*, pp. 369–374. New York, NY, USA: ACM Press/Addison-Wesley Publishing Co., 2000.

[Matusik et al. 02] Wojciech Matusik, Hanspeter Pfister, Addy Ngan, Paul Beardsley, Remo Ziegler, and Leonard McMillan. "Image-Based 3D Photography Using Opacity Hulls." *ACM Trans. Graph.* 21:3 (2002), 427–437.

[Matusik et al. 04] Wojciech Matusik, Matthew Loper, and Hanspeter Pfister. "Progressively-Refined Reflectance Functions from Natural Illumination." In *Proceedings of the Fifteenth Eurographics Conference on Rendering Techniques, EGSR'04*, pp. 299–308. Aire-la-Ville, Switzerland, Switzerland: Eurographics Association, 2004. Available online (http://dx.doi.org/10.2312/EGWR/EGSR04/299-308).

[McDonnell et al. 08] Rachel McDonnell, Michéal Larkin, Simon Dobbyn, Steven Collins, and Carol O'Sullivan. "Clone Attack! Perception of Crowd Variety." In *ACM Transactions on Graphics (TOG)*, 27, p. 26. ACM, 2008.

[McGlone 13] J. Chris McGlone, editor. *Manual of Photogrammetry.* Sixth Edition, American Society for Photogrammetry and Remote Sensing, 2013.

[McMillan and Bishop 95] Leonard McMillan and Gary Bishop. "Plenoptic Modeling: An Image-Based Rendering System." In *Proceedings of the 22nd Annual Conference on Computer Graphics and Interactive Techniques, SIGGRAPH '95*, pp. 39–46. New York, NY, USA: ACM, 1995.

[McNamara et al. 00] Ann McNamara, Alan Chalmers, Tom Troscianko, and Iain Gilchrist. "Comparing Real & Synthetic Scenes Using Human Judgements of Lightness." In *Proceedings of the Eurographics Workshop on Rendering Techniques 2000*, pp. 207–218. Springer-Verlag, 2000.

[Meyer et al. 86] Gary W. Meyer, Holly E. Rushmeier, Michael F. Cohen, Donald P. Greenberg, and Kenneth E. Torrance. "An Experimental Evaluation of Computer Graphics Imagery." *ACM Trans. Graph.* 5:1 (1986), 30–50.

[Meyer et al. 02] M. Meyer, M. Desbrun, P. Schroder, and A. H. Barr. "Discrete Differential-Geometry Operators for Triangulated 2-Manifolds." In *Proc. VisMath*, pp. 35–57, 2002.

[Meyer et al. 10] Quirin Meyer, Jochen Sümuth, Gerd Sussner, Marc Stamminger, and Günther Greiner. "On Floating-Point Normal Vectors." *Computer Graphics Forum* 29:4 (2010), 1405–1409.

[Micusik and Kosecka 10] Branislav Micusik and Jana Kosecka. "Multi-View Superpixel Stereo in Urban Environments." *International Journal of Computer Vision* 89:1 (2010), 106–119.

[Miguel et al. 12] Eder Miguel, Derek Bradley, Bernhard Thomaszewski, Bernd Bickel, Wojciech Matusik, Miguel A. Otaduy, and Steve Marschner. "Data-Driven Estimation of Cloth Simulation Models." *Comp. Graph. Forum* 31:2pt2 (2012), 519–528.

[Miguel et al. 13] Eder Miguel, Rasmus Tamstorf, Derek Bradley, Sara C. Schvartzman, Bernhard Thomaszewski, Bernd Bickel, Wojciech Matusik, Steve Marschner, and Miguel A. Otaduy. "Modeling and Estimation of Internal Friction in Cloth." *ACM Trans. Graph.* 32:6 (2013), 212:1–212:10.

[Mikolajczyk and Schmid 05] Krystian Mikolajczyk and Cordelia Schmid. "A Performance Evaluation of Local Descriptors." *TPAMI* 27:10.

[Mileva et al. 07] Yana Mileva, André Bruhn, and Joachim Weickert. "Illumination-Robust Variational Optical Flow with Photometric Invariants." In *DAGM Conference on Pattern Recognition*, pp. 152–162, 2007.

[Mitsunaga and Nayar 99] Tomoo Mitsunaga and Shree K. Nayar. "Radiometric Self Calibration." In *IEEE Conference on Computer Vision and Pattern Recognition (CVPR)*, pp. 374–380, 1999.

[Moeslund et al. 06] Thomas B. Moeslund, Adrian Hilton, and Volker Krüger. "A Survey of Advances in Vision-Based Human Motion Capture and Analysis." *Computer Vision and Image Understanding* 104:2 (2006), 90–126.

[Moffat et al. 95] A. Moffat, R. Neal, and Ian H. Witten. "Arithmetic Coding Revisited." In *Data Compression Conference, 1995. DCC '95. Proceedings*, pp. 202–211, 1995.

[Mohan et al. 08] Ankit Mohan, Ramesh Raskar, and Jack Tumblin. "Agile Spectrum Imaging: Programmable Wavelength Modulation for Cameras and Projectors." *Comput. Graph. Forum* 27:2 (2008), 709–717.

[Molina-Tanco and Hilton 00] Luis Molina-Tanco and Adrian Hilton. "Realistic Synthesis of Novel Human Movements from a Database of Motion Capture Examples." In *Workshop on Human Motion (HUMO)*, pp. 137–142, 2000.

[Moon and Marschner 06a] Jonathan T. Moon and Stephen R. Marschner. "Efficient Multiple Scattering in Hair Using Spherical Harmonics." *ACM Transactions on Graphics* 27:3.

[Moon and Marschner 06b] Jonathan T. Moon and Stephen R. Marschner. "Simulating Multiple Scattering in Hair Using a Photon Mapping Approach." *ACM Transactions on Graphics* 25:3.

[Moreno-Noguer et al. 07] Francesc Moreno-Noguer, Peter N. Belhumeur, and Shree K. Nayar. "Active Refocusing of Images and Videos." In *ACM SIGGRAPH 2007 Papers, SIGGRAPH '07*, pp. 67:1–67:9. New York, NY, USA: ACM, 2007.

[Mostafa et al. 99] Mostafa G.-H. Mostafa, Sameh M. Yamany, and Aly A. Farag. "Integrating Shape from Shading and Range Data Using Neural Networks." In *CVPR*, pp. 2015–2020, 1999.

[Mühlmann et al. 02] Karsten Mühlmann, Dennis Maier, Jürgen Hesser, and Reinhard Männer. "Calculating Dense Disparity Maps from Color Stereo Images, an Efficient Implementation." *International Journal of Computer Vision* 47:1-3 (2002), 79–88.

[Müller and Chentanez 10] Matthias Müller and Nuttapong Chentanez. "Wrinkle Meshes." In *Proceedings of the 2010 ACM SIGGRAPH/Eurographics Symposium on Computer Animation, SCA '10*, pp. 85–92. Aire-la-Ville, Switzerland, Switzerland: Eurographics Association, 2010.

[Müller et al. 04] Gero Müller, Jan Meseth, Mirko Sattler, Ralf Sarlette, and Reinhard Klein. "Acquisition, Synthesis and Rendering of Bidirectional Texture Functions." In *Eurographics 2004, State of the Art Reports*, edited by Christophe Schlick and Werner Purgathofer, pp. 69–94. INRIA and Eurographics Association, 2004.

[Müller et al. 08] K. Müller, A. Smolic, K. Dix, P. Kauff, and T. Wiegand. "Reliability-Based Generation and View Synthesis in Layered Depth Video." In *Multimedia Signal Processing, 2008 IEEE 10th Workshop on*, pp. 34–39. IEEE, 2008.

[Müller et al. 13] K. Müller, H. Schwarz, D. Marpe, C. Bartnik, S. Bosse, H. Brust, T. Hinz, H. Lakshman, P. Merkle, F.H. Rhee, G. Tech, M. Winken, and T. Wiegand. "3D High-Efficiency Video Coding for Multi-View Video and Depth Data." *Image Processing, IEEE Transactions on* 22:9 (2013), 3366–3378.

[Müller 09] Gero Müller. "Data-Driven Methods for Compression and Editing of Spatially Varying Appearance." Ph.D. thesis, University of Bonn, 2009.

[Mundermann et al. 07] Lars Mundermann, Stefano Corazza, and Thomas P. Andriacchi. "Accurately Measuring Human Movement Using Articulated ICP with Soft-joint Constraints and a Repository of Articulated Models." In *IEEE CVPR*, 2007.

[Murray et al. 94] Richard M. Murray, Zexiang Li, S. Shankar Sastry, and S. Shankara Sastry. *A Mathematical Introduction to Robotic Manipulation.* CRC Press, 1994.

[Mustafa et al. 12] Maryam Mustafa, Lea Lindemann, and Marcus Magnor. "EEG Analysis of Implicit Human Visual Perception." In *SIGCHI Conference on Human Factors in Computing Systems (CHI)*, pp. 513–516. ACM, 2012.

[Myronenko and Song 10] Andreiy Myronenko and Xubo Song. "Point Set Registration: Coherent Point Drift." *IEEE Transactions on PAMI* 32:12 (2010), 2262–2275.

[Myszkowski 98] Karol Myszkowski. "The Visible Differences Predictor: Applications to Global Illumination Problems." In *Rendering Techniques 98, Eurographics*, edited by George Drettakis and Nelson Max, pp. 223–236. Springer Vienna, 1998. Available online (http://dx.doi.org/10.1007/978-3-7091-6453-2_21).

[Nagel and Enkelmann 86] Hans-Helmut Nagel and Wilfried Enkelmann. "An Investigation of Smoothness Constraints for the Estimation of Displacement Vector Fields from Image Sequences." *IEEE Trans. Pattern Anal. Mach. Intell.* 8:5 (1986), 565–593.

[Nair et al. 12] Rahul Nair, Frank Lenzen, Stephan Meister, Henrik Schäfer, Christoph S. Garbe, and Daniel Kondermann. "High Accuracy TOF and Stereo Sensor Fusion at Interactive Rates." In *ECCV Workshops (2)*, pp. 1–11, 2012.

[Narain et al. 12] Rahul Narain, Armin Samii, and James F O'Brien. "Adaptive Anisotropic Remeshing for Cloth Simulation." *ACM Transactions on Graphics (TOG)* 31:6 (2012), 152.

[Narayanan et al. 98] P.J. Narayanan, Peter W. Rander, and Takeo Kanade. "Constructing Virtual Worlds Using Dense Stereo." In *Proceedings of the Sixth International Conference on Computer Vision*, pp. 3–11, 1998.

[Nayar and Mitsunaga 00] S.K. Nayar and T. Mitsunaga. "High Dynamic Range Imaging: Spatially Varying Pixel Exposures." In *IEEE Conference on Computer Vision and Pattern Recognition (CVPR)*, 1, 1, pp. 472–479, 2000.

[Nayar and Nakagawa 94] S.K. Nayar and Y. Nakagawa. "Shape from Focus." *IEEE Transactions on Pattern Analysis and Machine Intelligence* 16:8 (1994), 824–831.

[Nayar et al. 06] Shree K. Nayar, Gurunandan Krishnan, Michael D. Grossberg, and Ramesh Raskar. "Fast Separation of Direct and Global Components of a Scene Using High Frequency Illumination." *ACM Transactions on Graphics (Proc. SIGGRAPH 2006)* 25:3 (2006), 935–944.

[Nayar 97] Shree Nayar. "Catadioptric Omnidirectional Camera." In *Proc. CVPR*. San Juan, 1997.

[Nealen et al. 05] Andrew Nealen, Matthias Müller, Richard Keiser, Eddy Boxermann, and Mark Carlson. "Physically Based Deformable Models in Computer Graphics." In *Eurographics 2005. STAR—State of the Art Reports*, pp. 71–94, 2005.

[Nehab et al. 05] Diego Nehab, Szymon Rusinkiewicz, James Davis, and Ravi Ramamoorthi. "Efficiently Combining Positions and Normals for Precise 3D Geometry." *SIGGRAPH* 24:3, 2005.

[Neumann et al. 13a] T. Neumann, K. Varanasi, N. Hasler, M. Wacker, M. Magnor, and C. Theobalt. "Capture and Statistical Modeling of Arm-Muscle Deformations." *Computer Graphics Forum* 32:2pt3 (2013), 285–294.

[Neumann et al. 13b] T. Neumann, K. Varanasi, S. Wenger, M. Wacker, M. Magnor, and C. Theobalt. "Sparse Localized Deformation Components." *ACM Transactions on Graphics (Proc. SIGGRAPH Asia)* 32:6 (2013), Art.179.

[Newcombe et al. 11] R. Newcombe, S. Izadi, O. Hilliges, D. Molyneaux, D. Kim, A. Davison, P. Kohli, J. Shotton, S. Hodges, and A. Fitzgibbon. "KinectFusion: Real-Time Dense Surface Mapping and Tracking." In *Proc. IEEE Int. Symp. Mixed and Augm. Reality*, pp. 127–136, 2011.

[Ng et al. 02] A. Y. Ng, M. Jordan, and Y. Weiss. "On Spectral Clustering: Analysis and an Algorithm." In *Proc. NIPS*, 2002.

[Ng 06] R. Ng. "Digital Light Field Photography." PhD thesis, Stanford University, 2006. Note: thesis led to commercial light field camera, see also www.lytro.com.

[Nguyen et al. 12] H.M. Nguyen, B. Wünsche, P. Delmas, and C. Lutteroth. "3D Models from the Black Box: Investigating the Current State of Image-Based Modeling." *Journal of WSCG* 20:1 (2012), 1–10.

[Nießner et al. 13] M. Nießner, M. Zollhöfer, S. Izadi, and M. Stamminger. "Real-Time 3D Reconstruction at Scale Using Voxel Hashing." *ACM Transactions on Graphics (TOG)* 32:6 (2013), 169.

[Niquin et al. 10] Cédric Niquin, Stéphanie Prévost, and Yannick Rémion. "An Occlusion Approach with Consistency Constraint for Multiscopic Depth Extraction." *Int. J. Digital Multimedia Broadcasting* 2010 (2010), 1–8.

[Nistér and Stewenius 06] David Nistér and Henrik Stewenius. "Scalable Recognition with a Vocabulary Tree." In *Computer Vision and Pattern Recognition, 2006 IEEE Computer Society Conference on*, 2, pp. 2161–2168. IEEE, 2006.

[Nistér 04] David Nistér. "An Efficient Solution to the Five-Point Relative Pose Problem." *Pattern Analysis and Machine Intelligence, IEEE Transactions on* 26:6 (2004), 756–770.

[Nitsche 08] Michael Nitsche. "Experiments in the Use of Game Technology for Pre-Visualization." In *Proceedings of the 2008 Conference on Future Play: Research, Play, Share*, pp. 160–165. ACM, 2008.

[Oggier et al. 05] T. Oggier, B. Büttgen, F. Lustenberger, G. Becker, B. Rüegg, and A. Hodac. "SwissRanger SR3000 and First Experiences Based on Miniaturized 3D-ToF Cameras." In *Proc. of the First Range Imaging Research Day at ETH Zurich*, 2005.

[Oikonomidis et al. 11a] Iason Oikonomidis, Nikolaos Kyriazis, and Antonis A Argyros. "Efficient Model-Based 3D Tracking of Hand Articulations Using Kinect." In *BMVC*, pp. 1–11, 2011.

[Oikonomidis et al. 11b] Iasonas Oikonomidis, Nikolaos Kyriazis, and Antonis A Argyros. "Full dof Tracking of a Hand Interacting with an Object by Modeling Occlusions and Physical Constraints." In *Computer Vision (ICCV), 2011 IEEE International Conference on*, pp. 2088–2095. IEEE, 2011.

[Okutomi and Kanade 93] Masatoshi Okutomi and Takeo Kanade. "A Multiple-Baseline Stereo." *IEEE Trans. Pattern Anal. Mach. Intell.* 15:4 (1993), 353–363.

[O'Sullivan and Dingliana 01] Carol O'Sullivan and John Dingliana. "Collisions and Perception." *ACM Transactions on Graphics (TOG)* 20:3 (2001), 151–168.

[O'Toole and Kutulakos 10] Matthew O'Toole and Kiriakos N. Kutulakos. "Optical Computing for Fast Light Transport Analysis." *ACM Trans. Graph.* 29:6 (2010), 164:1–164:12.

[O'Toole et al. 12] Matthew O'Toole, Ramesh Raskar, and Kiriakos N. Kutulakos. "Primal-Dual Coding to Probe Light Transport." *ACM Trans. Graph.* 31:4 (2012), 39:1–39:11. Available online (http://doi.acm.org/10.1145/2185520.2185535).

[Ozawa et al. 12] Tomohiro Ozawa, Kris M. Kitani, and Hideki Koike. "Human-Centric Panoramic Imaging Stitching." In *Proceedings of the 3rd Augmented Human International Conference*, p. 20. ACM, 2012.

[Pajarola and Rossignac 00] Renato B. Pajarola and Jarek Rossignac. "SQUEEZE: Fast and Progressive Decompression of Triangle Meshes." In *Proc. CGI*, pp. 173–182, 2000.

[Park and Hodgins 06] S. I. Park and J. K. Hodgins. "Capturing and Animating Skin Deformation in Human Motion." *ACM Transactions on Graphics (Proceedings of SIGGRAPH)* 25:3.

[Park and Hodgins 08] S. I. Park and J. K. Hodgins. "Data-Driven Modeling of Skin and Muscle Deformation." *ACM Transactions on Graphics*, 27:3(2008), 96:1–96:6.

[Parker et al. 98] S. Parker, P. Shirley, Y. Livnat, Ch. Hansen, and P. Sloan. "Interactive Ray Tracing for Isosurface Rendering." In *Proc. Visualization*, pp. 233–238, 1998.

[Peers and Dutré 03] Pieter Peers and Philip Dutré. "Wavelet Environment Matting." In *Rendering Techniques 2003 (Proc. Eurographics Symposium on Rendering)*, pp. 157–166, 2003. Available online (http://www.eg.org/EG/DL/WS/EGWR/EGWR03/157-166.pdf).

[Peers and Dutré 05] Pieter Peers and Philip Dutré. "Inferring Reflectance Functions from Wavelet Noise." In *Rendering Techniques*, pp. 173–182, 2005.

[Peers et al. 09] Pieter Peers, Dhruv K. Mahajan, Bruce Lamond, Abhijeet Ghosh, Wojciech Matusik, Ravi Ramamoorthi, and Paul Debevec. "Compressive Light Transport Sensing." *ACM Trans. Graph.* 28:1 (2009), 3:1–3:18. Available online (http://doi.acm.org/10.1145/1477926.1477929).

[Peinsipp-Byma et al. 09] E. Peinsipp-Byma, N. Rehfeld, and R. Eck. "Evaluation of Stereoscopic 3D Displays for Image Analysis Tasks." In *Stereoscopic Displays and Applications XX*, edited by Andrew J. Woods, Nicolas S. Holliman, and John O. Merritt, p. 72370L. SPIE, 2009.

[Peleg et al. 99] Shmuel Peleg, Yael Pritch, and Moshe Ben-Ezra. "Stereo Panorama with a Single Camera." In *Proc. CVPR*. Fort Collins, USA, 1999.

[Peleg et al. 00] Shmuel Peleg, Yael Pritch, and Moshe Ben-Ezra. "Cameras for Stereo Panoramic Imaging." In *Proc. CVPR*, pp. 208–214. Hilton Head Island, 2000.

[Peleg et al. 01] Shmuel Peleg, Moshe Ben-Ezra, and Yael Pritch. "Omnistereo: Panoramic Stereo Imaging." *IEEE Trans. Pattern Analysis and Machine Intelligence* 23:3 (2001), 279–290.

[Pellacini et al. 00] Fabio Pellacini, James A. Ferwerda, and Donald P. Greenberg. "Toward a Psychophysically-Based Light Reflection Model for Image Synthesis." In *Proceedings of the 27th Annual Conference on Computer Graphics and Interactive Techniques, SIGGRAPH '00*, pp. 55–64. New York, NY, USA: ACM Press/Addison-Wesley Publishing Co., 2000.

[Peng et al. 05] Jingliang Peng, Chang-Su Kim, and C. C. Jay Kuo. "Technologies for 3D Mesh Compression: A Survey." *J. Vis. Commun. Image Represent.*, pp. 688–733.

[Perez and Luke 09] N. Perez and J. Luke. "Simultaneous Estimation of Super-Resolved Depth and All-in-Focus Images from a Plenoptic Camera." In *3DTV Conference: The True Vision-Capture, Transmission and Display of 3D Video*, pp. 1–4. IEEE, 2009.

[Perlin et al. 00] K. Perlin, S. Paxia, and J. S. Kollin. "An Autostereoscopic Display." In *ACM SIGGRAPH 2000 Conference Proceedings*, 33, 33, pp. 319–326, 2000.

[Perwass and Wietzke 12] C. Perwass and L. Wietzke. "Single Lens 3D-Camera with Extended Depth-of-Field." *SPIE Electronic Imaging*, pp. 22–26.

[Petit et al. 11] Benjamin Petit, Antoine Letouzey, Edmond Boyer, and Jean-Sébastien Franco. "Surface Flow from Visual Cues." In *VMV 2011—Vision, Modeling and Visualization Workshop*, pp. 1–8. Berlin, Germany, 2011.

[Pharr and Humphreys 10] Matt Pharr and Greg Humphreys. *Physically Based Rendering: From Theory To Implementation*, Second edition. Morgan Kaufmann Publishers Inc., 2010.

[Phong 75] Bui Tuong Phong. "Illumination for Computer Generated Pictures." *Commun. ACM* 18:6 (1975), 311–317.

[Pollefeys et al. 04] Marc Pollefeys, Luc Van Gool, Maarten Vergauwen, Frank Verbiest, Kurt Cornelis, Jan Tops, and Reinhard Koch. "Visual Modeling with a Hand-Held Camera." *International Journal of Computer Vision* 59:3 (2004), 207–232.

[Pool et al. 11] Jeff Pool, Anselmo Lastra, and Montek Singh. "Precision Selection for Energy-Efficient Pixel Shaders." In *Proceedings of the ACM SIGGRAPH Symposium on High Performance Graphics, HPG '11*, pp. 159–168. New York, NY, USA: ACM, 2011. Available online (http://doi.acm.org/10.1145/2018323.2018349).

[Popa et al. 09] Tiberiu Popa, Quan Zhou, Derek Bradley, Vladislav Kraevoy, Hongbo Fu, Alla Sheffer, and Wolfgang Heidrich. "Wrinkling Captured Garments Using Space-Time Data-Driven Deformation." *Comput. Graph. Forum* 28:2 (2009), 427–435.

[Potmesil 87] Michael Potmesil. "Generating Octree Models of 3D Objects from Their Silhouettes in a Sequence of Images." *Computer Vision, Graphics, and Image Processing* 40:1 (1987), 1–29.

[Press et al. 07] William Press, Saul Tuekolski, William Vetterling, and Brian Flannery. *Numerical Recipes: The Art of Scientific Computing, 3rd edition.* Cambridge University Press, 2007.

[Prévost et al. 13] Stéphanie Prévost, Cédric Niquin, Sylvie Chambon, and Guillaume Gales. "Multi- and Stereoscopic Matching, Depth and Disparity." In *In 3D Video: From Capture to Diffusion*, edited by Yannick Rémion Laurent Lucas and Cline Loscos, pp. 137–154. Wiley-ISTE, 2013.

[Prévoteau et al. 10] Jessica Prévoteau, Sylvia Chalençon-Piotin, Didier Debons, Laurent Lucas, and Yannick Remion. "Multi-View Shooting Geometry for Multiscopic Rendering with Controlled Distortion." *International Journal of Digital Multimedia Broadcasting (IJDMB), special issue Advances in 3DTV: Theory and Practice* 2010 (2010), 1–11.

[Pritchard and Heidrich 03] David Pritchard and Wolfgang Heidrich. "Cloth Motion Capture." *Comput. Graph. Forum* 22:3 (2003), 263–272.

[Provot 95] Xavier Provot. "Deformation Constraints in a Mass-Spring Model to Describe Rigid Cloth Behavior." In *In Graphics Interface*, pp. 147–154, 1995.

[Pujades et al. 14] S. Pujades, B. Goldlücke, and F. Devernay. "Bayesian View Synthesis and Image-Based Rendering Principles." In *Proc. International Conference on Computer Vision and Pattern Recognition*, 2014.

[Qi and Cooperstock 08] Zhi Qi and Jeremy R. Cooperstock. "Depth-Based Image Mosaicing for Both Static and Dynamic Scenes." In *Proc. Int. Conf. on Pattern Recognition (ICPR)*, pp. 1–4. Tampa, USA, 2008.

[Quan and Lan 99] Long Quan and Zhongdan Lan. "Linear N-Point Camera Pose Estimation." *IEEE Transactions on Pattern Analysis and Machine Intelligence* 21:7.

[Raguram et al. 11] Rahul Raguram, Changchang Wu, Jan-Michael Frahm, and Svetlana Lazebnik. "Modeling and Recognition of Landmark Image Collections Using Iconic Scene Graphs." *International Journal of Computer Vision* 95:3 (2011), 213–239.

[Ramamoorthi and Hanrahan 01a] Ravi Ramamoorthi and Pat Hanrahan. "An Efficient Representation for Irradiance Environment Maps." In *Proceedings of the 28th Annual Conference on Computer Graphics and Interactive Techniques*, pp. 497–500. ACM, 2001.

[Ramamoorthi and Hanrahan 01b] Ravi Ramamoorthi and Pat Hanrahan. "On the Relationship between Radiance and Irradiance: Determining the Illumination from Images of a Convex Lambertian Object." *Journal of the Optical Society of America* 18:10 (2001), 2448–2459.

[Ramamoorthi and Hanrahan 01c] Ravi Ramamoorthi and Pat Hanrahan. "A Signal-Processing Framework for Inverse Rendering." In *Proceedings of SIGGRAPH 2001, Computer Graphics Proceedings, Annual Conference Series*, edited by Eugene Fiume, pp. 117–128. ACM Press / ACM SIGGRAPH, 2001. ISBN 1-58113-292-1.

[Ramanarayanan et al. 07] Ganesh Ramanarayanan, James Ferwerda, Bruce Walter, and Kavita Bala. "Visual Equivalence: Towards a New Standard for Image Fidelity." *ACM Trans. Graph.* 26:3. Available online (http://doi.acm.org/10.1145/1276377.1276472).

[Raposo et al. 13] Carolina Raposo, Joao Pedro Barreto, and Urbano Nunes. "Fast and Accurate Calibration of a Kinect Sensor." In *Proceedings of the 2013 International Conference on 3D Vision, 3DV '13*, pp. 342–349. Washington, DC, USA: IEEE Computer Society, 2013.

[Read and Bohr 14] Jenny C.A. Read and Iwo Bohr. "User Experience while Viewing Stereoscopic 3D Television." *Ergonomics*: ahead-of-print (2014), 1–14.

[Reibel et al. 03] Y. Reibel, M. Jung, M. Bouhifd, B. Cunin, and C. Draman. "CCD or CMOS Camera Noise Characterisation." *The European Physical Journal Applied Physics* 21 (2003), 75–80.

[Reinhard et al. 08] Erik Reinhard, Erum Arif Khan, Ahmet Oguz Akyz, and Garrett M Johnson. *Color Imaging: Fundamentals and Applications.* A.K. Peters, Ltd., 2008.

[Reinhard et al. 10] Erik Reinhard, Greg Ward, Summant Pattanaik, Paul Debevec, Wolfgang Heidrich, and Karol Myszkowski. *High Dynamic Range Imaging: Acquisition, Display, and Image-Based Lighting*, Second edition. The Morgan Kaufmann series in Computer Graphics, Burlington, MA: Elsevier (Morgan Kaufmann), 2010.

[Ren et al. 10] Zhong Ren, Kun Zhou, Tengfei Li, Wei Hua, and Baining Guo. "Interactive Hair Rendering under Environment Lighting." *ACM Transactions on Graphics (Proc. SIGGRAPH 2010).*

[Rhemann et al. 11] Christoph Rhemann, Asmaa Hosni, Michael Bleyer, Carsten Rother, and Margrit Gelautz. "Fast Cost-Volume Filtering for Visual Correspondence and beyond." In *Computer Vision and Pattern Recognition (CVPR), 2011 IEEE Conference on*, pp. 3017–3024. IEEE, 2011.

[Richardt et al. 10] Christian Richardt, Douglas Orr, Ian Davies, Antonio Criminisi, and Neil A. Dodgson. "Real-Time Spatiotemporal Stereo Matching Using the Dual-Cross-Bilateral Grid." In *Computer Vision Conference on Computer Vision: Part III*, pp. 510–523, 2010.

[Richardt et al. 13] Christian Richardt, Yael Pritch, Henning Zimmer, and Alexander Sorkine-Hornung. "Megastereo: Constructing High-Resolution Stereo Panoramas." In *Proc. CVPR 2013*, pp. 1256–1263. Portland, USA, 2013.

[Ritschel et al. 12] Tobias Ritschel, Carsten Dachsbacher, Thorsten Grosch, and Jan Kautz. "The State of the Art in Interactive Global Illumination." *Computer Graphics Forum* 31:1 (2012), 160–188.

[Robinette et al. 99] K. Robinette, H. Daanen, and E. Paquet. "The CAESAR Project: A 3-D Surface Anthropometry Survey." In *Conference on 3D Digital Imaging and Modeling*, pp. 180–186, 1999.

[Rodriguez-Ramos et al. 11] J.M. Rodriguez-Ramos, J.G. Marichal-Hernandez, J.P. Luke, J. Trujillo-Sevilla, M. Puga, M. Lopez, J.J. Fernandez-Valdivia, C. Dominguez-Conde, J.C. Sanluis, F. Rosa, V. Guadalupe, H. Quintero, C. Militello, L.F. Rodriguez-Ramos, R. Lopez, I. Montilla, and B. Femenia. "New Developments at CAFADIS Plenoptic Camera." In *Information Optics (WIO), 2011 10th Euro-American Workshop on*, pp. 1 –3, 2011.

[Rogge et al. 14] Lorenz Rogge, Felix Klose, Michael Stengel, Martin Eisemann, and Marcus Magnor. "Garment Replacement in Monocular Video Sequences." *ACM Transactions on Graphics (TOG).* To appear.

[Rogmans et al. 09] Sammy Rogmans, Jiangbo Lu, Philippe Bekaert, and Gauthier Lafruit. "Real-Time Stereo-Based View Synthesis Algorithms: A Unified Framework and Evaluation on Commodity GPUs." *Image Communication* 24:1-2 (2009), 49–64.

[Rohmer et al. 10] Damien Rohmer, Tiberiu Popa, Marie-Paule Cani, Stefanie Hahmann, and Alla Sheffer. "Animation Wrinkling: Augmenting Coarse Cloth Simulations with Realistic-Looking Wrinkles." *ACM Trans. Graph.* 29:6 (2010), 157:1–157:8.

[Rose et al. 98] C. Rose, M. Cohen, and B. Bodenheimer. "Verbs and Adverbs: Multidimensional Motion Interpolation." *IEEE Computer Graphics and Applications* 18:5 (1998), 32–40.

[Rosenthal et al. 01] S. Rosenthal, D. Griffin, and M Sanders. "Real-Time Compter Graphics for On-set Visualization: "A.I." and "The Mummy Returns."" In *Siggraph 2001, Sketches and Applications*, 2001.

[Roth and Vona 12] Henry Roth and Marsette Vona. "Moving Volume KinectFusion." In *Proc. BMVC*, pp. 1–11, 2012.

[Rublee et al. 11] Ethan Rublee, Vincent Rabaud, Kurt Konolige, and Gary Bradski. "ORB: An Efficient Alternative to SIFT or SURF." In *Computer Vision (ICCV), 2011 IEEE International Conference on*, pp. 2564–2571. IEEE, 2011.

[Rucci 08] Michele Rucci. "Fixational Eye Movements, Natural Image Statistics, and Fine Spatial Vision." *Network: Computation in Neural Systems* 19:4 (2008), 253–285.

[Ruhl et al. 12a] Kai Ruhl, Benjamin Hell, Felix Klose, Christian Lipski, Sören Petersen, and Marcus Magnor. "Improving Dense Image Correspondence Estimation with Interactive User Guidance." In *ACM Multimedia*, pp. 1129–1132, 2012.

[Ruhl et al. 12b] Kai Ruhl, Felix Klose, Christian Lipski, and Marcus A. Magnor. "Integrating Approximate Depth Data into Dense Image Correspondence Estimation." In *European Conference on Visual Media Production (CVMP)*, pp. 26–31, 2012.

[Ruhl et al. 13] Kai Ruhl, Martin Eisemann, and Marcus Magnor. "Cost Volume-Based Interactive Depth Editing in Stereo Post-Processing." In *European Conference on Visual Media Production (CVMP)*, pp. 1–6, 2013.

[Rupprecht et al. 13] C. Rupprecht, O. Pauly, C. Theobalt, and S. Ilic. "3D Semantic Parameterization for Human Shape Modeling: Application to 3D Animation." In *3D Vision*, 2013.

[Rusinkiewicz and Levoy 01] Szymon Rusinkiewicz and Marc Levoy. "Efficient Variants of the ICP Algorithm." In *Third International Conference on 3D Digital Imaging and Modeling (3DIM)*, pp. 145–152, 2001.

[Rusinkiewicz et al. 02] S. Rusinkiewicz, O. Hall-Holt, and M. Levoy. "Real-Time 3D Model Acquisition." *ACM Trans. Graph* 21:3 (2002), 438–446.

[Ryan White 07] David Forsyth Ryan White, Keenan Crane. "Data Driven Cloth Animation." SIGGRAPH Technical Sketch, 2007.

[Sabov and Krüger 10] A. Sabov and J. Krüger. "Identification and Correction of Flying Pixels in Range Camera Data." In *Proc. Spring Conf. on Computer Graphics*, pp. 135–142, 2010.

[Sadeghi et al. 13] Iman Sadeghi, Oleg Bisker, Joachim De Deken, and Henrik Wann Jensen. "A Practical Microcylinder Appearance Model for Cloth Rendering." *ACM Transactions on Graphics (TOG)* 32:2 (2013), 14.

[Sand and Teller 06] Peter Sand and Seth J. Teller. "Particle Video: Long-Range Motion Estimation Using Point Trajectories." In *Computer Vision and Pattern Recognition*, pp. 2195–2202, 2006.

[Sander and Mitchell 05] Pedro V. Sander and Jason L. Mitchell. "Progressive Buffers: View-Dependent Geometry and Texture LOD Rendering." In *Proc. SGP*, pp. 129–138, 2005.

[Sato et al. 99] Imari Sato, Yoichi Sato, and Katsushi Ikeuchi. "Acquiring a Radiance Distribution to Superimpose Virtual Objects onto a Real Scene." *IEEE Transactions on Visualization and Computer Graphics* 5:1 (1999), 1–12.

[Sattler et al. 03] Mirko Sattler, Ralf Sarlette, and Reinhard Klein. "Efficient and Realistic Visualization of Cloth." In *Eurographics Symposium on Rendering 2003*, 2003.

[Schaefer and Yuksel 07] S. Schaefer and C. Yuksel. "Example-Based Skeleton Extraction." In *Proceedings of the Symposium on Geometry Processing (SGP)*, pp. 153–162, 2007.

[Scharstein and Szeliski 02] Daniel Scharstein and Richard Szeliski. "A Taxonomy and Evaluation of Dense Two-Frame Stereo Correspondence Algorithms." *International Journal of Computer Vision* 47:1/2/3 (2002), 7–42.

[Scharstein and Szeliski 03] Daniel Scharstein and Richard Szeliski. "High-Accuracy Stereo Depth Maps Using Structured Light." In *IEEE Computer Society Conference on Computer Vision and Pattern Recognition*, pp. 195–202, 2003.

[Schmeing and Jiang 11] Michael Schmeing and Xiaoyi Jiang. "Time-Consistency of Disocclusion Filling Algorithms in Depth Image-Based Rendering." In *3DTV Conference: The True Vision—Capture, Transmission and Display of 3D Video (3DTV-CON), 2011*, pp. 1–4, 2011.

[Schmidt and Jahne 11] M. Schmidt and B. Jahne. "Efficient and Robust Reduction of Motion Artifacts for 3D Time-of-Flight Cameras." In *Proc. Int. Conf. 3D Imaging (IC3D)*, pp. 1–8, 2011.

[Schneider and Eisert 12] David C. Schneider and Peter Eisert. "On User-Interaction in 3D Reconstruction." In *Eurographics (poster)*. Cagliari, Italy, 2012.

[Schödl et al. 00] Arno Schödl, Richard Szeliski, David H Salesin, and Irfan Essa. "Video Textures." In *Proceedings of the 27th Annual Conference on Computer Graphics and Interactive Techniques*, pp. 489–498. ACM Press/Addison-Wesley Publishing Co., 2000.

[Schoemake 94] Ken Schoemake. "Euler Angle Conversion." In *Graphics Gems, IV*, pp. 222–229. Academic Press, 1994.

[Scholz and Magnor 06] Volker Scholz and Marcus Magnor. "Texture Replacement of Garments in Monocular Video Sequences." In *Proc. Eurographics Symposium on Rendering (EGSR)*, pp. 305–312, 2006.

[Scholz et al. 05] Volker Scholz, Timo Stich, Michael Keckeisen, Markus Wacker, and Marcus Magnor. "Garment Motion Capture Using Color-Coded Patterns." *Computer Graphics Forum* 24:3 (2005), 439–448.

[Schreer et al. 12] Oliver Schreer, Peter Kauff, Peter Eisert, Christian Weissig, and Jean-Claude Rosenthal. "Geometrical Design Concept for Panoramic 3D Video Acquisition." In *Proc. 20th European Signal Processing Conference (EUSICPO 2012)*, pp. 2757–2761. Bucharest, Romania, 2012.

[Schreer et al. 13] Oliver Schreer, Ingo Feldmann, Christian Weissig, Peter Kauff, and Ralf Schäfer. "Ultrahigh-Resolution Panoramic Imaging for Format-Agnostic Video Production." *Proceedings of the IEEE* 101:1 (2013), 99–114.

[Schröder et al. 11] Kai Schröder, Reinhard Klein, and Arno Zinke. "A Volumetric Approach to Predictive Rendering of Fabrics." In *Computer Graphics Forum*, 30, 30, pp. 1277–1286. Wiley Online Library, 2011.

[Schröder et al. 12] Kai Schröder, Shuang Zhao, and Arno Zinke. "Recent Advances in Physically-Based Appearance Modeling of Cloth." In *SIGGRAPH Asia 2012 Courses, SA '12*, pp. 12:1–12:52, 2012.

[Schuon et al. 09] Sebastian Schuon, Christian Theobalt, James Davis, and Sebastian Thrun. "Lidarboost: Depth Superresolution for ToF 3D Shape Scanning." In *Proc. IEEE Conf. Computer Vision and Pattern Recognition*, pp. 343–350. IEEE, 2009.

[Schwartz et al. 10] Christopher Schwartz, Ruwen Schnabel, Patrick Degener, and Reinhard Klein. "PhotoPath: Single Image Path Depictions from Multiple Photographs." *Journal of WSCG* 18:1-3.

[Schwartz et al. 11a] Christopher Schwartz, Roland Ruiters, Michael Weinmann, and Reinhard Klein. "WebGL-Based Streaming and Presentation Framework for Bidirectional Texture Functions." In *The 12th International Symposium on Virtual Reality, Archeology and Cultural Heritage VAST 2011*, pp. 113–120. Eurographics Association, Eurographics Association, 2011.

[Schwartz et al. 11b] Christopher Schwartz, Michael Weinmann, Roland Ruiters, and Reinhard Klein. "Integrated High-Quality Acquisition of Geometry and Appearance for Cultural Heritage." In *The 12th International Symposium on Virtual Reality, Archeology and Cultural*

Heritage VAST 2011, pp. 25–32. Eurographics Association, Eurographics Association, 2011.

[Schwartz et al. 13] Christopher Schwartz, Ralf Sarlette, Michael Weinmann, and Reinhard Klein. "DOME II: A Parallelized BTF Acquisition System." In *Eurographics Workshop on Material Appearance Modeling: Issues and Acquisition*, edited by Holly Rushmeier and Reinhard Klein, pp. 25–31, 2013.

[Segal et al. 92] Marc Segal, Carl Korobkin, Rolf van Widenfelt, Jim Foran, and Paul Haeberli. "Fast Shadows and Lighting Effects Using Texture Mapping." *SIGGRAPH Comput. Graph.* 26:2 (1992), 249–252.

[Seitz and Dyer 96] Steven M. Seitz and Charles R. Dyer. "View Morphing." In *Proc. of ACM SIGGRAPH'96*, pp. 21–30. New York: ACM Press/ACM SIGGRAPH, 1996.

[Seitz and Dyer 99] Steven M. Seitz and Charles R. Dyer. "Photorealistic Scene Reconstruction by Voxel Coloring." *Int. J. Comput. Vision* 35:2 (1999), 151–173.

[Seitz et al. 05] Steven M. Seitz, Yasuyuki Matsushita, and Kiriakos N. Kutulakos. "A Theory of Inverse Light Transport." In *Proceedings of IEEE International Conference on Computer Vision*, pp. 1440–1447, 2005.

[Seitz et al. 06] Steven M. Seitz, Brian Curless, James Diebel, Daniel Scharstein, and Richard Szeliski. "A Comparison and Evaluation of Multi-View Stereo Reconstruction Algorithms." In *Computer Vision and Pattern Recognition, 2006 IEEE Computer Society Conference on*, 1, pp. 519–528, 2006.

[Sellent et al. 11] Anita Sellent, Martin Eisemann, Bastian Goldlücke, Daniel Cremers, and Marcus Magnor. "Motion Field Estimation from Alternate Exposure Images." *IEEE Transactions on Pattern Analysis and Machine Intelligence (PAMI)* 33:8 (2011), 1577–1589.

[Sellent et al. 12] Anita Sellent, Kai Ruhl, and Marcus Magnor. "A Loop-Consistency Measure for Dense Correspondences in Multi-View Video." *Journal of Image and Vision Computing* 30:9 (2012), 641–654.

[Sen and Darabi 09] Pradeep Sen and Soheil Darabi. "Compressive Dual Photography." *Computer Graphics Forum* 28:2 (2009), 609–618.

[Sen et al. 05] Pradeep Sen, Billy Chen, Gaurav Garg, Stephen R. Marschner, Mark Horowitz, Marc Levoy, and Hendrik P. A. Lensch. "Dual Photography." 24 (2005), 745755. Available online (http://dl.acm.org/citation.cfm?id=1073257).

[Shade et al. 98] J. Shade, S. Gortler, L. He, and R. Szeliski. "Layered Depth Images." In *Proceedings of the 25th Annual Conference on Computer Graphics and Interactive Techniques*, pp. 231–242. ACM, 1998.

[Shaheen et al. 09] Mohammed Shaheen, Jürgen Gall, Robert Strzodka, Luc J. Van Gool, and Hans-Peter Seidel. "A Comparison of 3D Model-Based Tracking Approaches for Human Motion Capture in Uncontrolled Environments." In *WACV*, 2009.

[Sharf et al. 07] A. Sharf, T. Lewiner, A. Shamir, and L. Kobbelt. "On-the-Fly Skeleton Computation for 3D Shapes." *Computer Graphics Forum (Proceedings of EUROGRAPHICS)*, pp. 323–328.

[Shi and Tomasi 94] Jianbo Shi and Carlo Tomasi. "Good Features to Track." In *Computer Vision and Pattern Recognition, 1994. Proceedings CVPR'94, 1994 IEEE Computer Society Conference on*, pp. 593–600. IEEE, 1994.

[Shotton et al. 11] Jamie Shotton, Andrew W. Fitzgibbon, Mat Cook, Toby Sharp, Mark Finocchio, Richard Moore, Alex Kipman, and Andrew Blake. "Real-Time Human Pose Recognition in Parts from Single Depth Images." In *IEEE Computer Vision and Pattern Recognition (CVPR)*, pp. 1297–1304, 2011.

[Shum and He 99] Heung-Yeung Shum and Li-Wei He. "Rendering with Concentric Mosaics." In *SIGGRAPH '99*, pp. 299–306. Los Angeles, USA, 1999.

[Shum and Kang 00] Heung-Yeung Shum and Sing Bing Kang. "A Review of Image-Based Rendering Techniques." *IEEE/SPIE Visual Communications and Image Processing (VCIP)* 213.

[Sigal et al. 12] Leonid Sigal, Michael Isard, Horst Haussecker, and Michael J. Black. "Loose-Limbed People: Estimating 3D Human Pose and Motion Using Non-Parametric Belief Propagation." *IJCV* 98:1 (2012), 15–48.

[Sinha et al. 11] Sudipta N. Sinha, Jan-Michael Frahm, Marc Pollefeys, and Yakup Genc. "Feature Tracking and Matching in Video Using Programmable Graphics Hardware." *Machine Vision and Applications* 22:1 (2011), 207–217.

[Sinha et al. 12] Sudipta N. Sinha, Drew Steedly, and Richard Szeliski. "A Multi-Stage Linear Approach to Structure from Motion." In *Trends and Topics in Computer Vision*, pp. 267–281. Springer, 2012.

[SizeGermany 07] SizeGermany, 2007. Available online (http://www.sizegermany.de).

[Slama 80] C. C. Slama. *Manual of Photogrammetry.* Americal Society of Photogrammetry, 1980.

[Smisek et al. 11] J. Smisek, M. Jancosek, and T. Pajdla. "3D with Kinect." In *Computer Vision Workshops (ICCV Workshops), 2011 IEEE International Conference on*, pp. 1154–1160, 2011.

[Smith and Cheeseman 86] Randall C. Smith and Peter Cheeseman. "On the Representation and Estimation of Spatial Uncertainly." *Int. J. Rob. Res.* 5:4 (1986), 56–68.

[Smolic et al. 09] A. Smolic, K. Mueller, P. Merkle, P. Kauff, and T. Wiegand. "An Overview of Available and Emerging 3D Video Formats and Depth Enhanced Stereo as Efficient Generic Solution." In *Picture Coding Symposium, 2009. PCS 2009*, pp. 1–4, 2009.

[Snavely et al. 06] N. Snavely, S.M. Seitz, and R. Szeliski. "Photo Tourism: Exploring Photo Collections in 3D." In *ACM Transactions on Graphics (Proc. SIGGRAPH)*, 25, 25, pp. 835–846, 2006.

[Snavely et al. 08a] Noah Snavely, Rahul Garg, Steven M. Seitz, and Richard Szeliski. "Finding Paths through the World's Photos." *ACM Transactions on Graphics (Proceedings of SIGGRAPH 2008)* 27:3 (2008), 11–21.

[Snavely et al. 08b] Noah Snavely, Steven M Seitz, and Richard Szeliski. "Modeling the World from Internet Photo Collections." *International Journal of Computer Vision* 80:2 (2008), 189–210.

[Snavely et al. 08c] Noah Snavely, Steven M Seitz, and Richard Szeliski. "Skeletal Graphs for Efficient Structure from Motion." In *CVPR*, 1, p. 2, 2008.

[Snavely 12] Noah Snavely. "Bundler: Structure from Motion (SfM) for Unordered Image Collections." http://phototour.cs.washington.edu/bundler/, 2012.

[Son et al. 07] Jung-Young Son, Yuri N. Gruts, Kae-Dal Kwack, Kyung-Hun Cha, and Sung-Kyu Kim. "Stereoscopic Image Distortion in Radial Camera and Projector Configurations." In *J. Opt. Soc. Am. A*, 24, pp. 643–650. OSA, 2007.

[Sorkine et al. 04] Olga Sorkine, Daniel Cohen-Or, Yaron Lipman, Marc Alexa, Christian Rssl, and Hans-Peter Seidel. "Laplacian Surface Editing." In *ACM Symposium on Geometry Processing*, pp. 175–184, 2004.

[Starck and Hilton 05] Jonathan Starck and Adrian Hilton. "Spherical Matching for Temporal Correspondence of Non-Rigid Surfaces." In *IEEE ICCV*, 2005.

[Starck and Hilton 07a] Jonathan Starck and Adrian Hilton. "Correspondence Labelling for Wide-Time Free-Form Surface Matching." In *IEEE ICCV*, 2007.

[Starck and Hilton 07b] Jonathan Starck and Adrian Hilton. "Surface Capture for Performance-Based Animation." *IEEE Computer Graphics and Application* 27 (2007), 21–31.

[Steinbrücker et al. 09] Frank Steinbrücker, Thomas Pock, and Daniel Cremers. "Large Displacement Optical Flow Computation without Warping." In *International Conference on Computer Vision, ICCV*, pp. 1609–1614, 2009.

[Steitz and Pannekamp 05] A. Steitz and J. Pannekamp. "Systematic Investigation of Properties of PMD-Sensors." In *Proc. 1st Range Imaging Research Day*, pp. 59–69. ETH Zurich, 2005.

[Stelmach et al. 03] Lew B Stelmach, Wa James Tam, Filippo Speranza, Ronald Renaud, and Taali Martin. "Improving the Visual Comfort of Stereoscopic Images." In *Electronic Imaging 2003*, pp. 269–282. International Society for Optics and Photonics, 2003.

[Stewart et al. 03] J. Stewart, J. Yu, S. J. Gortler, and L. McMillan. "A New Reconstruction Filter for Undersampled Light Fields." In *Proceedings of the 14th Eurographics Workshop on Rendering, EGRW '03*, pp. 150–156. Aire-la-Ville, Switzerland, Switzerland: Eurographics Association, 2003.

[Stich et al. 08] Timo Stich, Christian Linz, Georgia Albuquerque, and Marcus Magnor. "View and Time Interpolation in Image Space." *Computer Graphics Forum* 27:7 (2008), 1781–1787.

[Stich et al. 11] Timo Stich, Christian Linz, Christian Wallraven, Douglas Cunningham, and Marcus Magnor. "Perception-Motivated Interpolation of Image Sequences." *ACM Transactions on Applied Perception (TAP)* 8:2 (2011), 11:1–11:25.

[Stolfi 91] J. Stolfi. *Oriented Projective Geometry: A Framework for Geometric Computation.* Boston: Academic Press, 1991.

[Stoll et al. 10] Carsten Stoll, Juergen Gall, Edilson de Aguiar, Sebastian Thrun, and Christian Theobalt. "Video-Based Reconstruction of Animatable Human Characters." *ACM Transactions on Graphics (Proc. SIGGRAPH ASIA 2011)* 29:6 (2010), 139:1–139:10.

[Straka et al. 12] Matthias Straka, Stefan Hauswiesner, Matthias Rüther, and Horst Bischof. "Simultaneous Shape and Pose Adaption of Articulated Models Using Linear Optimization." In *ECCV*, pp. 724–737, 2012.

[Stumpfel et al. 04] Jessi Stumpfel, Chris Tchou, Andrew Jones, Tim Hawkins, Andreas Wenger, and Paul Debevec. "Direct HDR Capture of the Sun and Sky." In *Proceedings of the 3rd International Conference on Computer Graphics, Virtual Reality, Visualisation and Interaction in Africa, AFRIGRAPH '04*, pp. 145–149. New York, NY, USA: ACM, 2004.

[Sturm and Triggs 96] Peter Sturm and Bill Triggs. "A Factorization-Based Algorithm for Multi-Image Projective Structure and Motion." In *4th European Conference on Computer Vision*, Cambridge, England, pp. 709–720, 1996.

[Sturm et al. 11] Peter Sturm, Srikumar Ramalingam, Jean-Philippe Tardif, and Joao Barreto. "Camera Models and Fundamental Concepts Used in Geometric Computer Vision." *Foundations and Trends in Computer Graphics and Vision* 6:1-2.

[Sturm 01] Peter Sturm. "On Focal Length Calibration from Two Views." In *Computer Vision and Pattern Recognition, 2001. CVPR 2001. Proceedings of the 2001 IEEE Computer Society Conference on*, 2, pp. II–145. IEEE, 2001.

[Sumner and Popović 04] R. W. Sumner and J. Popović. "Deformation Transfer for Triangle Meshes." *ACM Transactions on Graphics (Proceedings of SIGGRAPH)* 23:3 (2004), 399–405.

[Sumner et al. 07] R. W. Sumner, J. Schmid, and M. Pauly. "Embedded Deformation for Shape Manipulation." *ACM Transactions on Graphics (Proceedings of SIGGRAPH)*, 2007.

[Sun and Abidi 01] Y. Sun and M. Abidi. "Surface Matching by 3D Point's Fingerprint." In *International Conference on Computer Vision*, 2, pp. 263–269, 2001.

[Sun et al. 03] Jian Sun, Nan-Ning Zheng, and Heung-Yeung Shum. "Stereo Matching Using Belief Propagation." *IEEE Trans. Pattern Anal. Mach. Intell.* 25:7 (2003), 787–800.

[Sun et al. 05] Jian Sun, Yin Li, Sing Bing Kang, and Heung-Yeung Shum. "Symmetric Stereo Matching for Occlusion Handling." In *Computer Vision and Pattern Recognition*, pp. 399–406, 2005.

[Sun et al. 10] Deqing Sun, Stefan Roth, and Michael J. Black. "Secrets of Optical Flow estimation and Their Principles." In *Computer Vision and Pattern Recognition*, pp. 2432–2439, 2010.

[Svarovsky 99] Jan Svarovsky. "Extreme Detail Graphics." In *Proc. Game Developers Conference*, pp. 889–904, 1999.

[Svoboda et al. 05] Tomáš Svoboda, Daniel Martinec, and Tomáš Pajdla. "A Convenient Multi-Camera Self-Calibration for Virtual Environments." *PRESENCE: Teleoperators and Virtual Environments* 14:4 (2005), 407–422.

[Szeliski and Golland 99] Richard Szeliski and Polina Golland. "Stereo Matching with Transparency and Matting." *Int. J. Comput. Vision* 32:1 (1999), 45–61.

[Szeliski et al. 08] Richard Szeliski, Ramin Zabih, Daniel Scharstein, Olga Veksler, Vladimir Kolmogorov, Aseem Agarwala, Marshall Tappen, and Carsten Rother. "A Comparative Study of Energy Minimization Methods for Markov Random Fields with Smoothness-Based Priors." *IEEE Trans. Pattern Anal. Mach. Intell.* 30:6 (2008), 1068–1080.

[Szeliski 96] Richard Szeliski. "Video-Mosaics for Virtual Environments." *IEEE Computer Graphics and Applications* 16:2 (1996), 22–30.

[Szeliski 99] Richard Szeliski. "A Multi-View Approach to Motion and Stereo." In *Computer Vision and Pattern Recognition, 1999. IEEE Computer Society Conference on*, 1, p. 163 Vol. 1, 1999.

[Szeliski 06] Richard Szeliski. "Image Alignment and Stitching: A Tutorial." *Foundations and Trends in Computer Graphics and Vision* 2:1 (2006), 1–104.

[Szeliski 11] Richard Szeliski. *Computer Vision: Algorithms and Applications.* Springer, 2011.

[Tagliasacchi et al. 09] A. Tagliasacchi, H. Zhang, and D. Cohen-Or. "Curve Skeleton Extraction from Incomplete Point Cloud." *ACM Transactions on Graphics (Proceedings of SIGGRAPH).*

[Taguchi et al. 08] Yuichi Taguchi, Bennett Wilburn, and C. Lawrence Zitnick. "Stereo Reconstruction with Mixed Pixels Using Adaptive Over-Segmentation." In *Computer Vision and Pattern Recognition*, 2008.

[Tam et al. 13] G. Tam, Z.-Q. Cheng, Y.-K. Lai, F. Langbein, Y. Liu, D. Marshall, R. Martin, X.-F. Sun, and P. Rosin. "Registration of 3D Point Clouds and Meshes: A Survey from Rigid to Non-Rigid." *Transactions on Visualization and Computer Graphics* 19:7 (2013), 1199–1217.

[Tan et al. 04] Kar-Han Tan, Hong Hua, and Narendra Ahuja. "Multiview Panoramic Cameras Using Mirror Pyramids." *IEEE Trans. on Pattern Analysis and Machine Intelligence* 26:7 (2004), 941–945.

[Tauber et al. 07] Z. Tauber, Ze-Nian Li, and M.S. Drew. "Review and Preview: Disocclusion by Inpainting for Image-Based Rendering." *Systems, Man, and Cybernetics, Part C: Applications and Reviews, IEEE Transactions on* 37:4 (2007), 527–540.

[Taylor 03] Camillo J. Taylor. "Surface Reconstruction from Feature Based Stereo." In *Proceedings of the Ninth IEEE International Conference on Computer Vision—Volume 2*, pp. 184–192, 2003.

[Teichman et al. 13] Alex Teichman, Stephen Miller, and Sebastian Thrun. "Unsupervised Intrinsic Calibration of Depth Sensors via SLAM." In *Proceedings of Robotics: Science and Systems*. Berlin, Germany, 2013.

[Teichmann and Teller 98] M. Teichmann and S. Teller. "Assisted Articulation of Closed Polygonal Models." *Computer Animation and Simulation*, pp. 87–102.

[Tejera et al. 13] M. Tejera, D. Casas, and A. Hilton. "Animation Control of Surface Motion Capture." *IEEE Transactions on Cybernetics* 43:6 (2013), 1532–1545.

[Tena et al. 11] Jose Rafael Tena, Fernando De la Torre, and Iain Matthews. "Interactive Region-Based Linear 3D Face Models." *ACM Transactions on Graphics (Proc. SIGGRAPH 2011)* 30:4.

[Terzopoulos et al. 87] Demetri Terzopoulos, John Platt, Alan Barr, and Kurt Fleischer. "Elastically Deformable Models." In *Proc. of the 14th Annual Conference on Computer Graphics and Interactive Techniques, SIGGRAPH '87*, pp. 205–214. New York, NY, USA: ACM, 1987.

[The Foundry 13] The Foundry. "Nuke." 2013. Http://www.thefoundry.co.uk/.

[Theobalt et al. 07] Christian Theobalt, Naveed Ahmed, Hendrik Lensch, Marcus Magnor, and H-P Seidel. "Seeing People in Different Light-Joint Shape, Motion, and Reflectance Capture." *IEEE Transactions on Visualization and Computer Graphics (TVCG)* 13:4 (2007), 663–674.

[Thomas et al. 97] Graham A Thomas, J Jin, T Niblett, and C Urquhart. "A Versatile Camera Position Measurement System for Virtual Reality TV Production."

[Thomas 07] Graham Thomas. "Real-Time Camera Tracking Using Sports Pitch Markings." *Journal of Real-Time Image Processing* 2:2-3 (2007), 117–132.

[Thomaszewski et al. 09] Bernhard Thomaszewski, Simon Pabst, and Wolfgang Straßer. "Continuum-Based Strain Limiting." *Comput. Graph. Forum* 28:2 (2009), 569–576.

[Tocci et al. 11] M. D. Tocci, C. Kiser, N. Tocci, and P. Sen. "A Versatile HDR Video Production System." *ACM Transactions on Graphics (TOG) (Proc. SIGGRAPH)* 30:4.

[Tompkin et al. 13] James Tompkin, Fabrizio Pece, Rajvi Shah, Shahram Izadi, Jan Kautz, and Christian Theobalt. "Video Collections in Panoramic Contexts." In *Proc. of Symposium on User Interface Software and Technology*, pp. 1256–1263. Portland, USA, 2013.

[Triggs et al. 00] Bill Triggs, Philip F. McLauchlan, Richard Hartley, and Andrew W. Fitzgibbon. "Bundle Adjustment Modern Synthesis." In *Vision Algorithms: Theory and Practice*, pp. 298–372. Springer, 2000.

[Tsai 87] R. Y. Tsai. "A Versatile Camera Calibration Technique for High-Accuracy 3D Machine Vision Metrology Using Off-the-Shelf TV Cameras and Lenses." *IEEE Journal of Robotics and Automation*, pp. 323–344.

[Tunwattanapong et al. 13] Borom Tunwattanapong, Graham Fyffe, Paul Graham, Jay Busch, Xueming Yu, Abhijeet Ghosh, and Paul Debevec. "Acquiring Reflectance and Shape from Continuous Spherical Harmonic Illumination." In *SIGGRAPH*. Anaheim, CA, 2013.

[Tuytelaers and Mikolajczyk 07] Tinne Tuytelaers and Krystian Mikolajczyk. "Local Invariant Feature Detectors: A Survey." *Foundations and Trends in Computer Graphics and Vision* 3:3 (2007), 177–280.

[Urey et al. 11] Hakan Urey, Kishore V Chellappan, Erdem Erden, and Phil Surman. "State of the Art in Stereoscopic and Autostereoscopic Displays." *Proceedings of the IEEE* 99:4 (2011), 540–555.

[Vaish and Adams 12] Vaibhav Vaish and Andrew Adams. "The (New) Stanford Light Field Archive." http://lightfield.stanford.edu/, 2012.

[Vaish et al. 04] V. Vaish, B. Wilburn, N. Joshi, and M. Levoy. "Using Plane + Parallax for Calibrating Dense Camera Arrays." In *Proc. International Conference on Computer Vision and Pattern Recognition*, 2004.

[Valgaerts et al. 12] Levi Valgaerts, Chenglei Wu, Andres Bruhn, Hans-Peter Seidel, and Christian Theobalt. "Lightweight Binocular Facial Performance Capture under Uncontrolled Lighting." *ACM Transactions on Graphics (Proc. SIGGRAPH ASIA 2012).*

[van den Hengel et al. 07] Anton van den Hengel, Anthony Dick, Thorsten Thormählen, Ben Ward, and Philip H. S. Torr. "VideoTrace: Rapid Interactive Scene Modelling from Video." *ACM Trans. Graph.* 26:3 (2007), 86:1–86:5.

[van Kaick et al. 11] O. van Kaick, H. Zhang, G. Hamarneh, and D. Cohen-Or. "A Survey on Shape Correspondence." *Computer Graphics Forum* 3:6 (2011), 1681–1707.

[Vangorp et al. 11] Peter Vangorp, Gaurav Chaurasia, Pierre-Yves Laffont, Roland Fleming, and George Drettakis. "Perception of Visual Artifacts in Image-Based Rendering of Façades." *Computer Graphics Forum (Proceedings of the Eurographics Symposium on Rendering)* 30:4 (2011), 1241–1250.

[Vangorp et al. 13] Peter Vangorp, Christian Richardt, Emily A. Cooper, Gaurav Chaurasia, Martin S. Banks, and George Drettakis. "Perception of Perspective Distortions in Image-Based Rendering." *ACM Trans. Graph.* 32:4 (2013), 58:1–58:12.

[Varanasi et al. 08] Kiran Varanasi, Andrei Zaharescu, Edmond Boyer, and Radu Horaud. "Temporal Surface Tracking Using Mesh Evolution." In *ECCV*, pp. 30–43, 2008.

[Veach and Guibas 94] Eric Veach and Leonidas Guibas. "Bidirectional Estimators for Light Transport." In *Proc. Eurographics Workshop on Rendering*, pp. 147–162, 1994.

[Veach and Guibas 97] Eric Veach and Leonidas J. Guibas. "Metropolis Light Transport." In *SIGGRAPH '97*, pp. 65–76, 1997.

[Veach 97] Eric Veach. "Robust Monte Carlo Methods for Light Transport Simulation." PhD thesis, Stanford University, 1997.

[Vedaldi and Fulkerson 08] Andrea Vedaldi and Brian Fulkerson. "VLFeat: An Open and Portable Library of Computer Vision Algorithms." http://www.vlfeat.org/, 2008.

[Vedula et al. 05] S. Vedula, S. Baker, and T. Kanade. "Image Based Spatio-Temporal Modeling and View Interpolation of Dynamic Events." *ACM Trans. on Graphics* 24:2 (2005), 240–261.

[Veksler 02] Olga Veksler. "Stereo Correspondence with Compact Windows via Minimum Ratio Cycle." *IEEE Trans. Pattern Anal. Mach. Intell.* 24:12 (2002), 1654–1660.

[Veksler 05] Olga Veksler. "Stereo Correspondence by Dynamic Programming on a Tree." In *Computer Vision and Pattern Recognition*, 2, pp. 384–390, 2005.

[Velten et al. 13] Andreas Velten, Di Wu, Adrian Jarabo, Belen Masia, Christopher Barsi, Chinmaya Joshi, Everett Lawson, Moungi Bawendi, Diego Gutierrez, and Ramesh Raskar. "Femto-Photography: Capturing and Visualizing the Propagation of Light." *ACM Trans. Graph.* 32:4 (2013), 44:1–44:8.

[Vergne et al. 10] Romain Vergne, Romain Pacanowski, Pascal Barla, Xavier Granier, and Christophe Schlick. "Radiance Scaling for Versatile Surface Enhancement." In *Proceedings of the 2010 ACM SIGGRAPH Symposium on Interactive 3D Graphics and Games*, pp. 143–150, 2010.

[Vetro et al. 11] A. Vetro, A.M. Tourapis, K. Muller, and Tao Chen. "3D-TV Content Storage and Transmission." *Broadcasting, IEEE Transactions on* 57:2 (2011), 384–394.

[Vlasic et al. 05] D. Vlasic, M. Brand, H. Pfister, and J. Popović. "Face Transfer with Multi-Linear Models." *ACM Transactions on Graphics (Proceedings SIGGRAPH)*, 2005.

[Vlasic et al. 08] D. Vlasic, I. Baran, W. Matusik, and J. Popović. "Articulated Mesh Animation from Multi-View Silhouettes." *ACM Transactions on Graphics (Proc. SIGGRAPH)* 27:3 (2008), 97:1–97:9.

[Vogiatzis et al. 06] George Vogiatzis, Carlos Hernández, and Roberto Cipolla. "Reconstruction in the Round Using Photometric Normals and Silhouettes." In *CVPR*, pp. 1847–1854, 2006.

[Volino et al. 09] Pascal Volino, Nadia Magnenat-Thalmann, and Francois Faure. "A Simple Approach to Nonlinear Tensile Stiffness for Accurate Cloth Simulation." *ACM Trans. Graph.* 28:4 (2009), 105:1–105:16.

[von der Pahlen et al. 14] Javier von der Pahlen, Jorge Jimenez, Etienne Danvoye, Paul Debevec, Graham Fyffe, and Hao Li. "Digital Ira and Beyond: Creating Photoreal Real-Time Digital Characters." In *SIGGRAPH Course Notes 2014*, 2014.

[Wade 00] L. Wade. "Automated Generation of Control Skeletons for Use in Animation." PhD thesis, Ohio State University, 2000.

[Wang et al. 01] Lifeng Wang, Sing Bing Kang, Richard Szeliski, and Heung-Yeung Shum. "Optimal Texture Map Reconstruction from Multiple Views." In *Computer Vision and Pattern Recognition, 2001. CVPR 2001. Proceedings of the 2001 IEEE Computer Society Conference on*, 1, pp. I–347. IEEE, 2001.

[Wang et al. 07] Huamin Wang, Mingxuan Sun, and Ruigang Yang. "Space-Time Light Field Rendering." *IEEE Transactions on Visualization and Computer Graphics* 13:4 (2007), 697–710.

[Wang et al. 08] Jiaping Wang, Shuang Zhao, Xin Tong, John Snyder, and Baining Guo. "Modeling Anisotropic Surface Reflectance with Example-Based Microfacet Synthesis." In *ACM Transactions on Graphics (TOG)*, 27, 27, p. 41. ACM, 2008.

[Wang et al. 09] Jiaping Wang, Yue Dong, Xin Tong, Zhouchen Lin, and Baining Guo. "Kernel Nyström Method for Light Transport." *ACM Transactions on Graphics (TOG)* 28:3 (2009), 29.

[Wang et al. 10] Huamin Wang, Florian Hecht, Ravi Ramamoorthi, and James O'Brien. "Example-Based Wrinkle Synthesis for Clothing Animation." *ACM Trans. Graph.* 29:4 (2010), 107:1–107:8.

[Wang et al. 11a] Huamin Wang, James F. O'Brien, and Ravi Ramamoorthi. "Data-Driven Elastic Models for Cloth: Modeling and Measurement." *ACM Trans. Graph.* 30:4 (2011), 71:1–71:12.

[Wang et al. 11b] Lili Wang, Kees Teunissen, Yan Tu, Li Chen, Panpan Zhang, Tingting Zhang, and Ingrid Heynderickx. "Crosstalk Evaluation in Stereoscopic Displays." *Display Technology, Journal of* 7:4 (2011), 208–214.

[Wang et al. 11c] O. Wang, M. Lang, M. Frei, A. Hornung, A. Smolic, and M. Gross. "StereoBrush: Interactive 2D to 3D Conversion Using Discontinuous Warps." In *Proceedings of the Eighth Eurographics Symposium on Sketch-Based Interfaces and Modeling, SBIM '11*, pp. 47–54. New York, NY, USA: ACM, 2011.

[Wang et al. 13a] J. Wang, K. Xu, L. Liu, J. Cao, S. Liu, Z. Yu, and X. D. Gu. "Consolidation of Low-Quality Point Clouds from Outdoor Scenes." *Computer Graphics Forum* 32:5 (2013), 207–216.

[Wang et al. 13b] Yangang Wang, Jianyuan Min, Jianjie Zhang, Yebin Liu, Feng Xu, Qionghai Dai, and Jinxiang Chai. "Video-Based Hand Manipulation Capture through Composite Motion Control." *ACM Transactions on Graphics (TOG)* 32:4 (2013), 43.

[Wanner and Goldlücke 12] S. Wanner and B. Goldlücke. "Spatial and Angular Variational Super-Resolution of 4D Light Fields." In *Proc. European Conference on Computer Vision*, 2012.

[Wanner and Goldlücke 13] S. Wanner and B. Goldlücke. "Reconstructing Reflective and Transparent Surfaces from Epipolar Plane Images." In *German Conference on Pattern Recognition (GCPR)*, 2013.

[Wanner and Goldlücke 14] S. Wanner and B. Goldlücke. "Variational Light Field Analysis for Disparity Estimation and Super-Resolution." *IEEE Transactions on Pattern Analysis and Machine Intelligence* 36:3 (2014), 606–619.

[Wanner et al. 11] S. Wanner, J. Fehr, and B. Jähne. "Generating EPI Representations of 4D Light Fields with a Single Lens Focused Plenoptic Camera." *Advances in Visual Computing*, pp. 90–101.

[Ward et al. 07] Kelly Ward, Florence Bertails, Tae-Yong Kim, Stephen R. Marschner, Marie-Paule Cani, and Ming Lin. "A Survey on Hair Modelling: Styling, Simulation and Rendering." *IEEE Transactions on Visualization and Computer Graphics TVCG* 13:2.

[Waschbüsch et al. 07] M. Waschbüsch, S. Würmlin, and M. Gross. "3D Video Billboard Clouds." In *Computer Graphics Forum*, 26, 26, pp. 561–569. Wiley Online Library, 2007.

[Watson et al. 00] Benjamin Watson, Alinda Friedman, and Aaron McGaffey. "Using Naming Time to Evaluate Quality Predictors for Model Simplification." In *Proceedings of the SIGCHI Conference on Human Factors in Computing Systems, CHI '00*, pp. 113–120. New York, NY, USA: ACM, 2000.

[Weber et al. 07] O. Weber, O. Sorkine, Y. Lipman, and C. Gotsman. "Context-Aware Skeletal Shape Deformation." *Computer Graphics Forum (Proceedings of Eurographics)* 26:3 (2007), 267–273.

[Weber et al. 09] O. Weber, M. Ben-Chan, and C. Gotsman. "Complex Barycentric Coordinates with Applications to Planar Shape Deformations." *Computer Graphics Forum (Proceedings EUROGRAPHICS)* 28:2.

[Wei et al. 12] Xiaolin Wei, Peizhao Zhang, and Jinxiang Chai. "Accurate Realtime Full-Body Motion Capture Using a Single Depth Camera." *ACM Transactions on Graphics (TOG)* 31:6 (2012), 188.

[Weise et al. 09] Thibaut Weise, Hao Li, Luc Van Gool, and Mark Pauly. "Face/Off: Live Facial Puppetry." In *ACM/Eurographics Symposium on Computer Animation*, 2009.

[Weise et al. 11] Thibaut Weise, Sofien Bouaziz, Hao Li, and Mark Pauly. "Realtime Performance Based Facial Animation." *ACM Transactions on Graphics (Proc. SIGGRAPH 2011)* 30:4.

[Weiss et al. 11] A. Weiss, D. Hirshberg, and M. Black. "Home 3D Body Scans from Noisy Image and Range Data." In *International Conference on Computer Vision*, pp. 1951–1958, 2011.

[Weissig et al. 12] Christian Weissig, Oliver Schreer, Peter Eisert, and Peter Kauff. "The Ultimate Immersive Experience: Panoramic 3D Video Acquisition." In *Proc. 18th Int. Conf. on MultiMedia Modelling*, pp. 671–681. Klagenfurt, Austria, 2012.

[Wendland 95] H. Wendland. "Piecewise Polynomial, Positive Definite and Compactly Supported Radial Functions of Minimal Degree." *Adv. Comput. Math.* 4:4 (1995), 389–396.

[Wenger et al. 05] Andreas Wenger, Andrew Gardner, Chris Tchou, Jonas Unger, Tim Hawkins, and Paul Debevec. "Performance Relighting and Reflectance Transformation with Time-Multiplexed Illumination." *ACM Transactions on Graphics (Proc. SIGGRAPH)* 24:3 (2005), 756–764.

[Werlberger et al. 09] Manuel Werlberger, Werner Trobin, Thomas Pock, Andreas Wedel, Daniel Cremers, and Horst Bischof. "Anisotropic Huber-L1 Optical Flow." In *British Machine Vision Conference*, pp. 108.1–108.11, 2009.

[Wetzstein et al. 07] Gordon Wetzstein, Oliver Bimber, et al. "Radiometric Compensation through Inverse Light Transport." In *Pacific Conference on Computer Graphics and Applications*, pp. 391–399, 2007.

[Wetzstein et al. 11] G. Wetzstein, I. Ihrke, D Lanman, and W. Heidrich. "State of the Art in Computational Plenoptic Imaging." In *STAR Proceedings of Eurographics*, 2011.

[Weyrich et al. 08] Tim Weyrich, Jason Lawrence, Hendrik Lensch, Szymon Rusinkiewicz, and Todd Zickler. "Principles of Appearance Acquisition and Representation." *Foundations and Trends in Computer Graphics and Vision* 4:2 (2008), 75–191.

[White and Forsyth 06] Ryan White and David A. Forsyth. "Retexturing Single Views Using Texture and Shading." In *Proc. Europ. Conf. on Computer Vision (ECCV)*, pp. 70–81. Springer, 2006.

[White et al. 07] Ryan White, Keenan Crane, and David A. Forsyth. "Capturing and Animating Occluded Cloth." *ACM Trans. Graph.* 26:3.

[Wietzke 12] Lennart Wietzke. "Raytrix 3D Lightfield Camera Technology." http://www.raytrix.de/, 2012.

[Wilburn et al. 05] Bennett Wilburn, Neel Joshi, Vaibhav Vaish, Eino-Ville Talvala, Emilio Antunez, Adam Barth, Andrew Adams, Mark Horowitz, and Marc Levoy. "High Performance Imaging Using Large Camera Arrays." *ACM Transactions on Graphics* 24 (2005), 765–776.

[Wilson and Snavely 13] Kyle Wilson and Noah Snavely. "Network Principles for SfM: Disambiguating Repeated Structures with Local Context." In *International Conference on Computer Vision*. IEEE, 2013.

[Witkin and Popovic 95] A. Witkin and Z. Popovic. "Motion Warping." In *Proceedings of ACM SIGGRAPH*, pp. 105–108, 1995.

[Wood et al. 00] Daniel N. Wood, Daniel I. Azuma, Ken Aldinger, Brian Curless, Tom Duchamp, David H. Salesin, and Werner Stuetzle. "Surface Light Fields for 3D Photography." In *Proceedings of the 27th Annual Conference on Computer Graphics and Interactive Techniques*, pp. 287–296. ACM Press/Addison-Wesley Publishing Co., 2000.

[Woodford et al. 09] Oliver Woodford, Philip Torr, Ian Reid, and Andrew Fitzgibbon. "Global Stereo Reconstruction under Second-Order Smoothness Priors." *IEEE Trans. Pattern Anal. Mach. Intell.* 31:12 (2009), 2115–2128.

[Woodham 80] R. J. Woodham. "Photometric Method for Determining Surface Orientation from Multiple Images." In *Optical Engineering*, 19(1), 19(1), pp. 139–144, 1980.

[Wu et al. 11] Changchang Wu, Sameer Agarwal, Brian Curless, and Steven M. Seitz. "Multicore Bundle Adjustment." In *Computer Vision and Pattern Recognition (CVPR), 2011 IEEE Conference on*, pp. 3057–3064. IEEE, 2011.

[Wu et al. 12a] Chenglei Wu, Kiran Varanasi, and Christian Theobalt. "Full Body Performance Capture under Uncontrolled and Varying Illumination: A Shading-Based Approach." In *Computer Vision–ECCV 2012*, pp. 757–770. Springer, 2012.

[Wu et al. 12b] Di Wu, M. O'Toole, A. Velten, A. Agrawal, and R. Raskar. "Decomposing Global Light Transport Using Time of Flight Imaging." In *Computer Vision and Pattern Recognition (CVPR), 2012 IEEE Conference on*, pp. 366–373, 2012.

[Wu et al. 13] Chenglei Wu, Carsten Stoll, Levi Valgaerts, and Christian Theobalt. "On-Set Performance Capture of Multiple Actors with a Stereo Camera." *ACM Trans. Graph. (Proc. of SIGGRAPH Asia)* 32:6.

[Wu 11] Changchang Wu. "VisualSFM: A Visual Structure from Motion System." http://ccwu.me/vsfm/, 2011.

[Wuhrer and Shu 13] S. Wuhrer and C. Shu. "Estimating 3D Human Shapes from Measurements." *Machine Vision and Applications* 24:6 (2013), 1133–1147.

[Wuhrer et al. 10] S. Wuhrer, Z. Ben Azouz, and C. Shu. "Semi-Automatic Prediction of Landmarks on Human Models in Varying Poses." In *Canadian Conference on Computer and Robot Vision*, pp. 136–142, 2010.

[Wuhrer et al. 11] S. Wuhrer, C. Shu, and P. Xi. "Landmark-Free Posture Invariant Human Shape Correspondence." *The Visual Computer* 27:9 (2011), 843–852.

[Wuhrer et al. 12] S. Wuhrer, C. Shu, and P. Xi. "Posture-Invariant Statistical Shape Analysis Using Laplace Operator." *Computers & Graphics* 36:5 (2012), 410–416. *Proceedings of Shape Modeling International.*

[Xiao et al. 06] Jiangjian Xiao, Hui Cheng, Harpreet S. Sawhney, Cen Rao, and Michael A. Isnardi. "Bilateral Filtering-Based Optical Flow Estimation with Occlusion Detection." In *European Conference on Computer Vision*, 3951, 3951, pp. 211–224, 2006.

[Xu et al. 98] Z. Xu, R. Schwarte, H. Heinol, B. Buxbaum, and T. Ringbeck. "Smart Pixel—Photonic Mixer Device (PMD)." In *Proc. Int. Conf. on Mechatron. & Machine Vision*, pp. 259–264, 1998.

[Xu et al. 11] F. Xu, Y. Liu, C. Stoll, J. Tompkin, G. Bharaj, Q. Dai, H.-P. Seidel, J. Kautz, and C. Theobalt. "Video-Based Characters—Creating New Human Performances from a Multi-View Video Database." *Proc. ACM SIGGRAPH*, 2011.

[Y. Sahin and Backhaus 14] S. Spielmann Y. Sahin and M. Backhaus. "Dark Matter—A Tale of Virtual Production." *In ACM SIGGRAPH 2014 Talks (SIGGRAPH '14)*, ACM. New York, NY, USA, Article 25, 1 page DOI=10.1145/2614106,2614195 http://doi.acm.org/1145/2614106.2614195.

[Yahav et al. 07] G. Yahav, G. J. Iddan, and D. Mandelbaum. "3D Imaging Camera for Gaming Application." In *Digest of Technical Papers of Int. Conf. on Consumer Electronics*, 2007. DOI: 10.1109/ICCE.2007.341537.

[Yamanoue 06] H. Yamanoue. "The Differences between Toed-in Camera Configurations and Parallel Camera Configurations in Shooting Stereoscopic Images." *Multimedia and Expo, IEEE International Conference on* 0 (2006), 1701–1704.

[Yamauchi et al. 05] H. Yamauchi, S. Gumhold, R. Zayer, and Seidel H.-P. "Mesh Segmentation Driven by Gaussian Curvature." *The Visual Computer* 21 (2005), 649–658.

[Yamazaki et al. 02] Shuntaro Yamazaki, Ryusuke Sagawa, Hiroshi Kawasaki, Katsushi Ikeuchi, and Masao Sakauchi. "Microfacet Billboarding." In *Proceedings of the 13th Eurographics Workshop on Rendering, EGRW '02*, pp. 169–180. Aire-la-Ville, Switzerland, Switzerland: Eurographics Association, 2002.

[Yamazoe et al. 12] H. Yamazoe, H. Habe, I. Mitsugami, and Y. Yagi. "Easy Depth Sensor Calibration." In *Pattern Recognition (ICPR), 2012 21st International Conference on*, pp. 465–468, 2012.

[Yang and Pollefeys 03] Ruigang Yang and Marc Pollefeys. "Multi-Resolution Real-Time Stereo on Commodity Graphics Hardware." In *Computer Vision and Pattern Recognition, 2003. Proceedings. 2003 IEEE Computer Society Conference on*, 1, pp. I–211. IEEE, 2003.

[Yang et al. 93] Yibing Yang, Alan Yuille, and Jie Lu. "Local, Global, and Multilevel Stereo Matching." In *Proceedings of the IEEE Computer Society Conference on Computer Vision and Pattern Recognition, CVPR '93*, pp. 274–279. IEEE, 1993.

[Yang et al. 06a] Qingxiong Yang, Liang Wang, Ruigang Yang, Henrik Stewénius, and David Nistér. "Stereo Matching with Color-Weighted Correlation, Hierarchical Belief Propagation and Occlusion Handling." In *Computer Vision and Pattern Recognition*, pp. 2347–2354, 2006.

[Yang et al. 06b] Qingxiong Yang, Liang Wang, Ruigang Yang, Shengnan Wang, Miao Liao, and David Nistér. "Real-Time Global Stereo Matching Using Hierarchical Belief Propagation." In *British Machine Vision Conference*, 2006.

[Yang et al. 06c] R. Yang, D. Guinnip, and L. Wang. "View-Dependent Textured Splatting." *The Visual Computer* 22:7 (2006), 456–467.

[Yang et al. 10] Qingxiong Yang, Kar-Han Tan, W. Bruce Culbertson, and John G. Apostolopoulos. "Fusion of Active and Passive Sensors for Fast 3D Capture." In *MMSP*, pp. 69–74, 2010.

[Yaraş et al. 10] Fahri Yaraş, Hoonjong Kang, and Levent Onural. "State of the Art in Holographic Displays: A Survey." *Journal of Display Technology* 6:10 (2010), 443–454.

[Ye et al. 11] Mao Ye, Xianwang Wang, Ruigang Yang, Liu Ren, and Marc Pollefeys. "Accurate 3D Pose Estimation from a Single Depth Image." In *Proceedings of the 2011 International Conference on Computer Vision, ICCV '11*, pp. 731–738. Washington, DC, USA: IEEE Computer Society, 2011.

[Ye et al. 12] Genzhi Ye, Yebin Liu, Nils Hasler, Xiangyang Ji, Qionghai Dai, and Christian Theobalt. "Performance Capture of Interacting Characters with Handheld Kinects." In *ECCV*, pp. 828–841, 2012.

[Yoon and Kweon 05] Kuk-Jin Yoon and In-So Kweon. "Locally Adaptive Support-Weight Approach for Visual Correspondence Search." In *Computer Vision and Pattern Recognition*, pp. 924–931, 2005.

[Yuksel et al. 12] Cem Yuksel, Jonathan M. Kaldor, Doug L. James, and Steve Marschner. "Stitch Meshes for Modeling Knitted Clothing with Yarn-Level Detail." *ACM Trans. Graph.* 31:4 (2012), 37:1–37:12.

[Zabih and Woodfill 94] Ramin Zabih and John Woodfill. "Non-Parametric Local Transforms for Computing Visual Correspondence." In *Proceedings of the Third European Conference on Computer Vision (Vol. II), ECCV '94*, pp. 151–158. Secaucus, NJ, USA: Springer-Verlag New York, Inc., 1994.

[Zach et al. 07] C. Zach, T. Pock, and H. Bischof. "A Duality Based Approach for Realtime TV-L 1 Optical Flow." In *Proc. German Conference on Pattern Recognition (DAGM)*, pp. 214–223. Springer, 2007.

[Zach et al. 08] Christopher Zach, David Gallup, Jan-Michael Frahm, and Marc Niethammer. "Fast Global Labeling for Real-Time Stereo Using Multiple Plane Sweeps." In *Vision, Modeling and Visualization*, pp. 243–252, 2008.

[Zalevsky et al. 07] Zeev Zalevsky, Alexander Shput, Aviad Maizels, and Javier Garcia. "Method and System for Object Reconstruction." 2007.

[Zhang and Huang 04] Song Zhang and Peisen Huang. "High-Resolution, Real-Time 3D Shape Acquisition." In *Proceedings of the 2004 Conference on Computer Vision and Pattern Recognition Workshop (CVPRW'04)*, 3, 3, pp. 28–36. Washington, DC, USA: IEEE Computer Society, 2004.

[Zhang and Zhang 11] Cha Zhang and Zhengyou Zhang. "Calibration between Depth and Color Sensors for Commodity Depth Cameras." In *Proceedings of the 2011 IEEE International Conference on Multimedia and Expo, ICME '11*, pp. 1–6. Washington, DC, USA: IEEE Computer Society, 2011.

[Zhang et al. 02] Li Zhang, Brian Curless, and Steven M. Seitz. "Rapid Shape Acquisition Using Color Structured Light and Multi-Pass Dynamic Programming." In *The 1st IEEE International Symposium on 3D Data Processing, Visualization, and Transmission*, pp. 24–36, 2002.

[Zhang et al. 06] Qingshan Zhang, Zicheng Liu, Baining Guo, and Harry Shum. "Geometry-Driven Photorealistic Expression Synthesis." *IEEE Trans. Vis. Comput. Graphics* 12:1 (2006), 48–60.

[Zhang et al. 08] Yilei Zhang, Minglun Gong, and Yee-Hong Yang. "Local Stereo Matching with 3D Adaptive Cost Aggregation for Slanted Surface Modeling and Sub-Pixel Accuracy." In *International Conference on Pattern Recognition*, pp. 1–4, 2008.

[Zhang et al. 10a] Ke Zhang, Gauthier Lafruit, Rudy Lauwereins, and Luc J. Van Gool. "Joint Integral Histograms and Its Application in Stereo Matching." In *International Conference on Image Processing*, pp. 817–820, 2010.

[Zhang et al. 10b] Song Zhang, Daniel Van Der Weide, and James Oliver. "Superfast Phase-Shifting Method for 3-D Shape Measurement." *Optics Express* 18:9 (2010), 9684–9689.

[Zhang et al. 11] Lei Zhang, Xiaolin Wu, Antoni Buades, and Xin Li. "Color Demosaicking by Local Directional Interpolation and Nonlocal Adaptive Thresholding." *Journal of Electronic Imaging* 20:2.

[Zhang et al. 12] Qing Zhang, Mao Ye, Ruigang Yang, Yasuyuki Matsushita, Bennett Wilburn, and Huimin Yu. "Edge-Preserving Photometric Stereo via Depth Fusion." In *Proceedings of the 2012 IEEE Conference on Computer Vision and Pattern Recognition (CVPR), CVPR '12*, pp. 2472–2479, 2012.

[Zhang 00] Z. Zhang. "A Flexible New Technique for Camera Calibration." *IEEE Transactions on Patterns Analysis and Machine Intelligence* 22:11 (2000), 1330–1334.

[Zhao et al. 11] Shuang Zhao, Wenzel Jakob, Steve Marschner, and Kavita Bala. "Building Volumetric Appearance Models of Fabric Using Micro CT Imaging." In *ACM Transactions on Graphics (TOG)*, 30, 30, p. 44. ACM, 2011.

[Zheng et al. 09] K.C. Zheng, A. Colburn, A. Agarwala, M. Agrawala, D. Salesin, B. Curless, and M.F. Cohen. "Parallax Photography: Creating 3D Cinematic Effects from Stills." In *Proceedings of Graphics Interface 2009*, pp. 111–118. Canadian Information Processing Society, 2009.

[Zhou et al. 10] S. Zhou, H. Fu, L. Liu, D. Cohen-Or, and X. Han. "Parametric Reshaping of Human Bodies in Images." *ACM Transactions on Graphics* 29 (2010), 126:1–10. Proceedings of SIGGRAPH.

[Zhou et al. 12] Zhenglong Zhou, Bo Shu, Shaojie Zhuo, Xiaoming Deng, Ping Tan, and Stephen Lin. "Image-Based Clothes Animation for Virtual Fitting." In *ACM SIGGRAPH Asia 2012 Technical Briefs*, pp. 33:1–33:4. ACM, 2012.

[Zhu et al. 11] Jiejie Zhu, Liang Wang, Ruigang Yang, James E. Davis, and Zhigeng Pan. "Reliability Fusion of Time-of-Flight Depth and Stereo Geometry for High Quality Depth Maps." *IEEE Trans. Pattern Anal. Mach. Intell.* 33:7 (2011), 1400–1414.

[Zilly et al. 10] Frederik Zilly, M Muller, Peter Eisert, and Peter Kauff. "The Stereoscopic Analyzer—An Image-Based Assistance Tool for Stereo Shooting and 3D Production." In *Image Processing (ICIP), 2010 17th IEEE International Conference on*, pp. 4029–4032. IEEE, 2010.

[Zimmer et al. 11] Henning Zimmer, Andrés Bruhn, and Joachim Weickert. "Optic Flow in Harmony." *Int. J. Comput. Vision* 93:3 (2011), 368–388.

[Zitnick et al. 04] C Lawrence Zitnick, Sing Bing Kang, Matthew Uyttendaele, Simon Winder, and Richard Szeliski. "High-Quality Video View Interpolation Using a Layered Representation." *ACM Transactions on Graphics (Proc. SIGGRAPH)* 23:3 (2004), 600–608.

[Zongker et al. 99] Douglas E. Zongker, Dawn M. Werner, Brian Curless, and David H. Salesin. "Environment Matting and Compositing." In *SIGGRAPH '99: Proceedings of the 26th Annual Conference on Computer Graphics and Interactive Techniques*, pp. 205–214. New York, NY, USA: ACM Press/Addison-Wesley Publishing Co., 1999.

[Zurdo et al. 13] Javier S. Zurdo, Juan P. Brito, and Miguel A. Otaduy. "Animating Wrinkles by Example on Non-Skinned Cloth." *IEEE Trans. on Visualization and Computer Graphics* 19:1 (2013), 149–158.

Index